Lecture Notes in Mathematics 1559

Editors:
A. Dold, Heidelberg
B. Eckmann, Zürich
F. Takens, Groningen

Subseries:
LOMI and Euler International
Mathematical Institute, St. Petersburg

Adviser:
L. D. Faddeev

Vladimir G. Sprindžuk

Classical
Diophantine Equations

Springer-Verlag

Berlin Heidelberg New York
London Paris Tokyo
Hong Kong Barcelona
Budapest

Author

Vladimir G. Sprindžuk †

Translation Editors

Ross Talent †

Alf van der Poorten
Centre for Number Theory Research
Macquarie University
NSW 2109, Australia

Title of the original Russian edition: Klassicheskie diofantovy uravneniya ot dvukh neizvestnykh. "Nauka". Moscow 1982.

Mathematics Subject Classification (1991): 11J86, 11D00, 11J68, 12E25

ISBN 3-540-57359-3 Springer-Verlag Berlin Heidelberg New York
ISBN 0-387-57359-3 Springer-Verlag New York Berlin Heidelberg

Library of Congress Cataloging-in-Publication Data. Sprindzhuk, V. G. (Vladimir Gennadievich) [Klassicheskie diofantovy uravneniia ot dvukh neizvestnykh. English] Classical diophantine equations / Vladimir G. Sprindzhuk. p. cm. – (Lecture notes in mathematics; 1559) Includes bibliographical references and index.
ISBN 3-540-57359-3 (Berlin: acid-free) – ISBN 0-387-57359-3 (New York: acid-free) 1. Diophantine equations. I. Title. II. Series: Lecture notes in mathematics (Springer-Verlag); 1559. QA3.L28 no. 1559 [QA242] 512'.74–dc20 93-33443

© Springer-Verlag Berlin Heidelberg 1993
Printed in Germany

2146/3140-543210 - Printed on acid-free paper

Foreword

The author had initiated a revision and translation of this volume prior to his death.

Given the rapid advances in transcendence theory and diophantine approximation over recent years, one might fear that the present monograph, which is essentially a translation of a work originally published in the then USSR in 1982, is mostly superseded. That is not so. There is in any event a certain amount of updating inserted by the author. However, the author's emphasis remains original and almost unique, and well warrants study now that this work appears in the mathematical *lingua franca** thus making it easily accessible to the majority of mathematicians.

Most research mathematicians will be familiar with the eccentricities of Russian style — in this case I should correct that to Byelorussian style — in mathematical writing. There is quite an amount of repetitive detail and little assumption about notation, exemplified by a great deal more 'letting' in enunciations of lemmata and theorems than now seems customary; and the natural logarithm remains ln, just as on the engineer's calculator. Notwithstanding that, Sprindžuk maintains a pleasant and chatty approach, full of wise and interesting remarks. His emphases well warrant emulation.

I had the pleasure of meeting the author at several Oberwolfach meetings. Indeed, it was his instruction 'You will walk with me,' that led to the one and only time that I have allowed myself to be subjected to the post-breakfast perambulation all the way down and then, worse, back up the drive. I was a little surprised to find that Sprindžuk's spoken English was rather better than I had been led to expect given his apparent reticence at tea and dinner. But that may have been a function of the bad old days.

Nonetheless, the translation from which the present volume is derived was just from Russian to 'Russlish'. I am indebted, in the first instance to the late Ross Talent who commenced TEXing and 'translating' the translation prior to his death in a car accident in September 1991, and then to Sam Williams and Dr Deryn Griffiths who assisted with preliminary typing of the remainder of the manuscript. I owe special and extensive thanks to Dr Chris Pinner who carefully read all that preliminary typescript and carefully annotated it with corrections both to its TEX and to its mathematics. Incidentally, Chris Pinner's efforts make it clear that at least some of the detail provided must be

---

* I cannot resist using this phrase and irritating my French colleagues.

taken flexibly. What is presented here is entirely correct in spirit; that is, in its principal parameters. In applying it one should, as always, rework the details to the purpose at hand. That will be all the more so given the errors I will inadvertently have introduced, notwithstanding all the efforts of my minders.

I have gone to some pains to translate from the Russlish to English, but restrainedly, if only so as not to hide Sprindžuk's style and personality. That may mean the retention of some eccentric phrasing. I hope that I have not done so to such an extent as to hide important meaning. However, once or twice, I should confess, I had no idea what was intended, even after retreating to the original Russian. So it goes.

I began by saying that much of this monograph remains fresh, interesting and useful. The reader should notice the unusual emphases in the first seven chapters; I am confident that there is much yet to be usefully done along the lines there delineated. I am not aware of any other place that a reader can find a congenial entry to the ideas of the final two chapters and am certain that the present volume will spark a great deal of useful thought and fascinating work.

Alfred J. van der Poorten
ceNTRe for Number Theory Research
Macquarie University
alf@mpce.mq.edu.au

Sydney, Australia
May 1993

Afterword: In mid-1993, a volume on diophantine equations seems incomplete if it fails to allude to the surprising announcement by Andrew Wiles of his proof of the Shimura-Taniyama-Weil conjecture for semi-stable elliptic curves, and its spectacular consequence. As it happens, Fermat's Last Theorem gets barely a mention in the present volume; the one oasis is the concluding remarks of Chapter VII. Thus to bring this volume up to date in this respect it suffices just to eliminate mention of a paper of Inkeri and mine! Of course it is no longer totally out of the question that the work on elliptic curves be extended to prove the abc conjecture; that will warrant a rather more significant revision.

July 1993

Preface

The theory of diophantine equations has a long history, and like human culture as a whole, has had its ups and downs. This monograph aims to show that the last 10 to 15 years were a period of uplift, at least in the field of diophantine equations in two integral unknowns, a part of the subject which has intrigued and attracted researchers throughout its history.

Even a cursory acquaintance with the work preceding the papers of Runge [166] in 1887 and Thue [229] in 1909 will impress with the dramatic search for general laws for the behaviour of solutions of diophantine equations, and the realisation of the peculiar difficulties of attaining this aim (see, for example, [56], vol. 2). It was Runge who obtained the first general theorem on the finiteness of the number of integer points on a wide variety of algebraic curves. After nearly a century it is difficult to judge the influence of Runge's work on his contemporaries. Certainly it is evident in Hilbert's proof of his irreducibility theorem [98], which initiated research on the inverse problem of Galois theory. It is possible that Thue was stimulated by Runge's arguments to investigate the representation of numbers by irreducible binary forms, a closely related problem not covered by Runge's theorem. However, the peculiar virtue of Runge's methods – the possibility of making them effective and obtaining explicit bounds for the solutions – was lost in both cases.

Thue's work initiated a most fruitful period of development of the theory of diophantine equations in two unknowns – the golden age of ineffective methods! Two monumental results of that period are widely known: Siegel [193] proved that curves of genus greater than zero have only finitely many integer points, and Roth [165], in the problem of representation of numbers by irreducible binary forms (the Thue equation), obtained the best possible exponent estimate for the unknowns in terms of the number represented. Both results were achieved by a thorough development and enrichment of Thue's method, and on the way to these results many specific facts were obtained, special methods were worked out, and phenomena arising from these two general theorems were discovered. The monographs by Skolem [196], Lang [120], Mahler [136] and Mordell [145] give a good idea of the variety of the results obtained.

One of the above-mentioned special methods is among the most beautiful in the theory of diophantine equations: Skolem's method. Though Skolem himself and his adherents achieved much by this method, and were for a long time the leaders in questions of number representation by norm forms in three or more variables, the ascendancy was finally won by Thue's method (Schmidt's theory of representation of numbers by norm forms is a fine testament to that [185]). Nevertheless, the fundamental idea of Skolem's method, the reduction of algebraic diophantine equations to exponential equations, has shown exceptional vitality in recent episodes of the theory of diophantine equations.

In 1952 Gelfond [77] suggested that non-trivial effective estimates for linear forms in the logarithms of three or more algebraic numbers would make it possible to obtain explicit bounds for the solutions of exponential diophantine equations, in particular those to which Thue's equation reduces, thereby yielding an effective bound for the solutions of this equation. By that time the necessary estimates were known in the case of two logarithms, but the transition to three logarithms presented considerable difficulty and had not been carried through. In 1966 Baker [8] obtained such estimates for forms in logarithms of any number of algebraic numbers, and later applied them to diophantine equations. Baker's work had a stimulating effect on his close colleagues, and during the next decade the theory of diophantine equations was enriched by results of a qualitatively new type, which will occupy a considerable portion of this monograph.

This book covers all the main types of diophantine equations in two unknowns for which the solutions are to be integers or S-integers or rationals or algebraic numbers from a fixed field. Such a broad notion of solution domain makes available a wider arsenal of arithmetic facts than would be possible if only the classical case of rational integer solutions (which, of course, remains the main case here as well) were considered. In particular, by transcending the rational integer domain, we are able to analyse certain classes of diophantine equations in several unknowns (for example, representations of numbers by certain norm forms). Special attention is given to the influence of the parameters of the equation on the magnitude of its solutions, and to the construction (in principle) of best possible bounds for the solutions. Here an interesting general phenomenon is observed which formerly revealed itself in very special cases only: the regulator of some algebraic number field connected with the equation has a preeminent influence on the magnitude of the equation's solutions. (In virtue of the Siegel-Brauer formula, this amounts to preeminence of the class number.) We use this phenomenon to describe parametric construction of algebraic number fields with large class number. Further work in this direction may lead to major improvements to known bounds for solutions of diophantine equations in terms of the height of the equation, or to a proof that such an improvement is impossible (which seems more probable). Not all results concerning the value of class numbers are directly connected with the theory of linear forms in logarithms of algebraic numbers, but they were inspired by the above-mentioned relationship between class numbers and the value of solutions of diophantine equations. Chapter IX is altogether independent of the theory of logarithms.

The theory of algebraic units, the theory of ideals in algebraic number fields, and the concepts and techniques of p-adic analysis in both arithmetic and analytic form predominate in this monograph. The informed reader will notice that p-adic analysis makes some quite unexpected appearences. Many of the results can be obtained without the use of p-adic analysis, but there are some which cannot even be formulated without reference to p-adic metrics (See Chap. IX).

There is also another approach to the investigation of integer points on algebraic curves which uses parametrisation of curves and the Mordell-Weil theorem on the group of rational points on the curve. We do not touch upon this approach, because the main results obtained in this way are still ineffective. Besides, this topic is treated in a recent monograph by Lang [124].

I have often seen the admiration felt for modern diophantine analysis by older mathematicians who have worked in number theory or taken an interest in its development; for what is done now was in their youth just a pleasant dream. Younger mathematicians will take its achievements for granted, and will feel that its deficiencies should be criticised. If this monograph should stimulate them to creative work or offer clues to new discoveries, its aims will be more than fulfilled.

As this work was nearing completion, it became clear that for many readers it will make an impression much as the one tourists in Paris feel on seeing the Pompidou Centre: all the main lines, informative and logical, are extremely plain and to the fore. It is, of course, easier to construct a building or write a book in the 'good old style', but then inevitably a great deal will be hidden for the sake of a favourable external impression. Extreme frankness, whether in art or science, imposes much more on our time.

Central themes of this monograph were the subject of my lectures at the Institut Henri Poincaré (Paris, May-June 1980) by invitation of the Université de Paris VI. Namely, (1) generalisations and effective improvements to Liouville's inequality, (2) a connection between bounds for the solutions of diophantine equations and class numbers, and also the manner in which the class number varies, (3) effective versions of Hilbert's irreducibility theorem and rational points on algebraic curves. The audience's interest in and attention to these topics helped to finalise their presentation in this monograph. Michel Waldschmidt and Daniel Bertrand contributed most of all. I am obliged to Alan Baker for the exceptional stimulus which his works gave me in the late sixties, and to Andrzej Schinzel for information given to me during previous investigations of Hilbert's theorem. I am heartily grateful to all the above-mentioned persons.

Minsk   V. Sprindžuk
September 1980

Table of Contents

I. Origins 1
   I.1 Runge's Theorem 1
   I.2 Liouville's Inequality; the Theorem and Method of Thue 3
   I.3 Exponential Equations and Skolem's Method 7
   I.4 Hilbert's Seventh Problem and its Development 10

II. Algebraic Foundations 14
   II.1 Height and Size of an Algebraic Number 14
   II.2 Bounds for Units, Regulators and Class Numbers 17
   II.3 Analytic Functions in p-adic Fields 22

III. Linear Forms in the Logarithms of Algebraic Numbers 30
   III.1 Direct Bounds and Connections
      Between Bounds in Different Metrics 30
   III.2 Preliminary Facts 32
   III.3 The Auxiliary System of Linear Equations 36
   III.4 Transition to the Auxiliary Function 39
   III.5 Analytic-arithmetical Extrapolation 43
   III.6 Completion of the Proof of Lemma 1.1 46
   III.7 Connection Between Bounds
      in Different Non-archimedean Metrics 53
   III.8 Lemmas on Direct Bounds 58

IV. The Thue Equation 61
   IV.1 Existence of a Computable Bound for Solutions 61
   IV.2 Dependence on the Number Represented by the Form 64
   IV.3 Exceptional Forms 70
   IV.4 Estimates for the Solutions in terms of the Main Parameters 73
   IV.5 Norm Forms with Two Dominating Variables 75
   IV.6 Equations in Relative Fields 80

V. The Thue-Mahler Equation 85
   V.1 Solution of the Thue Equation in S-integers 85

V.2 Rational Approximation to Algebraic Numbers
in Several Metrics 92
V.3 The Greatest Prime Factor of a Binary Form 94
V.4 The Generalised Thue-Mahler Equation 98
V.5 Approximations to Algebraic
Numbers by Algebraic Numbers of Fixed Field 105

VI. Elliptic and Hyperelliptic Equations 111
VI.1 The Simplest Elliptic Equations 111
VI.2 The General Hyperelliptic Equation 116
VI.3 Linear Dependence of Three Algebraic Units 118
VI.4 Main Theorem 120
VI.5 Estimate for the Number of Solutions 125
VI.6 Linear Dependence of Three S-units 126
VI.7 Solutions in S-integers 131
VI.8 S-integer Points on Elliptic Curves 133

VII. Equations of Hyperelliptic Type 138
VII.1 Equations with Fixed Exponent 138
VII.2 Equations with indefinite exponent 144
VII.3 The Catalan Equation 148

VIII. The Class Number Value Problem 155
VIII.1 Influence of the Value of the Class Number
on the Size of the Solutions 155
VIII.2 Real Quadratic Fields 159
VIII.3 Fields of Degree 3,4 and 6 163
VIII.4 Superposition of Polynomials 167
VIII.5 The Ankeny-Brauer-Chowla Fields 170
VIII.6 A Statistical Approach 176
VIII.7 Conjectures and Perspectives 184

IX. Reducibility of Polynomials and Diophantine Equations 188
IX.1 An Irreducibility Theorem of Hilbert's Type 188
IX.2 Main argument 191
IX.3 Details and Sharpenings 197
IX.4 Theorems on Reducibility 205
IX.5 Proofs of the Reducibility Theorems 208
IX.6 Further Results and Remarks 213

References 219

Notation

The following notation, mainly standard, is frequently used.

$\mathbb{Q}$	the field of rational numbers
$\mathbb{C}$	the field of complex numbers
$\overline{\mathbb{Q}}$	the field of all algebraic numbers
$\mathbb{K}, \mathbb{L}, \ldots$	algebraic number fields of finite degree over $\mathbb{Q}$
$[\mathbb{L} : \mathbb{Q}]$	the degree of the field $\mathbb{L}$
$[\mathbb{L} : \mathbb{K}]$	the degree of $\mathbb{L}$ over $\mathbb{K}$
$\mathbb{Z}$	the ring of rational integers
$I_{\mathbb{K}}$	the ring of integers of $\mathbb{K}$
$\mathbb{K}(x, y, \ldots)$	the field of rational functions in $x, y, \ldots$ over $\mathbb{K}$
$\mathbb{K}[x, y, \ldots]$	the ring of polynomials in $x, y, \ldots$ over $\mathbb{K}$
$E_{\mathbb{K}}$	the group of units of the field $\mathbb{K}$
$D_{\mathbb{K}}$	the discriminant of $\mathbb{K}$
$R_{\mathbb{K}}$	the regulator of $\mathbb{K}$
$h_{\mathbb{K}}$	the number of ideal classes of $\mathbb{K}$
$N(\mathfrak{a})$	the absolute norm of an ideal $\mathfrak{a}$
$\mathrm{Nm}(\alpha)$	the absolute norm of an algebraic number α
$\mathrm{Nm}_{\mathbb{L}/\mathbb{K}}(\alpha)$	the absolute norm from $\mathbb{L}$ to $\mathbb{K}$ of an algebraic number α
$h(\alpha)$	the height of an algebraic number α
$\lceil \alpha \rceil$	the size of an algebraic number (the maximum modulus of the conjugates of α)
$\deg \alpha$	the degree of an algebraic number α
$\mathrm{ord}_{\mathfrak{p}} \alpha$	the exponent of the power to which a prime ideal $\mathfrak{p}$ divides α
$\lvert\ \rvert_p$	the p-adic metric, normalised so that $\lvert p \rvert_p = p^{-1}$
$\mathbb{Q}_p$	the field of p-adic numbers
$\mathbb{Z}_p$	the ring of p-adic integers
$\overline{\mathbb{Q}}_p$	the algebraic closure of $\mathbb{Q}_p$
Ω_p	the completion in $\lvert\ \rvert_p$ of an algebraic closure of $\mathbb{Q}_p$
$\lceil F \rceil$ or H_F	the height of a polynomial F
$\deg F$	the degree of a polynomial F
$\deg_x F$	the degree of a polynomial F with respect to x
$D(F)$	the discriminant of a polynomial F
$R(F, G)$	the resultant of polynomials F and G
$R_x(F, G)$	the resultant of polynomials F and G with respect to x
$c(n), c(n, \epsilon) \ldots$	positive quantities depending only on the indicated parameters
$\ln x$	the 'natural logarithm' of x, the logarithm to base e
$\lfloor a \rfloor$	the integer part of a real number a.

I. Origins

This chapter reviews the origin and development of the fundamental principles of the contemporary analysis of diophantine equations, from the perspective of the theory of diophantine approximation.

1. Runge's Theorem

Let $F(x, y)$ be an integral polynomial irreducible in $\mathbb{Q}[x, y]$. We suppose as we may without loss of generality that its degree in y is at least its degree in x and set $\deg_y F(x, y) = n \geq 2$. We consider solutions in integers x, y of the equation

$$F(x, y) = 0. \tag{1.1}$$

Although Fermat and Euler had analysed special equations of this form (for example, $x^2 - Dy^2 = 1$ with square-free D), results of a more or less general nature were for a long time elusive. In 1887 Runge devised the general approach whose essence is described below (see also [145], p.262).

Equation (1.1) determines an algebraic function $y(x)$ which takes integral values at integer solutions of the equation. Suppose there are infinitely many solutions. One can find $y(x)$ numerically for sufficiently large x by expansion in a power series about the point at infinity.

Let

$$F(x, y) = f_n(x, y) + f_{n-1}(x, y) + \ldots + f_0(x, y)$$

where $f_j(x, y)$ is a binary form of degree j. Put $x = t^{-1}$ and $y = st^{-1}$, and write

$$
\begin{aligned}
G(t, s) = t^n F(t^{-1}, st^{-1}) &= \\
&= t^n f_n(t^{-1}, st^{-1}) + t \cdot t^{n-1} f_{n-1}(t^{-1}, st^{-1}) + \ldots \\
&= g_n(s) + g_{n-1}(s)t + \cdots
\end{aligned}
$$

where $g_n(s)$, $g_{n-1}(s)$, $\ldots$ are polynomials in s. Suppose that $g_n(s) = f_n(1, s)$ has no multiple roots. The equation $G(t, s) = 0$ following from (1.1) defines n power series expansions

$$s = \alpha + \alpha_0 t + \alpha_1 t^2 + \ldots ,$$

one for each root α of the polynomial $g_n(s)$, the numbers α_i being in the field $\mathbb{K} = \mathbb{Q}(\alpha)$. Consequently we have n expansions

$$y = \alpha x + \alpha_0 + \alpha_1 x^{-1} + \dots \qquad (1.2)$$

corresponding to the roots α of $g_n(s)$.

Let $\phi(x)$ denote one of the n power series (1.2). The principal idea of Runge consists in a choice of integral polynomials $A_i(x)$, $(0 \le i \le n - 1)$ of degree not exceeding some bound h, such that a power series expansion of the function

$$\Phi(x) = \sum_{i=0}^{n-1} A_i(x)\phi^i(x)$$

about the point at infinity has only negative powers of x:

$$\Phi(x) = \beta_1 x^{-1} + \beta_2 x^{-2} + \dots \qquad (1.3)$$

When can this be done? Writing $\Phi(x)$ in the form

$$\Phi(x) = \sum_{j=-h-n+1}^{\infty} \beta_j x^{-j},$$

observe that each β_j is a linear form in the unknown integer coefficients of the polynomials $A_0(x), \dots, A_{n-1}(x)$. We will have (1.3) when

$$\beta_j = 0 \qquad (j = -h - n + 1, -h - n, \dots, 0). \qquad (1.4)$$

Each β_j lies in $\mathbb{K}$, and may therefore be represented by its coordinates with respect to a basis of $\mathbb{K}$ as a $\mathbb{Q}$-vector space. Then the system of equations (1.4) becomes a system of $d(h+n)$ linear homogeneous equations with rational coefficients, where $d = \deg \mathbb{K}$, in the $n(h + 1)$ unknown integer coefficients of the polynomials $A_0(x), \dots, A_{n-1}(x)$. Provided $n(h + 1) > d(h + n)$ we can guarantee the existence of a non-zero set of integers satisfying the system. If $d = n$ this cannot be done, but for $d < n$ it suffices to take $h = n^2 - n + 1$. Thus we can find a non-zero set of integral polynomials of degree not exceeding $n^2 - n + 1$ for which (1.3) will hold, provided that the polynomial $f_n(1, s)$ is reducible in $\mathbb{Q}[s]$.

We now substitute in (1.3) the integer values of x for which there exists an integer y satisfying (1.1) and (1.2). For such x, y, with $|x|$ sufficiently large, we obtain

$$\sum_{i=0}^{n-1} A_i(x)y^i = 0, \qquad (1.5)$$

since it follows from (1.3) that $|\Phi(x)| < 1$ for sufficiently large $|x|$, and so $\Phi(x)$, being a rational integer, is zero. We have obtained an equation (1.5) which is independent of (1.1). The polynomial $F(x, y)$ is irreducible in $\mathbb{Q}[x, y]$ and its degree with respect to y is n, while the left hand side of (1.5) has degree in y not exceeding $n - 1$. Writing the resultant of these polynomials with respect to y, we obtain an equation only for x, which completes the proof of the finiteness of the number of solutions to (1.1) under the assumption that the polynomial $f_n(1, y)$ is reducible.

Clearly the above argument is effective, and may be used in concrete cases to determine all solutions of (1.1). Its further development yields very strong bounds on the solutions (such as a power of the height of $F(x, y)$).

Certainly the requirement that $f_n(1, y)$ be reducible is a serious restriction. Even the case $F(x, y) = f_n(x, y) + f_0(x, y)$, with f_n irreducible, is of interest, being the problem of representation of numbers by irreducible binary forms. For $n = 2$ the finiteness or otherwise of the number of solutions is easily resolved, but even for $n = 3$ significant difficulties arise. The general case was solved by Thue, using a method which has influenced the development of the whole of this branch of number theory.

2. Liouville's Inequality; the Theorem and Method of Thue

In 1844 Liouville [128] observed that algebraic numbers do not admit 'very strong' approximation by rational numbers, and was thereby able to give the first construction of transcendental numbers. Since then the approximation estimate he obtained has been so frequently and widely applied that it has acquired a proper name: Liouville's Inequality.

Let α be a real algebraic number of degree $n \geq 2$ and let p, q be integers. Then Liouville's inequality is

$$|\alpha - p/q| > c_1 q^{-n}, \tag{2.1}$$

where $c_1 = c_1(\alpha) > 0$ is a value depending explicitly on α. The proof is immediate from the upper bound for the absolute value of $\mathrm{Nm}(\alpha q - p)$ and the observation that it is a non-zero rational number with denominator dividing a^n, where a is an integer such that $a\alpha$ is an algebraic integer. For $n = 2$ it is impossible to improve on (2.1) by replacing c_1 by some positive function $\lambda(q)$ increasing monotonically to infinity, for it is known from the theory of continued fractions that, for any quadratic irrational α, the reverse of (2.1) has infinitely many solutions in integers p, q when c_1 is replaced by $\sqrt{5}$ (see [40], Ch. II). For $n \geq 3$, however, a sharpening of the (2.1) of the type

$$|\alpha - p/q| > \lambda(q)/q^n, \qquad \lambda(q) \uparrow \infty \tag{2.2}$$

is of great interest for the study of diophantine equations.

Indeed, let $f(x, y)$ be an integral irreducible binary form of degree $n \geq 3$, and suppose that $A \neq 0$ is an integer. If the inequality (2.1) admits a sharpening of the form (2.2) for some $\lambda(q)$, then the diophantine equation

$$f(x, y) = A \tag{2.3}$$

has only finitely many solutions.

If $f(x, 1)$ is a polynomial without real roots, it is obvious that (2.3) has only a finite number of solutions. Suppose instead that α is a real root of $f(x, 1)$ and $\alpha^{(i)}, i = 1, 2, \ldots n$ its conjugates. It follows from (2.3) and $y \neq 0$ that

$$\prod_{i=1}^{n} |\alpha^{(i)} - x/y| = A/(|a||y|^n) \tag{2.4}$$

where a is the leading coefficient of the polynomial $f(x, 1)$. Assuming the equation (2.3) has integer solutions with arbitrarily large $|y|$ we see that the product on the left of (2.4) takes arbitrarily small values for solutions x, y of (2.3). As all the $\alpha^{(i)}$ are different, x/y must be correspondingly close to one of the real numbers $\alpha^{(i)}$, say α.

Thus we obtain

$$|\alpha - x/y| < c_2/|y|^n$$

where c_2 depends only on a, n, and $\prod_{i \neq j} |\alpha^{(i)} - \alpha^{(j)}|^{-1} A$ (see Ch. IV, §1). Comparison of this inequality with (2.2) shows that $|y|$ cannot be arbitrarily large, and so the number of solutions of (2.3) is finite.

It is not difficult to see that the arguments are effective, and that an explicit bound can be constructed for solutions of (2.3) once an effective inequality (2.2) is known. The sharpening of the Liouville inequality (2.1), however, especially in effective form, proved to be very difficult.

In 1909 Thue published a proof [229] that

$$|\alpha - p/q| < q^{-\frac{n}{2}-1-\varepsilon} \tag{2.5}$$

has only finitely many solutions in integers $p, q > 0$ for all algebraic numbers α of degree $n \geq 3$ and any $\varepsilon > 0$. In essence, he obtained the inequality (2.2) with $\lambda(q)$ of the form $c_3 q^{\frac{1}{2}n-1-\varepsilon}$, where $c_3 > 0$ depends on α and ε. But Thue's arguments do not allow one to find a bound for the greatest q satisfying (2.5), so it is impossible to exhibit the dependence of c_3 on α and ε, and so the bound for the number of solutions to (2.3) cannot be given in explicit form either: it is *ineffective*.

We shall show that the inequality (2.5) has just finitely many solutions following the arguments of Thue himself (see also [51]). Obviously, one may suppose that $(p, q) = 1$ in (2.5) and that α is an algebraic integer. Suppose that $h > 0$ is an integer, δ satisfies $0 < \delta < 1$, and

$$m = \lfloor \tfrac{1}{2}(n - 2)(1 + \delta)h \rfloor . \tag{2.6}$$

For each h we will construct auxiliary polynomials $P(x)$, $Q(x)$ of minimal degree and height such that $P(x) - \alpha Q(x)$ is divisible by $(x - \alpha)^h$. In more detail, put

$$P(x) - \alpha Q(x) = (x - \alpha)^h \left\{ R_0 + R_1(x)\alpha + \ldots + R_{n-1}(x)\alpha^{n-1} \right\} \tag{2.7}$$

where the integral polynomials $R_0(x), \ldots, R_{n-1}(x)$ are chosen so that their degrees do not exceed m and not all of them are zero. Then we have $n(m + 1)$

unknown integer coefficients of these polynomials which are to satisfy the $(n-2)(h+m+1)$ linear homogeneous equations implied by the representation of the right-hand side of (2.7) in the form

$$S_0(x) + \alpha S_1(x) + \ldots + \alpha^{n-1} S_{n-1}(x)$$

and by the vanishing of all coefficients of the polynomials $S_2(x), \ldots, S_{n-1}(x)$. Condition (2.6) implies

$$\frac{n(m+1)}{(n-2)(h+m+1)} > \frac{n(n-2)(1+\delta)}{n(n-2)(1+\delta) - 4\delta + 2(n-2)/h} > 1$$

if $h > (n/2-1)\delta^{-1}$, and so the number of unknowns is greater than the number of equations. It is easy to obtain a bound of the type c_4^h for coefficients of these equations, where c_4 depends only on α. Then for the unknown coefficients of the polynomials $R_0(x), \ldots, R_{n-1}(x)$ one obtains an estimate of the same form c_5^h with $c_5 = c_5(\alpha, \delta)$. This shows that the heights of the polynomials $P(x)$, $Q(x)$ satisfy

$$\max(\overline{|P(x)|}, \overline{|Q(x)|}) < c_5^h. \tag{2.8}$$

Equation (2.7) shows that the polynomial

$$W(x) = P(x)Q'(x) - P'(x)Q(x) \tag{2.9}$$

cannot be identically zero, since otherwise $P(x)$ and $Q(x)$ differ only by a numerical factor, and are divisible by $(x-\alpha)^h$ and hence by the h-th power of the minimal polynomial of α; so their degrees are not less than nh. But that is impossible, as in fact their degrees do not exceed

$$h + m \le \left(\frac{n}{2} + \left(\frac{n}{2} - 1\right)\delta\right)h < nh.$$

Now (2.8) and (2.9) imply

$$\overline{|W(x)|} < c_6^h. \tag{2.10}$$

Let p, q be coprime integers, $q > 0$, and suppose that $W(x)$ has a zero of order r at the point $x = p/q$. Then, by Gauss's lemma, $(qx - p)^r$ divides $W(x)$. Comparison of the leading coefficients of these two polynomials together with (2.10) shows that

$$q^r < c_6^h. \tag{2.11}$$

Since $W^{(r)}(p/q) \ne 0$, there exist indices i, j such that $0 \le i < j \le r + 1$ and

$$P^{(i)}(p/q)Q^{(j)}(p/q) \ne P^{(j)}(p/q)Q^{(i)}(p/q), \tag{2.12}$$

as is easily established by considering $W(x)$ as a determinant

$$W(x) = \begin{vmatrix} P(x) & Q(x) \\ P'(x) & Q'(x) \end{vmatrix}$$

and differentiating r times.

Assume p, q and p_1, q_1 satisfy (2.5), with $(p_1, q_1) = 1$ and $q < q_1$. It follows from (2.12) that we have either

$$q_1 P^{(j)}(p/q) - p_1 Q^{(j)}(p/q) \neq 0 \qquad (2.13)$$

or the analogous inequality with j replaced by i. Put

$$U = \frac{q^{m+h-j}}{j!} P^{(j)}(p/q), \qquad V = \frac{q^{m+h-j}}{j!} Q^{(j)}(p/q).$$

Then U, V are integers, not both zero, with

$$|V| < c_7^h q^{m+h-j}$$

from (2.8) and (2.11). Differentiating (2.7) j times, we obtain

$$|U - \alpha V| < c_8^h q^{m+h-j} |\alpha - p/q|^{h-j}.$$

Now we find from (2.13) and the latter estimates that

$$1 \leq |U q_1 - V p_1| \leq q_1 |U - \alpha V| + |V| |p_1 - \alpha q_1| <$$
$$< c_9^h (q_1 q^{m+h-j-\lambda(h-j)} + q^{m+h-j} q_1^{-\lambda+1}),$$

where $\lambda = \frac{1}{2}n + 1 + \varepsilon$. Recalling that $0 \leq j \leq r+1$, and that (2.11) holds, we find

$$1 < c_{10}^h (q_1 q^{m+h-\lambda h+\lambda-1} + q^{m+h} q_1^{-\lambda+1} =$$
$$= c_{10}^h (q_1 q^{m-(\lambda-1)(h-1)} + q^{(m+h)} q_1^{-\lambda+1}). \qquad (2.14)$$

So far we have not nominated a number h. Take $h = \lfloor \ln q_1 / \ln q \rfloor + 2$. Then from (2.14)

$$1 < 2 c_{10}^2 q_1^{-\varepsilon + (\frac{1}{2}n-1)\delta} q_1^{\ln c_{10} / \ln q} q^{2n-2}.$$

Assuming $\delta = \varepsilon / 2(n-2)$, observe that the existence of any solution of (2.5) with $(p, q) = 1$ and $q > c_{10}^{4/\varepsilon}$ implies $q_1 < (2 c_{10} q^{n-1})^{4/\varepsilon}$ for every other solution p_1, q_1 with $(p_1, q_1) = 1$. Hence there are only finitely many solutions, which is the assertion.

The source of the ineffectiveness of the above arguments is clear: having assumed that a solution of the inequality (2.5) with sufficiently large q exists, we bounded all other large solutions by a value depending on q. The conclusions are wholly uninformative concerning existence and magnitude of possible solutions of the inequality. At the same time it should not be forgotten that the conclusions on the number of solutions are quite satisfactory. In particular, one can find a computable value $c_{11} = c_{11}(\alpha, \varepsilon)$ such that (2.5) will have at most one solution p, q with $(p, q) = 1$ and $q > c_{11}$.

Thue's arguments were extended significantly by Siegel [191], who introduced a different construction of the auxiliary polynomials and proved that (2.5) has only finitely many solutions when $\frac{1}{2}n + 1$ in the exponent is replaced with $2\sqrt{n}$. Gelfond [77] and Dyson [59] independently improved this exponent to $\sqrt{2n}$. Finally, in 1955 Roth [155] proved that

$$|\alpha - p/q| < q^{-2-\varepsilon}$$

has only finitely many solutions for any $\varepsilon > 0$.

Siegel [191] also extended Thue's method to the analysis of approximations of algebraic numbers by algebraic numbers from a fixed field. He thereby proved the finiteness theorem for the number of integer points on algebraic curves of genus greater than zero. Mahler [129] brought Siegel's arguments to bear on p-adic approximation, allowing him to prove a finiteness theorem for solutions of Thue's equation in rational numbers having denominators with only fixed prime factors, and to similarly extend Siegel's result for integer points on curves [132].

A recent extension of Thue's method by Schmidt has enabled him to solve the problem of simultaneous rational approximation of any number of algebraic numbers ([179], [180]; see also [185]). From there he was able to settle questions on the representation of numbers by norm forms, a considerable generalisation of (2.3).

These offshoots of Thue's original idea all inherit its shortcoming: like Thue's own theorems, they remain ineffective.

3. Exponential Equations and Skolem's Method

Skolem [196] devised an essentially different method to prove the finiteness of the number of solutions of (2.3) in the case of polynomials $f(x, 1)$ having at least one complex root. This method is based on the theory of analytic functions in p-adic fields and the theory of local analytic varieties.

Let p be a prime number. For every integer $a \neq 0$ we define its p-adic norm $|a|_p$ by

$$|a|_p = p^{-\operatorname{ord}_p a},$$

where $\operatorname{ord}_p a$ is the exponent to which p divides a (for $a = 0$ we put $|0|_p = 0$). For every rational $r = a/b$ let $|r|_p = |a|_p/|b|_p$. The function $|r|_p$ so defined for all rational numbers is called a p-adic metric on the field $\mathbb{Q}$ of rationals, and has useful properties: for any two numbers $r_1, r_2 \in \mathbb{Q}$,

$$|r_1 r_2|_p = |r_1|_p |r_2|_p, \qquad |r_1 + r_2|_p \leq \max(|r_1|_p, |r_2|_p). \tag{3.1}$$

Just as the real numbers can be constructed as Cauchy's fundamental sequences of rationals in the ordinary metric induced by the absolute value, the field of p-adic numbers, $\mathbb{Q}_p$, is constructed with respect to the metric $| \ |_p$ on $\mathbb{Q}$. Thus every p-adic number ω is a limit of rationals r_n with respect to $| \ |_p$, the norm $|\omega|_p$ being defined as the limit of the norms $|r_n|_p$ $(n \to \infty)$. The relations (3.1) remain valid for elements of $\mathbb{Q}_p$. The metric $| \ |_p$ is uniquely continued onto finite algebraic extensions of $\mathbb{Q}_p$, and hence onto the algebraic closure $\overline{\mathbb{Q}}$ of $\mathbb{Q}_p$. Constructing the completion of $\overline{\mathbb{Q}}$, once again by fundamental sequences with respect to $|\ |_p$, we obtain a complete, algebraically closed field Ω_p, on which $|\ |_p$ is defined and satisfies (3.1).

The theory of analytic functions on Ω_p is constructed like that for analytic functions of complex variables (see Chap. II, §3 below). In particular, the power series

$$\exp x = 1 + x + \frac{x^2}{2!} + \ldots + \frac{x^n}{n!} + \ldots , \qquad x \in \Omega_p$$

defines the exponential function in the p-adic disc $|x|_p < p^{-1/(p-1)}$, and the power series

$$\log(1 + x) = x - \frac{x^2}{2} + \frac{x^3}{3} - \ldots + (-1)^{n-1}\frac{x^n}{n} + \ldots , \qquad x \in \Omega_p$$

defines the logarithmic function in the disc $|x|_p < 1$. For $z, t \in \Omega_p$, the function $z^t = \exp(t \log z)$ is defined provided $|z - 1|_p < p^{-1/(p-1)}$ and $|t|_p < 1$.

The principle behind Skolem's idea of applying the theory of analytic functions in Ω_p to diophantine equations becomes apparent from his proof of the finiteness of the number of solutions of the diophantine equation

$$\lambda_1 \alpha_1^x + \lambda_2 \alpha_2^x + \ldots + \lambda_n \alpha_n^x = 0 \tag{3.2}$$

in which $\lambda_1, \ldots, \lambda_n$, $\alpha_1, \ldots, \alpha_n$ are algebraic numbers, none of the quotients α_i/α_j ($i \neq j$) is a root of unity, and x is an unknown integer.

Set $\mathbb{K} = \mathbb{Q}(\lambda_1, \ldots, \lambda_n, \alpha_1, \ldots, \alpha_n)$, let p be an odd rational prime relatively prime to each α_i, $\mathfrak{p}$ a prime ideal of $I_{\mathbb{K}}$ dividing p, $|\ |_\mathfrak{p}$ the metric on $\mathbb{K}$ defined by this ideal, and $\mathbb{K}_\mathfrak{p}$ a completion of $\mathbb{K}$ with respect to the metric $|\ |_\mathfrak{p}$. Then $\mathbb{K}_\mathfrak{p}$ contains $\mathbb{Q}_p$ and has finite degree over it, hence $\mathbb{K}_\mathfrak{p}$ may be isomorphically embedded into Ω_p and the metric $|\ |_\mathfrak{p}$ coincides with the metric $|\ |_p$ induced by this embedding.

Suppose $\mathfrak{p}^e \| p$, and set $N = N(\mathfrak{p})^e - N(\mathfrak{p})^{e-1}$, where $N(\mathfrak{p})$ is the norm of the ideal $\mathfrak{p}$. Then for any $\alpha \in \mathbb{K}$ with $(\alpha, \mathfrak{p}) = 1$ we have $\alpha^N \equiv 1 \pmod{\mathfrak{p}^e}$, hence $|\alpha^N - 1|_p < 1/p$. Therefore both the p-adic logarithm $\log(\alpha^N)$ and, for $t \in \mathbb{Q}_p$ with $|t|_p \leq 1$, the p-adic function of t

$$\alpha^{Nt} = \exp(t \log \alpha^N)$$

are defined.

Suppose that infinitely many integers x satisfy equation (3.2). Every such x may be represented in the form $x = Nt + x_0$, $0 \leq x_0 < N$, $t, x_0 \in \mathbb{Z}$, and the same x_0 will occur for infinitely many x. Hence the equation

$$\mu_1 \beta_1^t + \mu_2 \beta_2^t + \ldots + \mu_n \beta_n^t = 0, \tag{3.3}$$

in which $\mu_i = \lambda_i \alpha_i^{x_0}$ and $\beta_i = \alpha_i^N$, has infinitely many solutions in integers t. The left hand side of (3.3) may be regarded as an analytic function of the p-adic argument $t \in \Omega_p$ with $|t|_p \leq 1$, having infinitely many zeros. As for analytic functions of a complex variable, a p-adic analytic function cannot have infinitely many zeros in a bounded disc unless it is identically zero. But (3.3) cannot hold identically on the unit disc $|t|_p \leq 1$, as can be seen by reducing it to

$$\mu_1 + \mu_2 \left(\frac{\beta_2}{\beta_1}\right)^t + \ldots + \mu_n \left(\frac{\beta_n}{\beta_1}\right)^t = 0,$$

differentiating with respect to t, and employing induction on n, taking into account that the p-adic logarithms $\log(\beta_i/\beta_j)$ $(i \neq j)$ are all nonzero, since none of the β_i/β_j $(i \neq j)$ is a root of unity.

By similar arguments Skolem and his adherents proved the finiteness of the number of solutions of systems of exponential diophantine equations in several unknowns. In particular Thue's equation (2.3) reduces to this form. Indeed, for any solution x, y of this equation, the principal ideal $(x - \alpha y)$ belongs to a finite set of ideals which depends on the number A. (Here α is a root of the polynomial $f(x, 1)$ and the binary form $f(x, y)$ is fixed). Hence $x - \alpha y = \beta \varepsilon$, where β lies in the field $\mathbb{L} = \mathbb{Q}(\alpha)$ and belongs to a fixed finite set of numbers, and ε is a unit of $\mathbb{L}$. Together with $x - \alpha y = \beta \varepsilon$ we have conjugate equalities

$$x - \alpha^{(i)}y = \beta^{(i)}\varepsilon^{(i)} \qquad (i = 1, 2, \ldots, n).$$

Eliminating x and y from these equalities, we obtain for any indices i, j, k $(1 \leq i < j < k \leq n)$

$$\beta^{(i)}\varepsilon^{(i)}(\alpha^{(j)} - \alpha^{(k)}) + \beta^{(j)}\varepsilon^{(j)}(\alpha^{(k)} - \alpha^{(i)}) + \beta^{(k)}\varepsilon^{(k)}(\alpha^{(i)} - \alpha^{(j)}) = 0. \quad (3.4)$$

By Dirichlet's unit theorem

$$\varepsilon = \zeta \varepsilon_1^{x_1} \ldots \varepsilon_r^{x_r}, \tag{3.5}$$

where $\varepsilon_1, \ldots, \varepsilon_r$ are the fundamental units of $\mathbb{L}$, ζ is a root of unity, and $x_1, \ldots, x_r$ are integers. If we substitute the expressions for the conjugates of ε from (3.5) in (3.4), we obtain a system of exponential equations with respect to $x_1, \ldots x_r$. When the α_i are not all real, one can show that (3.4) has only finitely many solutions in integers $x_1, \ldots x_r$ by going over to the p-adic exponential function and analysing the corresponding analytic variety [33]. In particular, if $n = 3$ then $r = 1$, and then the system (3.4) is reduced to a single equation of the form (3.2), and the finite number of solutions of (2.3) is guaranteed by the arguments above.

The finiteness of the solution set for norm form equations in three unknowns has also been proved by this method [196]. As all the constructions are explicit, it is possible not only to estimate the number of solutions of the equations that arise, but also to compute all the p-adic points which "cover" the set of integer solutions. However, there is no algorithm to determine in a finite number of operations, bounded by the parameters of the equation, whether the existence of a p-adic solution guarantees the existence of a rational integer solution. So in this case also we have no effective method for determining the solutions of the equation.

An interesting feature of Skolem's method is the transition from algebraic diophantine equations to exponential equations, with an accompanying transition from algebraic functions to transcendental ones (for example, on the left of (3.2)). In this very way effective methods for the analysis of diophantine equations were eventually devised, and the influence of parameters of the equations on the magnitude of solutions was evaluated. But there were considerable advances in transcendental number theory preceding the achievement of these results.

4. Hilbert's Seventh Problem and its Development

In his famous lecture at the International Congress of Mathematicians in 1900, Hilbert had recognised the great importance of investigation of the arithmetic nature of the values of transcendental functions at algebraic points. He drew special attention to the conjecture on transcendence of numbers α^β for algebraic $\alpha \neq 0, 1$ and irrational algebraic β. He subsequently propagandised this problem in his lectures, and considered it so difficult that its solution could be expected only in the remote future.

In 1934 A. O. Gelfond [76] and Schneider [186] independently gave a complete solution of the problem, by proving the transcendence of any irrational quotient of logarithms of algebraic numbers. Gelfond later extended his arguments to prove the inequality

$$|\beta \ln \alpha_1 - \ln \alpha_2| > e^{-(\ln h(\beta))^{5+\epsilon}}, \qquad (4.1)$$

where α_1, α_2 are algebraic numbers other than 0 or 1, $\ln \alpha_1 / \ln \alpha_2$ is an irrational number, β is an algebraic number, and $h(\beta)$ is the height of β and is taken to exceed some computable value depending on α_1, α_2, the degree of β, and $\epsilon > 0$. Later still he improved the exponent $5 + \epsilon$ to $2 + \epsilon$ [78].

Let us examine the main ideas which Gelfond applied towards these goals. He begins by constructing an auxiliary transcendental function of a complex variable z:

$$f(z) = \sum_{\lambda_1=0}^{L} \sum_{\lambda_2=0}^{L} p(\lambda_1, \lambda_2)(\alpha_1^{\lambda_1} \alpha_2^{\lambda_2})^z \qquad (4.2)$$

with zero derivatives $f^{(s)}(\ell)$ for all integers ℓ, s in the range $0 \leq \ell \leq \ell_0$, $0 \leq s \leq s_0$, where ℓ_0, s_0 are arbitrary natural numbers but L must be consistent with ℓ_0 and s_0. Since

$$(\ln \alpha_1)^{-s} f^{(s)}(\ell) = \sum_{\lambda_1=0}^{L} \sum_{\lambda_2=0}^{L} p(\lambda_1, \lambda_2)(\lambda_1 + \eta \lambda_2)^s (\alpha_1^{\lambda_1} \alpha_2^{\lambda_2})^\ell \qquad (4.3)$$

is an algebraic number when $\eta = \ln \alpha_1 / \ln \alpha_2$ is algebraic, we can assure the equalities $f^{(s)} = 0$ by choosing integers $p(\lambda_1, \lambda_2)$ not too large and not all zero, provided that the number of resulting linear constraints on the $p(\lambda_1, \lambda_2)$

is less than the number of unknowns, that is when $(s_0 + 1)(\ell_0 + 1) < (L+1)^2$.
The large number of zeros of the function $f(z)$ and the integral representation

$$f^{(s)}(\ell) = \frac{s!}{(2\pi i)^2} \int_{|z|=R_1} \frac{dz}{(z-\ell)^{s+1}} \int_{|\xi|=R_2} \prod_{r=0}^{s_0} \left(\frac{z-r}{\xi-r}\right)^{\ell_0} \frac{f(\xi)}{\xi-z} d\xi$$

with suitably chosen values R_1, R_2 $(R_1 \geq 2\ell_0, R_2 \geq 2R_1 + 3s_0)$ show that
the numbers $f^{(s)}(\ell)$ are very small in absolute value for values of s, ℓ in an
expanded range $0 \leq \ell \leq \ell_1, 0 \leq s \leq s_1$ $(\ell_1 > \ell_0, s_1 > s_0)$. Following from
(4.3), algebraicity of the numbers $(\ln \alpha_1)^{-s} f^{(s)}(\ell)$ now implies

$$f^{(s)}(\ell) = 0 \qquad (0 \leq \ell \leq \ell_1, 0 \leq s \leq s_1).$$

The argument may be repeated anew and the range of ℓ, s extended once more,
and so on. Gelfond called the procedure *analytic-arithmetical extrapolation*,
and it is the primary mechanism of the method ([77], p.131).

To disprove the algebraicity of η it suffices to show

$$f^{(s)}(0) = 0 \qquad (0 \leq s \leq (L+1)^2),$$

giving a system of linear homogeneous equations in $p(\lambda_1, \lambda_2)$ with, by the
irrationality of η, nonzero Vandermonde determinant. Consequently, all the
numbers $p(\lambda_1, \lambda_2)$ are zero, contradicting the stipulation that they be not all
zero. This proves the transcendence of η.

It is easily seen that the arguments can be improved to yield a lower bound
for the difference between η and any algebraic number β, as a function of the
height of β. The assumption that $|\eta - \beta|$ is sufficiently small (violating (4.1),
say) implies that $(\ln \alpha_1)^{-s} f^{(s)}(\ell)$ is close to the algebraic number

$$\Phi_s(\ell) = \sum_{\lambda_1=0}^{L} \sum_{\lambda_2=0}^{L} p(\lambda_1, \lambda_2)(\lambda_1 + \beta\lambda_2)^s (\alpha_1^{\lambda_1} \alpha_2^{\lambda_2})^\ell.$$

Now one applies analytic-arithmetical extrapolation to $(\ln \alpha_1)^{-s} f^{(s)}(\ell)$ and
$\Phi_s(\ell)$ simultaneously, bounding the values $(\ln \alpha_1)^{-s} f^{(s)}(\ell)$ from above by analytic considerations, and bounding the $\Phi_s(\ell)$ from below by arithmetic considerations. These bounds are only compatible when $\Phi_s(\ell) = 0$ for ℓ, s in
certain definite ranges, of which one or the other is extended at each step of
the procedure. In this way one can obtain, for example, a system of equations

$$\Phi_0(\ell) = 0 \qquad (0 \leq \ell \leq (L+1)^2),$$

whence, given multiplicative independence of α_1 and α_2, one finds that all the
numbers $p(\lambda_1, \lambda_2)$ are zero. Since they were chosen not all zero, the contradiction follows.

Gelfond [77] obtained by different means an estimate of the form (4.1) with
exponent $2+\varepsilon$ instead of $5+\varepsilon$, under the assumption that $\ln \alpha_1$ and $\ln \alpha_2$ are
linearly independent over $\mathbb{Q}$, that is, that their quotient is irrational. Given
the proven transcendence of $\eta = \ln \alpha_1 / \ln \alpha_2$ and the fact that $(\ln \alpha_1)^{-s} f^{(s)}(\ell)$

is a polynomial in η with algebraic coefficients, it is possible to construct many different integral polynomials taking small values at the point η, but always with certain inequalities relating their degrees and heights so as to exclude the possibility of their taking very small values at η.

Gelfond considered the extension of inequalities of the type (4.1) to handle any number of logarithms of algebraic numbers to be a matter of exceptional importance, and made attempts on the problem in the case of three logarithms ([78], p.155). He mentioned as possible applications norm form and exponential diophantine equations. By way of example, he considered the equation of Thue

$$\text{Nm}(\alpha x + \beta y) = 1$$

where α and β are integers of a cubic field with negative discriminant and

$$\frac{\alpha^{(3)}\beta^{(2)} - \alpha^{(2)}\beta^{(3)}}{\alpha^{(1)}\beta^{(2)} - \alpha^{(2)}\beta^{(1)}} = \zeta, \tag{4.4}$$

ζ being a root of unity, and showed that an inequality like (4.1) gives an effective bound for the solutions x, y ([77], Chap.III, §5).

In 1966 Baker [8] generalised (4.1) to any number of logarithms of algebraic numbers: assuming that the algebraic numbers $\alpha_1, \ldots \alpha_n$ are multiplicatively independent, he proved

$$|\beta_1 \ln \alpha_1 + \ldots + \beta_{n-1} \ln \alpha_{n-1} - \ln \alpha_n| > e^{-(\ln H)^{n+1+\epsilon}}, \tag{4.5}$$

where the β_j are arbitrary algebraic numbers of heights not exceeding H, and $H > H_0$, H_0 being effectively determined by $\alpha_1, \ldots, \alpha_n$, together with the bound for the degrees of the β_j and $\epsilon > 0$. His proof works by introducing a new interpolation method into Gelfond's scheme which enables the analytic-arithmetical extrapolation to be carried out in the case of an arbitrary number of logarithms of algebraic numbers.

In place of the auxiliary function (4.2) Baker uses a function in several complex variables:

$$\Phi(z_1, \ldots, z_{n-1}) = \sum_\lambda p(\lambda) \alpha_1^{(\lambda_1 + \lambda_n \beta_1)z_1} \ldots \alpha_{n-1}^{(\lambda_{n-1} + \lambda_n \beta_{n-1})z_{n-1}},$$

where the summation is over all integer vectors $\lambda = (\lambda_1, \ldots, \lambda_n)$ satisfying $0 \leq \lambda_i \leq L$ $(i = 1, \ldots, n)$, and L is an integer parameter chosen in accordance with H. Even for $n = 2$ this function, being independent of α_2, differs from (4.2). Numbers $p(\lambda)$ are chosen as non-trivial integer solutions of the system of linear equations with algebraic coefficients

$$\Phi_{s_1, \ldots, s_{n-1}}(\ell, \ldots, \ell) =$$
$$= \sum_\lambda p(\lambda)(\lambda_1 + \lambda_n \beta_1)^{s_1} \ldots (\lambda_{n-1} + \lambda_n \beta_{n-1})^{s_{n-1}}(\alpha_1^{\lambda_1} \ldots \alpha_n^{\lambda_n})^\ell = 0$$

$$(0 \leq \ell \leq \ell_0, \ s_1 + \ldots s_{n-1} = s \leq S_0), \tag{4.6}$$

where $(\ell_0 + 1)(S_0 + 1)^{n-1}$ is somewhat less than $(L + 1)^n$ so that the number of equations is less than then number of unknowns and a suitable estimate for $|p(\lambda)|$ may be obtained. By assuming that the opposite to (4.5) holds, and by choosing L, ℓ_0, S_0 judiciously, the system (4.6) implies that the partial derivatives

$$\frac{\partial^s}{\partial z_1^{s_1} \dots \partial z_{n-1}^{s_{n-1}}} \Phi(\ell, \dots, \ell) \qquad (0 \le \ell \le \ell_0, \ s_1 + \dots + s_{n-1} = s \le S_0)$$

are small. Application of the Hermite interpolation formula allows this smallness to be demonstrated for a wider range of ℓ under some restriction on the order of the derivatives: $0 \le \ell \le \ell_1, 0 \le s \le S_1$ $(\ell_1 > \ell_0, S_1 < S_0)$. It follows, given the reverse of (4.5), that the corresponding algebraic numbers (4.6) are so small that they must vanish, and this gives the system (4.6) a new range: $0 \le \ell \le \ell_1, 0 \le s \le S_1$. The procedure is iterated, and eventually yields the system

$$\Phi_{0,\dots,0}(\ell, \dots \ell) = \sum_{\lambda} p(\lambda)(\alpha_1^{\lambda_1} \dots \alpha_n^{\lambda_n})^\ell = 0 \qquad (0 \le \ell \le (L + 1)^n).$$

Since the numbers $\alpha_1, \dots, \alpha_n$ are assumed multiplicatively independent, one finds that all the numbers $p(\lambda)$ vanish, contrary to their choice.

Baker and those who followed him subsequently introduced new analytic and arithmetic tools for the investigation of linear forms in logarithms of algebraic numbers. Today the multitude of results and their applications to various problems in number theory constitute a vast and ever-expanding field (see [25], [222], where a survey of results is given). Some of the most recent achievements of this field underlie many of the results of the present monograph (see Chap. III). Baker's work has ushered in a new era in the theory of diophantine equations in two unknowns.

II. Algebraic Foundations

This chapter covers some general lemmas on algebraic numbers, algebraic number fields and analytic functions in p-adic fields. More specialised lemmas will be adduced as needed in subsequent chapters.

1. Height and Size of an Algebraic Number

Let α be an algebraic number. There is a polynomial of minimal degree, with integer coefficients having no common divisor, and with positive leading coefficient, which has α as a root. This polynomial is called the *minimal polynomial* of α, and the maximum modulus of its coefficients is called the *height* of α and denoted $h(\alpha)$. Many estimates in the theory of diophantine approximation with algebraic numbers are connected with behaviour of the value $h(\alpha)$ as a function of α.

Let $\mathbb{K}$ be an algebraic number field of finite degree over $\mathbb{Q}$. On the elements x of the field we define the function

$$\lceil x \rceil = \max |x^{(i)}| \qquad (i = 1, 2, \ldots, d),$$

where the $x^{(i)}$ are the conjugates of x, and $d = [\mathbb{K} : \mathbb{Q}]$. Obviously, the roots of the minimal polynomial of x determine $\lceil x \rceil$, so $\lceil x \rceil$ is independent of the field in which it is defined. We shall call this value the *size* of the algebraic number x.

The size possesses two important properties. First, for any $\alpha, \lambda \in \mathbb{K}$,

$$\lceil \alpha + \lambda \rceil \leq \lceil \alpha \rceil + \lceil \lambda \rceil, \qquad \lceil \alpha\lambda \rceil \leq \lceil \alpha \rceil \lceil \lambda \rceil. \tag{1.1}$$

This follows immediately from the definition of the function $\lceil \alpha \rceil$. Second, if $\alpha \neq 0$ is an integer, then

$$\lceil \alpha^{-1} \rceil \leq \lceil \alpha \rceil^{d-1}. \tag{1.2}$$

For, writing $\alpha = \alpha^{(1)}$, we have $\mathrm{Nm}_{\mathbb{K}/\mathbb{Q}}(\alpha) = \alpha^{(1)} \ldots \alpha^{(d)}$, and

$$\lceil \alpha^{-1} \rceil = \lceil \alpha^{(2)} \ldots \alpha^{(d)} \rceil \Big/ |\mathrm{Nm}_{\mathbb{K}/\mathbb{Q}}(\alpha)| \leq \lceil \alpha^{(2)} \rceil \ldots \lceil \alpha^{(d)} \rceil = \lceil \alpha \rceil^{d-1},$$

since $\lceil \alpha^{(j)} \rceil = \lceil \alpha \rceil$. Now for any rational integer v, (1.1) and (1.2) yield

$$\overline{|\alpha^v|} \leq \overline{|\alpha|}^{(d-1)|v|}. \tag{1.3}$$

There is a relationship between the height and the size of an algebraic number. Let $p(x) = a_0 x^n + a_1 x^{n-1} + \ldots + a_n$ be the minimal polynomial of α. Since every conjugate of α satisfies

$$\alpha^{(i)} = -\frac{a_1}{a_0} - \frac{a_2}{a_0}(\alpha^{(i)})^{-1} - \ldots - \frac{a_n}{a_0}(\alpha^{(i)})^{-n+1}$$

we see that

$$|\alpha^{(i)}| \leq \min(1, nh(\alpha)|a_0|^{-1}) \leq nh(\alpha),$$

and so

$$\overline{|\alpha|} \leq nh(\alpha). \tag{1.4}$$

On the other hand, each number $\pm a_j / a_0$ is a symmetric function of the roots $\alpha^{(i)}$ of $p(x)$, and hence can be estimated in terms of $\max|\alpha^{(i)}|$:

$$h(\alpha) = \max|a_j| \leq 2^n |a_0| \max(1, \overline{|\alpha|}^n). \tag{1.5}$$

Comparison of (1.4) and (1.5) shows a considerable difference between the upper and lower bounds for $h(\alpha)$ in terms of $\overline{|\alpha|}$. We will have occasion to compare $\ln h(\alpha)$ and $\ln \overline{|\alpha|}$, with α of finite degree, and prominent in that comparison will be the 'denominator' $a(\alpha) = a_0$ of α, the leading coefficient of the minimal polynomial of α, which we assume > 0.

Let α, β be nonzero algebraic numbers of degrees n and m respectively. We will often need to estimate $h(\alpha + \beta)$ and $h(\alpha\beta)$ in terms of $h(\alpha)$ and $h(\beta)$. It follows from (1.5) that

$$h(\alpha\beta) \leq 2^{nm} a(\alpha\beta) \max(1, \overline{|\alpha\beta|}^{nm}),$$

where $a(\alpha\beta)$ is the leading coefficient of the minimal polynomial of $\alpha\beta$. This coefficient is bounded by $a^m(\alpha)a^n(\beta)$, since the polynomial

$$a^m(\alpha)a^n(\beta) \prod_{\substack{1 \leq i \leq n \\ 1 \leq j \leq m}} (x - \alpha^{(i)}\beta^{(j)})$$

is integral by the theorem on symmetric polynomials, and is divisible by the minimal polynomial of $\alpha\beta$, and, by Gauss's Lemma, its leading coefficient is divisible by the leading coefficient of the minimal polynomial of $\alpha\beta$. Now applying (1.1), (1.4) and (1.5) we obtain

$$h(\alpha\beta) \leq 2^{nm} a^m(\alpha)a^n(\beta) \max(1, \overline{|\alpha|}^{nm}) \max(1, \overline{|\beta|}^{nm}) \leq$$
$$\leq 2^{nm} h^m(\alpha)h^n(\beta)(nh(\alpha))^{nm}(mh(\beta))^{nm} =$$
$$= (2nm)^{nm} h^{(n+1)m}(\alpha)h^{(m+1)n}(\beta).$$

Similarly, by (1.5) we have

$$h(\alpha + \beta) \leq 2^{nm} a(\alpha + \beta) \max(1, \overline{|\alpha + \beta|}^{nm}).$$

Here $a(\alpha + \beta)$ is the leading coefficient of the minimal polynomial of $\alpha + \beta$, and is bounded by $a^m(\alpha)a^n(\beta)$ because the polynomial

$$a^m(\alpha)a^n(\beta) \prod_{\substack{1 \le i \le n \\ 1 \le j \le m}} (x - \alpha^{(i)} - \beta^{(j)})$$

is integral. Again from (1.1) and (1.4), we obtain

$$h(\alpha + \beta) \le 2^{nm} a^m(\alpha) a^n(\beta) \max\left(1, (\lceil\alpha\rceil + \lceil\beta\rceil)^{nm}\right) \le$$
$$\le 4^{nm} a^m(\alpha) a^n(\beta) \max\left(1, \lceil\alpha\rceil^{nm}\right) \max\left(1, \lceil\beta\rceil^{nm}\right) \le$$
$$\le (4nm)^{nm} h^{(n+1)m}(\alpha) h^{(m+1)n}(\beta).$$

We will use such inequalities frequently without special explanation. In our applications, a most vital role will be played by the inequalities

$$\ln h(\alpha\beta) < c_1(\ln h(\alpha) + \ln h(\beta)) + c_2, \tag{1.6}$$
$$\ln h(\alpha + \beta) < c_1(\ln h(\alpha) + \ln h(\beta)) + c_2; \tag{1.7}$$

where c_1, c_2 depend only on the degrees of α, β. Thus for algebraic numbers of bounded degree one can bound the order of $\ln h(\alpha\beta)$ or $\ln h(\alpha + \beta)$ by the sum of the orders of $\ln h(\alpha)$ and $\ln h(\beta)$. Since the derivation yields an explicit dependence on the degrees of α and β, the inequalities can also be applied in situations where the degrees are allowed to grow, but it would then be advisable instead to introduce notions of height and size of an algebraic number in another way (see, e.g. [242], [243]).

We conclude this section with a lemma on a lower bound for the size of algebraic integers. By a well-known theorem of Kronecker, if an algebraic integer α satisfies $\lceil\alpha\rceil = 1$ then α is a root of unity. A slight revision of the usual proof of this theorem yields somewhat more:

Lemma 1.1 *The inequalities*

$$\lceil\alpha\rceil \ge c_3 > 1 \tag{1.8}$$

hold for any algebraic integer α which is not a root of unity, where c_3 is determined explicitly by the degree of α.

Proof. Let N be the number of integers of the field $\mathbb{K} = \mathbb{Q}(\alpha)$ which have size less than 2. Then c_3 may be taken as $2^{1/N}$. For, assume contrary to (1.8) that $\lceil\alpha\rceil < 2^{1/N}$ and consider the series $1, \alpha, \ldots, \alpha^N$. These $N + 1$ numbers all lie in $\mathbb{K}$ and have size less than 2, and so at least two of them coincide. Hence α is a root of unity, contrary to assumption.

A conjecture exists that one can take $c_3 = 1 + c_4/\deg\alpha$ with an absolute constant $c_4 > 0$. This is unproved, but it is known that $c_3 = 1 + (30d^2 \ln 6d)^{-1}$, $d = \deg\alpha \ge 3$, is admissible ([31], see also [57]).

2. Bounds for Units, Regulators and Class Numbers

The lemmas proved below will be used repeatedly in the analysis of the magnitude of solutions of diophantine equations.

Let $\mathbb{L}$ be an algebraic number field of degree ℓ over $\mathbb{Q}$, with r the rank of the group of units of $\mathbb{L}$, and $R = R_{\mathbb{L}}$ the regulator of $\mathbb{L}$. The following important lemma is due to Siegel [195].

Lemma 2.1 $\mathbb{L}$ *contains independent units* $\eta_1, \ldots, \eta_r$ *such that*

$$\prod_{i=1}^{r} \ln |\overline{\eta_i}| < c_5 R, \tag{2.1}$$

where c_5 *is determined explicitly by* ℓ.

Proof. Suppose that $\mathbb{L}$ has exactly s real units and $2t$ complex isomorphisms into $\mathbb{C}$, and enumerate the conjugate fields $\mathbb{L}^{(j)}$ in such a way that

$\mathbb{L}^{(j)}$	$(j = 1, \ldots, s)$	are real,
$\mathbb{L}^{(j)}$	$(j = s+1, \ldots, s+2t)$	are complex,
$\mathbb{L}^{(j)}$ and $\mathbb{L}^{(j+t)}$	$(j = s+1, \ldots, s+t)$	are complex conjugate.

Take any basis $\varepsilon_1, \ldots \varepsilon_r$ of the group of units of infinite order in $\mathbb{L}$ and consider the system of linear forms

$$L_j(\mathbf{x}) = \sum_{i=1}^{r} x_i \ln |\varepsilon_i^{(j)}|^{\ell_i} \qquad (j = 1, 2, \ldots, r) \tag{2.2}$$

where $\ell_i = 1$ (respectively 2) for real (respectively complex) isomorphisms. Obviously, the determinant of this system of forms is equal to R. Let λ_1 be the minimal real number λ such that the system of inequalities

$$\max_{1 \leq j \leq r} |L_j(\mathbf{x})| \leq \lambda \tag{2.3}$$

has a non-trivial integer solution $\mathbf{x}_{11}$; let λ_2 be the minimal real λ for which the system has two linearly independent integer solutions $\mathbf{x}_{21}, \mathbf{x}_{22}$; let λ_3 be the minimal real λ for which the system has three linearly independent integer solutions $\mathbf{x}_{31}, \mathbf{x}_{32}, \mathbf{x}_{33}$; and so on up to λ_r. By Minkowski's theorem on successive minima, the product $\lambda_1 \lambda_2 \cdots \lambda_r$ does not exceed the absolute value of the determinant of the system (2.3); that is,

$$\lambda_1 \lambda_2 \cdots \lambda_r \leq R. \tag{2.4}$$

It is possible to choose $i_1, i_2, \ldots, i_r$ $(i_1 = 1, 1 \leq i_2 \leq 2, \ldots, 1 \leq i_r \leq r)$ so that the vectors $y_1 = \mathbf{x}_{1i_1}, \ldots, y_r = \mathbf{x}_{ri_r}$ will be linearly independent, and then by (2.4) we have

$$\prod_{k=1}^{r} \max_{1 \leq j \leq r} |L_j(y_k)| \leq R. \tag{2.5}$$

Setting $y_k = (y_{k1}, \ldots, y_{kr})$, introduce units $\eta_k = \varepsilon_1^{y_{k1}} \ldots \varepsilon_r^{y_{kr}}$ $(k = 1, 2, \ldots, r)$. Substituting y_k for $\mathbf{x}$ in (2.2) and applying (2.5) gives

$$\prod_{k=1}^{r} \max_{1 \le j \le r} \left| \ln |\eta_k^{(j)}| \right| \le R. \tag{2.6}$$

Since $r = s + t - 1$ and

$$\sum_{j=1}^{s} \ln |\eta_k^{(j)}| + 2 \sum_{j=s+1}^{s+t} \ln |\eta_k^{(j)}| = 0, \tag{2.7}$$

we find that $\ln \overline{|\eta_k|} < \ell \max_{(j)} \left| \ln |\eta_k^{(j)}| \right|$ $(j, k = 1, 2, \ldots, r)$, which by (2.6) implies (2.1).

Given a number in $\mathbb{L}$, we use Lemma 2.1 to transform it into another number having all its conjugates in a sufficiently narrow domain, by multiplying it by a special unit.

Lemma 2.2 *Let $U = U_\mathbb{L}$ be the group of units of the field $\mathbb{L}$ generated by the units $\eta_1, \ldots, \eta_r$ determined in Lemma 2.1, and let $\alpha \ne 0$ be a number in $\mathbb{L}$. Then there exists a unit $\eta \in U$ such that*

$$\overline{|\alpha \eta^{-1}|} = |\mathrm{Nm}(\alpha)|^{1/\ell} e^{\theta c_6 R}, \qquad |\theta| < 1,$$

where c_6 is determined explicitly by ℓ.

Proof. Assume the enumeration of the conjugate fields $\mathbb{L}^{(j)}$ given in the proof of the previous lemma, and consider a system of equations

$$\sum_{i=1}^{r} x_i \ln |\eta_i^{(j)}| = \ln \frac{|\alpha^{(j)}|}{N^{1/\ell}} \qquad (j = 1, 2, \ldots, r),$$

where $N = |\mathrm{Nm}(\alpha)|$. Since the determinant of this system is not zero, because the units $\eta_1, \ldots, \eta_r$ are independent, it has a solution $(x_{10}, \ldots, x_{r0})$. Replacing each x_{i0} by the nearest integer a_i we obtain

$$\left| \sum_{i=1}^{r} a_i \ln |\eta_i^{(j)}| - \ln \frac{|\alpha^{(j)}|}{N^{1/\ell}} \right| \le \frac{r}{2} \max_{1 \le i \le r} \left| \ln |\eta_i^{(j)}| \right|. \tag{2.8}$$

Since $1 \le |\mathrm{Nm}_{\mathbb{L}^{(j)}/\mathbb{Q}}(\eta_i^{(j)})| \le |\eta_i^{(j)}| \overline{|\eta_i|}^{\ell-1}$, we have $\overline{|\eta_i|}^{-\ell+1} \le |\eta_i^{(j)}| \le \overline{|\eta_i|}$ $(1 \le i \le r, 1 \le j \le \ell)$, and hence

$$\left| \ln |\eta_i^{(j)}| \right| \le (\ell - 1) \ln \overline{|\eta_i|}.$$

In addition, Lemma 1.1 and (2.1) show that

$$\ln \overline{|\eta_i|} \ge c_7 > 0, \qquad \ln \overline{|\eta_i|} < c_5 c_7^{-r+1} R \qquad (i = 1, \ldots, r). \tag{2.9}$$

Therefore the left-hand side of inequality (2.8) does not exceed

$$\frac{r}{2}(\ell - 1)c_5 c_7^{-r+1} R = c_8 R. \tag{2.10}$$

Due to the chosen enumeration of the conjugate fields, we have

$$|\eta_i^{(j)}| = |\eta_i^{(j+t)}|, \quad |\alpha_i^{(j)}| = |\alpha_i^{(j+t)}| \quad (j = s+1, \ldots, s+t),$$

and so (2.8) holds for all $j \neq s+t, s+2t$; but it holds for these exceptions as well when the right hand side is replaced by ℓ times the value of (2.10), as is shown by (2.7) and a similar identity for the values $|\alpha^{(j)}| N^{-1/\ell}$. Thus

$$\left| \sum_{i=1}^{r} a_i \ln|\eta_i^{(j)}| - \ln \frac{|\alpha^{(j)}|}{N^{1/\ell}} \right| < c_6 R \quad (j = 1, \ldots, \ell),$$

where $c_6 = \ell c_8$.

Hence we obtain

$$\left| \alpha^{(j)} (\eta_1^{(j)})^{-a_1} \ldots (\eta_r^{(j)})^{-a_r} \right| = N^{1/\ell} e^{\theta c_6 R}, \quad |\theta| < 1.$$

In particular, we see that the unit $\eta = \eta_1^{a_1} \ldots \eta_r^{a_r}$ verifies the assertion of the lemma.

Now we show by virtue of inequality (2.5) above that the regulator of every subfield of the field $\mathbb{L}$ can be bounded in terms of the regulator of $\mathbb{L}$. As a preliminary, note that (2.6) implies

$$\prod_{k=1}^{r} \max_{1 \leq j \leq \ell} \left| \ln|\eta_k^{(j)}| \right| \leq \left(\frac{\ell}{2} - 1\right)^r R. \tag{2.11}$$

For, from (2.7) we obtain

$$2 \left| \ln|\eta_k^{(s+t)}| \right| \leq (\ell - 2) \max_{1 \leq j \leq r} \left| \ln|\eta_k^{(j)}| \right|,$$

hence

$$\max_{1 \leq j \leq \ell} \left| \ln|\eta_k^{(j)}| \right| \leq \left(\frac{\ell}{2} - 1\right) \max_{1 \leq j \leq r} \left| \ln|\eta_k^{(j)}| \right|,$$

which gives (2.11) because of (2.6).

Lemma 2.3 *Let $\mathbb{K}$ be a subfield of $\mathbb{L}$. Then the regulator of this subfield satisfies*

$$R_{\mathbb{K}} < c_9 R_{\mathbb{L}}, \tag{2.12}$$

where c_9 is depends explicitly on ℓ.

Proof. Set $n = [\mathbb{K} : \mathbb{Q}]$, $m = [\mathbb{L} : \mathbb{K}]$. For $(j = 1, 2, \ldots n; k = 1, 2, \ldots m)$ denote by $\mathbb{K}^{(j)}$ and $\mathbb{L}^{(j,k)}$ subfields of $\mathbb{C}$ isomorphic to $\mathbb{K}$ and $\mathbb{L}$ respectively, where the notation is to mean that the $\mathbb{L}^{(j,k)}$ with fixed j are conjugate fields over $\mathbb{K}^{(j)}$.

(Any embedding of $\mathbb{K}$ into $\mathbb{C}$ has exactly m continuations to the embedding of $\mathbb{L}$ into $\mathbb{C}$.) Hence, for example, if $\eta \in \mathbb{L}$ then we write

$$\mathrm{Nm}_{\mathbb{L}^{(j)}/\mathbb{K}}(\eta) = \eta^{(j,1)} \ldots \eta^{(j,m)} \qquad (j = 1, 2, \ldots n).$$

Suppose that $\eta_1, \ldots, \eta_r$ is a system of independent units in $\mathbb{L}$ satisfying

$$\prod_{i=1}^{r} \max_{(j,k)} \left| \ln |\eta_i^{(j,k)}| \right| < c_{10} R, \qquad (2.13)$$

corresponding to (2.11). We set

$$\theta_i^{(j)} = \mathrm{Nm}_{\mathbb{L}^{(j)}/\mathbb{K}}(\eta_i) \qquad (i = 1, 2, \ldots, r; \; j = 1, 2, \ldots n).$$

The units $\theta_i = \mathrm{Nm}_{\mathbb{L}/\mathbb{K}}(\eta_i)$ $(i = 1, 2, \ldots, r)$ lie in $\mathbb{K}$ and form a system of the same rank as that of the group of all units in $\mathbb{K}$. Concretely, if $\varepsilon_1, \ldots, \varepsilon_q$ is a system of fundamental units (of infinite order) in $\mathbb{K}$, then for every unit ε_u there exist integers $a_u \neq 0, a_{1u}, \ldots, a_{ru}$ with

$$\varepsilon_u^{a_u} = \eta_1^{a_{1u}} \ldots \eta_r^{a_{ru}} \qquad (u = 1, 2, \ldots, q).$$

Taking the norm from $\mathbb{L}$ to $\mathbb{K}$ in this equation, we find that the units $\varepsilon_u^{ma_u}$ can be expressed multiplicatively in terms of $\theta_1, \ldots, \theta_r$. If M is the least common multiple of the numbers $ma_1, \ldots, ma_q$ then the Mth power of each unit $\varepsilon_1, \ldots, \varepsilon_q$ is given by a multiplicative expression in $\theta_1, \ldots, \theta_r$. Hence the rank of the system $\theta_1, \ldots, \theta_r$ is not less than q, and so equals q. This means that the rank of the matrix

$$\left(\ln |\theta_i^{(j)}| \right)_{\substack{i=1,2,\ldots,r \\ j=1,2,\ldots,n}}$$

is q. Let

$$\det \left(\ln |\theta_i^{(j)}| \right) \neq 0 \qquad (i, j = 1, 2, \ldots, q).$$

Then we find

$$R_{\mathbb{K}} \leq \left| \det \left(\ln |\theta_i^{(j)}| \right) \right| = \left| \det \left(\ln |\mathrm{Nm}_{\mathbb{L}^{(j)}/\mathbb{K}}(\eta_i)| \right) \right| =$$

$$= \left| \det \left(\sum_{k=1}^{m} \ln |\eta_i^{(j,k)}| \right) \right| \leq \sum_{(j_1, \ldots j_q)} \prod_{i=1}^{q} \sum_{k=1}^{m} \left| \ln |\eta_i^{(j_i,k)}| \right| =$$

$$= \sum_{(j_1, \ldots j_q)} \sum_{k=1}^{m} \left| \ln |\eta_1^{(j_1,k)}| \right| \ldots \left| \ln |\eta_q^{(j_q,k)}| \right|,$$

where $(j_1, \ldots j_q)$ runs over all permutations of $1, 2, \ldots, q$. Since

$$\max_{(j,k)} \left| \ln |\eta_i^{(j,k)}| \right| \geq c_{11} > 0 \qquad (i = 1, 2, \ldots, r),$$

where c_{11} depends on ℓ (Lemma 1.1), it follows from (2.13) that

$$\left|\ln|\eta_1^{(j_1,k)}|\right|\cdots\left|\ln|\eta_q^{(j_q,k)}|\right| < c_{12}R$$

for any set $(j_1,\ldots,j_q)$. This proves inequality (2.12).

From (2.9) we see that the regulator is bounded from below by a value which depends only on the degree of the field. In fact stronger bounds are known (for example, if $\mathbb{L}$ is not a purely imaginary quadratic extension of a purely real field, then $R > c_{13}\ln|D_{\mathbb{L}}|$, where $D_{\mathbb{L}}$ is the discriminant of $\mathbb{L}$, and c_{13} depends only on the degree of $\mathbb{L}$; see [163]), but we do not need them. On the other hand, we will often make use of upper bounds for the regulator.

Lemma 2.4 *The class number h, the regulator R and the discriminant D of a field $\mathbb{L}$ are connected by the inequality*

$$hR < c_{14}|D|^{1/2}(\ln|D|)^{\ell-1}, \tag{2.14}$$

where c_{14} is determined explicitly by ℓ.

Proof. The proof of this lemma is well-known: it follows from an upper bound for the residue of the Dedekind zeta function of $\mathbb{L}$. It was probably first obtained by Landau [119]. Expressions for c_{14} are given in various works (e.g. [125], [195]).

Since $h \geq 1$ and R is bounded from below by a value depending only on ℓ, it follows that h and R are each independently bounded from above by expressions of the form (2.14).

An upper bound for the residue of the zeta function of $\mathbb{L}$ implies the inequality

$$hR > c_{15}|D|^{\frac{1}{2}-\epsilon} \tag{2.15}$$

for any $\epsilon > 0$, where $c_{15} = c_{15}(\epsilon,\ell)$ depends on ϵ and ℓ, but an explicit form of this dependence is unknown. The proof of (2.15) is considerably more complicated than that of (2.14). The consequence $hR < c_{16}|D|^{\frac{1}{2}+\epsilon}$ of (2.14) combines with (2.15) to give the single assertion known as the *Siegel-Brauer theorem* ([194], [34]).

In concrete cases, to bound the regulator or the class number from above using (2.14) requires a bound for the discriminant of the field in terms of values associated with a generator of the field (its height, size or discriminant). To do so we shall use the *theory of differents* of algebraic number fields, relying on three fundamental facts: the discriminant of a field as the norm of the field different, the multiplicative property of differents in relative extensions, and divisibility of the differents of all integers of a field by the field different ([245], [96]).

Note that if $\mathbb{K}$ is a subfield of $\mathbb{L}$, then $h_{\mathbb{K}} \leq mh_{\mathbb{L}}$ where $m = [\mathbb{L} : \mathbb{K}]$. This follows immediately from the existence of an absolute class field for $\mathbb{L}$. Comparison with (2.12) shows an analogy between the behaviours of regulators and class numbers in field extensions: in extensions of bounded degree the order of magnitude of neither regulator nor class number decreases. A similar, stronger phenomenon exists for discriminants: $D_{\mathbb{K}}^m | D_{\mathbb{L}}$. Evidently for problems connected with upper bounds for regulators or class numbers or discriminants, the arguments are most expedient in a minimal field. The construction of effective bounds for solutions of diophantine equations is precisely such a problem, but carrying out the arguments in a minimal field is not always convenient. Besides, although the value of the discriminant is quite controllable, the same cannot be said for the regulator or the class number, and we shall see some unexpected peculiarities of their behaviour revealed by the analysis of diophantine equations (Chap. VIII).

3. Analytic Functions in p-adic Fields

We shall frequently apply p-adic analysis in both its arithmetic and analytic variants.

We denote by Ω_p the completion in the metric $|\ |_p$ of the algebraic closure $\overline{\mathbb{Q}}$ of the field $\mathbb{Q}_p$ of p-adic numbers. It is a complete and algebraically closed field [107].

A function $f : \Omega_p \longrightarrow \Omega_p$ is *analytic* (or *regular*) at a point $a \in \Omega_p$ if it may be defined as a power series

$$f(z) = \sum_{k=0}^{\infty} a_k(z-a)^k, \qquad a_k, a \in \Omega_p, \tag{3.1}$$

converging in the metric of Ω_p. In view of the non-archimedean nature of the metric, the convergence of (3.1) is equivalent to the condition

$$|a_k|_p|z-a|_p^k \to 0 \quad (k \to \infty).$$

If the series converges at a point $z_0 \neq a$, it converges in the disc $|z-a|_p \leq |z_0-a|_p$ as well, so that the domain of convergence includes a disc $|z-a|_p < \rho$. The derivative $f'(z)$ defined formally by the series

$$f'(z) = \sum_{k=1}^{\infty} ka_k(z-a)^{k-1}$$

makes sense in any disc $|z-a|_p < \rho$ where $f(z)$ exists. Defining the successive derivatives $f^{(n)}(z)$ in this way we obtain, in particular, $f^{(n)}(a) = n!a_n$, and hence arrive at the Taylor expansion

$$f(z) = \sum_{n=0}^{\infty} \frac{f^{(n)}(a)}{n!}(z-a)^n, \qquad |z-a|_p < \rho.$$

If $P(z)$ is a polynomial over Ω_p, then $1/P(z)$ is an analytic function at all points other than zeros of P. The quotient $f(z)/P(z)$ is an analytic function at any point in the disc $|z-a|_p < \rho$ other than zeros of P. The poles and residues of this function may be determined by local expansion in Laurent series.

L. G. Schnirelman [187] introduced an analogue of the Cauchy contour integral which allows many arguments from complex analysis to be imported. Let $(n, p) = 1$ and $p_n(x) = x^n - 1 = (x - \zeta_{1n}) \ldots (x - \zeta_{nn})$ (factorisation in Ω_p). Since ζ_{in} are algebraic units, we have $|\zeta_{in}|_p = 1$ for all i, n. If for fixed numbers $a, r \in \Omega_p$ ($r \neq 0$) the function $f(z)$ is defined at all the points $a + r\zeta_{in}$, form the sum

$$\frac{1}{n} \sum_{i=1}^{n} f(a + r\zeta_{in}), \qquad (n, p) = 1,$$

and consider the limit (in the metric of Ω_p) of the sums as $n \to \infty$. This limit, when it exists, is called the *Schnirelman integral*, and denoted

$$\int_{a,r} f(z) dz. \tag{3.2}$$

[Schnirelman himself denoted it by $\int_{a,r}^{z} f(z)$, but we shall use the now widespread notation (3.2)]. In the following presentation we follow [2], [126].

Directly from the definition we see that

$$\left| \int_{a,r} f(z) dz \right|_p \leq \max |f(z)|_p, \tag{3.3}$$

where the maximum is taken over all z with $|z - a|_p = |r|_p$. Therefore any convergent series of analytic functions, defined and uniformly bounded on a circle in Ω_p, can be integrated term by term.

There is a certain analogy between the Schnirelman integral and the Cauchy contour integral, but it is far from straightforward. To convince ourselves, let us compute the integrals

$$I_k = \int_{0,r} z^k dz \qquad (r \in \Omega_p; k = 0, \pm 1, \pm 2, \ldots)$$

directly from the definition. The integral sums are

$$\frac{1}{n} \sum_{i=1}^{n} (r\zeta_{in})^k = \frac{1}{n} r^k \sum_{i=1}^{n} \zeta_{in}^k,$$

and since for $n > |k|$ one has

$$\sum_{i=1}^{n} \zeta_{in}^k = \begin{cases} 0, & \text{if } k \neq 0 \\ n & \text{if } k = 0, \end{cases}$$

we have $I_0 = 1$, $I_k = 0$ ($k \neq 0$).

Since the definition of the integral implies

$$\int_{a,r} f(z) dz = \int_{0,r} f(z + a) dz,$$

we obtain

$$\int_{a,r} z^k dz = a^k \qquad (k = 0, 1, 2, \ldots). \tag{3.4}$$

Lemma 3.1 *Suppose that the series*

$$g(z) = \sum_{k=0}^{\infty} b_k z^k, \quad b_k \in \Omega_p \tag{3.5}$$

converges for $|z|_p < \rho$ *($\rho > 0$). Let* $a, r \in \Omega_p$, $|a|_p < \rho$, $|r|_p < \rho$. *Then*

$$\int_{a,r} g(z) dz = g(a).$$

Proof. Integrate the series (3.5) term by term and apply (3.4).

Lemma 3.2 *Let* $x, r \in \Omega_p$ *and let* $n > 0$ *be an integer. Then*

$$\int_{0,r} \frac{dz}{(z-x)^n} = \begin{cases} 0 & \text{if } |x|_p < |r|_p, \\ (-1)^n / x^n & \text{if } |x|_p > |r|_p. \end{cases}$$

Proof. The series

$$(-x)^{-n} \sum_{k=0}^{\infty} \binom{-n}{k} \left(-\frac{z}{x}\right)^k$$

converges for $|z|_p < |x|_p$ and represents a regular function of z (for fixed x), namely the function $(z-x)^{-n}$. Hence for $|x|_p > |r|_p$ the integral is $(-x)^{-n}$ by Lemma 3.1. When $|x|_p < |r|_p$ the integrand has the Laurent series expansion

$$(z-x)^{-n} = z^{-n} \left(1 - \frac{x}{z}\right)^{-n} = z^{-n} \sum_{k=0}^{\infty} \binom{-n}{k} \left(-\frac{x}{z}\right)^k.$$

Since 'on the circle of integration' $|z|_p = |r|_p$, the series may be integrated term by term, which gives

$$\sum_{k=0}^{\infty} \binom{-n}{k} (-x)^k I_{n+k},$$

which is zero.

Lemma 3.3 *With the same setting as Lemma 3.1, assume further that* $x \in \Omega_p$, $|x|_p < \rho$. *Then*

$$\int_{a,r} \frac{g(z)(z-a)}{z-x} dz = \begin{cases} g(x) & \text{if } |x-a|_p < |r|_p, \\ 0 & \text{if } |x-a|_p > |r|_p. \end{cases}$$

Proof. On replacing z by $z+a$ and x by $x+a$, we see that it suffices to consider the case $a = 0$. First note that for any integer $k \geq 0$

$$\int_{0,r} \frac{z^{k+1}}{z-x} dz = \int_{0,r} \frac{z^{k+1} - x^{k+1} + x^{k+1}}{z-x} dz$$

$$= \int_{0,r} (z^k + xz^{k-1} + \ldots + x^k) dz + x^{k+1} \int_{0,r} \frac{dz}{z-x}.$$

Integrating the first integral term by term and applying Lemma 3.2 to the second one, we obtain

$$\int_{0,r} \frac{z^{k+1}}{z-x} dz = \begin{cases} x^k & \text{if } |x|_p < |r|_p, \\ 0 & \text{if } |x|_p > |r|_p. \end{cases}$$

Since

$$\int_{0,r} \frac{g(z)z}{z-x} dz = \sum_{k=0}^{\infty} b_k \int_{0,r} \frac{z^{k+1}}{z-x} dz,$$

we obtain the assertion of the lemma.

Lemma 3.4 *Under the hypotheses of the previous lemma we have for x in the disc $|x - a|_p < |r|_p$*

$$g^{(n)}(x) = n! \int_{a,r} \frac{g(z)(z-a)}{(z-x)^{n+1}} dz \qquad (n = 0, 1, 2, \ldots).$$

Proof. We again start by computing some special integrals, namely

$$\int_{a,r} \frac{z-a}{(z-x)^k} dz \qquad (k = 1, 2, \ldots).$$

Since $|x - a|_p < |r|_p$, the integrand expands in the series

$$\frac{z-a}{(z-x)^k} = \frac{1}{(z-a)^{k-1}} \left(\sum_{q=0}^{\infty} \left(\frac{x-a}{z-a} \right)^q \right)^k = \frac{1}{(z-a)^{k-1}} \sum_{s=0}^{\infty} e_s \left(\frac{x-a}{z-a} \right)^s,$$

say. We know that

$$\int_{a,r} \frac{dz}{(z-a)^{k+s-1}} = \begin{cases} 0 & \text{if } k+s-1 \neq 0, \\ 1 & \text{if } k+s-1 = 0. \end{cases}$$

Therefore

$$\int_{a,r} \frac{z-a}{(z-x)^k} dz = \sum_{s=0}^{\infty} e_s (x-a)^s \int_{a,r} \frac{dz}{(z-a)^{k+s-1}} = \begin{cases} 0 & \text{if } k > 1, \\ 1 & \text{if } k = 1. \end{cases} \qquad (3.6)$$

Now we note that

$$g(z) = \sum_{k=0}^{\infty} \frac{g^{(k)}(x)}{k!} (z - x)^k.$$

Hence

$$\int_{a,r} \frac{g(z)(z - a)}{(z - x)^{n+1}} dz = \sum_{k=0}^{\infty} \frac{g^{(k)}(x)}{k!} \int_{a,r} \frac{(z - a)}{(z - x)^{n+1-k}} dz = \frac{g^{(n)}(x)}{n!},$$

where we apply (3.6) if $k \le n$, and Lemma 3.1 if $k > n$.

We now turn to lemmas on residues of functions of the type $g(z)/P(z)$, where $g(z)$ is regular in some disc and $P(z)$ is a polynomial over Ω_p. If z_0 is a pole of such a function, then $\operatorname{Res}_{z_0} g(z)/P(z)$ will denote the residue at the point z_0. We begin with the simplest case, the quotient of polynomials.

Lemma 3.5 *Suppose that $G(z)$ and $P(z) = (z - z_1)^{k_1} \ldots (z - z_n)^{k_n}$ are polynomials over Ω_p, with $z_i \ne z_j$ ($i \ne j$). Let $r \in \Omega_p$, with $|z_j|_p \ne |r|_p$. Then*

$$\int_{0,r} \frac{G(z)z}{P(z)} dz = \sum_{|z_j|_p < |r|_p} \operatorname{Res}_{z_j} \frac{G(z)}{P(z)}. \tag{3.7}$$

That is, the integral is equal to the sum of the residues of $G(z)/P(z)$ at those poles z_j with $|z_j|_p < |r|_p$.

Proof. The integrand has a partial fraction expansion

$$\frac{G(z)}{P(z)} = G_1(z) + \frac{\lambda_{11}}{z - z_1} + \frac{\lambda_{12}}{(z - z_1)^2} + \ldots + \frac{\lambda_{1k_1}}{(z - z_1)^{k_1}}$$

$$+ \ldots$$

$$+ \frac{\lambda_{n1}}{z - z_n} + \frac{\lambda_{n2}}{(z - z_n)^2} + \ldots + \frac{\lambda_{nk_n}}{(z - z_n)^{k_n}},$$

where $G_1(z)$ is a polynomial over Ω_p, $\lambda_{ij} \in \Omega_p$. By Lemma 3.1, and by taking $a = 0$, $x = z_j$, and $|z_j|_p < |r|_p$ in (3.6), we obtain (3.7).

Lemma 3.6 *Suppose that the hypotheses of Lemma 3.5 hold, and further that $|z_j|_p < |r|_p$ ($j = 1, 2, \ldots, n$), and let $t \in \Omega_p$ with $|t|_p < |z_i - z_j|_p$ ($i \ne j$; $i, j = 1, 2, \ldots n$). Then*

$$\int_{0,r} \frac{G(z)z}{P(z)} dz = \int_{z_1,t} \frac{G(z)(z - z_1)}{P(z)} dz + \ldots + \int_{z_n,t} \frac{G(z)(z - z_n)}{P(z)} dz.$$

Proof. By Lemma 3.5

$$\int_{z_i,t} \frac{G(z)(z - z_i)}{P(z)} dz = \operatorname{Res}_{z_i} \frac{G(z)}{P(z)},$$

and again by this lemma the residue sum is equal to the initial integral.

Lemma 3.7 *Let $g(z)$ be the power series* (3.5), *convergent in the disc* $|z|_p < \rho$; *let* $r \in \Omega_p$, $|r|_p < \rho$; *let* $P(z) = (z - z_1)^{k_1} \ldots (z - z_n)^{k_n}$ *be a polynomial over* Ω_p, *with* $|z_i|_p < \rho$ $(i = 1, 2, \ldots, n)$; *and let* $t \in \Omega_p$, $|z_i - z_j|_p > |t|_p$ $(i \neq j;\ i, j = 1, 2, \ldots n)$. *Then*

$$\int_{0,r} \frac{g(z)z}{P(z)}\,dz = \int_{z_1,t} \frac{g(z)(z - z_1)}{P(z)}\,dz + \ldots + \int_{z_n,t} \frac{g(z)(z - z_n)}{P(z)}\,dz.$$

Proof. We proceed by induction on n. For $n = 1$,

$$\int_{0,r} \frac{g(z)z}{P(z)}\,dz = \int_{z_1,t} \frac{g(z)(z - z_1)}{P(z)}\,dz,$$

since $P(z) = (z - z_1)^{k_1}$ and both integrals represent $g^{(k_1 - 1)}(z_1)/(k_1 - 1)!$ by Lemma 3.4.

Introduce an expansion of $g(z)$ in a neighbourhood of z_1:

$$g(z) = g_0 + g_1(z - z_1) + g_2(z - z_1)^2 + \ldots ,$$

and put

$$G(z) = g_0 + g_1(z - z_1) + \ldots + g_{k_1 - 1}(z - z_1)^{k_1 - 1},$$
$$g(z) = G(z) + (z - z_1)^{k_1} h(z).$$

Then

$$\int_{0,r} \frac{g(z)z}{P(z)}\,dz = \int_{0,r} \frac{G(z)z}{P(z)}\,dz + \int_{0,r} \frac{h(z)z}{P_1(z)}\,dz,$$

where $P_1(z)$ is defined by $P_1(z)(z - z_1)^{k_1} = P(z)$. The first integral is

$$\sum_{i=1}^{n} \int_{z_i,t} \frac{G(z)(z - z_i)}{P(z)}\,dz$$

by Lemma 3.6, while the second integral is

$$\sum_{i=2}^{n} \int_{z_i,t} \frac{h(z)(z - z_i)}{P_1(z)}\,dz$$

by the induction assumption. Since

$$\int_{z_i,t} \frac{h(z)(z - z_i)}{P_1(z)}\,dz = \int_{z_i,t} \frac{g(z)(z - z_i)}{P(z)}\,dz - \int_{z_i,t} \frac{G(z)(z - z_i)}{P(z)}\,dz,$$

we obtain

$$\int_{0,r} \frac{g(z)z}{P(z)}\,dz = \int_{z_1,t} \frac{G(z)(z - z_1)}{P(z)}\,dz + \sum_{i=2}^{n} \int_{z_i,t} \frac{g(z)(z - z_i)}{P(z)}\,dz.$$

Finally,

$$\int_{z_1,t} \frac{G(z)(z-z_1)}{P(z)}\,dz = \int_{z_1,t} \frac{g(z)(z-z_1)}{P(z)}\,dz,$$

for, by Lemma 3.1,

$$\int_{z_1,t} \frac{(z-z_1)^{k_1} h(z)(z-z_1)}{P(z)}\,dz = 0$$

because of the regularity of the function

$$(z-z_1)^{k_1} h(z)/P(z) = h(z)/P_1(z)$$

in the disc $|z - z_1|_p < |t|_p$.

Lemma 3.8 *Suppose that the hypotheses of Lemma 3.7 hold, and that $a \in \Omega_p$, $|a|_p < \rho$, $|z_i - a|_p \neq |r|_p$ $(i = 1, 2, \ldots, n)$. Then*

$$\int_{a,r} \frac{g(z)(z-a)}{P(z)}\,dz = \sum_{|z_j - a|_p < |r|_p} \mathrm{Res}_{z_j} \frac{g(z)}{P(z)}.$$

Proof. This follows from the previous lemma.

We see that the theory of residues is analogous to the complex version. The main lemma of this section can now be proved.

Lemma 3.9 *Suppose $f(z)$ is a function regular in the disc $|z|_p < \rho$, $\rho > 0$; let $z_0 = 0$, $z_1, \ldots, z_m, t, r \in \Omega_p$, with $t \neq 0$, $r \neq 0$, $|t|_p < |z_i|_p < |r|_p < \rho$, and $|z_i - z_j|_p > |t|_p$ $(i \neq j;\ i, j = 1, 2, \ldots, m)$; let $n > 0$ be an integer, and*

$$F(z) = [z(z-z_1)\ldots(z-z_m)]^{n+1}.$$

Then for every $z \in \Omega_p$ with $|z - z_i|_p > |t|_p$ $(i = 1, 2, \ldots, m)$ we have

$$f(z) = \int_{0,r} \frac{F(z)}{F(\xi)} \frac{f(\xi)\xi}{\xi - z}\,d\xi - \sum_{k=0}^{n}\sum_{\ell=0}^{m} \frac{f^{(k)}(z_\ell)}{k!} \int_{z_\ell,t} \frac{F(z)}{F(\xi)} \frac{(\xi - z_\ell)^{k+1}}{\xi - z}\,d\xi.$$

Proof. Put

$$I = \int_{0,r} \frac{f(\xi)\xi}{F(\xi)(\xi - z)}\,d\xi.$$

Using Lemma 3.7,

$$I = \int_{z,t} \frac{f(\xi)}{F(\xi)}\,d\xi + \sum_{i=0}^{m} \int_{z_i,t} \frac{f(\xi)(\xi - z_i)}{F(\xi)(\xi - z)}\,d\xi.$$

In the first integral the integrand is regular in the disc $|\xi - z|_p < \lambda|t|_p$, where $\lambda = \min |z_i|_p |t|_p^{-1} > 1$. So by Lemma 3.1 this integral is equal to $f(z)/F(z)$. Each integral

$$I_i = \int_{z_i,t} \frac{f(\xi)(\xi - z_i)}{F(\xi)(\xi - z)} d\xi$$

is easily calculated by the substitution

$$f(\xi) = \sum_{s=0}^{\infty} \frac{f^{(s)}(z_i)}{s!}(\xi - z_i)^s,$$

which gives

$$I_i = \sum_{s=0}^{n} \frac{f^{(s)}(z_i)}{s!} \int_{z_i,t} \frac{(\xi - z_i)^{s+1}}{F(\xi)(\xi - z)} d\xi,$$

because by Lemma 3.1

$$\int_{z_i,t} \frac{(\xi - z_i)^{s+1}}{F(\xi)(\xi - z)} d\xi = 0 \qquad (s = n+1, n+2, \ldots),$$

due to the regularity of the function $(\xi - z_i)^{n+1}/F(\xi)$ at the point $z - z_i$. Thus

$$I = \frac{f(z)}{F(z)} + \sum_{i=0}^{m} \sum_{s=0}^{n} \frac{f^{(s)}(z_i)}{s!} \int_{z_i,t} \frac{(\xi - z_i)^{s+1}}{F(\xi)(\xi - z)} d\xi$$

and we obtain the desired formula.

We conclude this section with a mention of three special functions:

$$\log(1 + z) = \sum_{k=1}^{\infty} \frac{(-1)^{k-1}}{k} z^k, \qquad |z|_p < 1;$$

$$\exp z = \sum_{k=0}^{\infty} \frac{z^k}{k!}, \qquad |z|_p < p^{-1/(p-1)};$$

$$(1 + z)^t = \sum_{k=0}^{\infty} \binom{t}{k} z^k, \qquad |z|_p < p^{-1/(p-1)}, \quad |t|_p \leq 1;$$

where $z, t \in \Omega_p$. We record the main properties of these functions [93]:

a) $|\log(1 + z)|_p = |z|_p$ if $|z|_p < p^{-1/(p-1)}$;
b) $\log z_1 z_2 = \log z_1 + \log z_2$ if $|z_1 - 1|_p < 1$ and $|z_2 - 1|_p < 1$;
c) $|\exp z - 1|_p = |z|_p$ if $|z|_p < p^{-1/(p-1)}$;
d) $\exp(z_1 + z_2) = \exp z_1 \cdot \exp z_2$ if $|z_1|_p < p^{-1/(p-1)}$ and $|z_2|_p < p^{-1/(p-1)}$;
e) $\exp \log(1 + z) = 1 + z$ and $\log \exp z = z$, if $|z|_p < p^{-1/(p-1)}$;
f) $(1 + z)^t = \exp(t \log(1 + z))$ if $|z|_p < p^{-1/(p-1)}$ and $|t|_p \leq 1$.

We note in addition that $\log z = 0$ only for those z which are p^s-th roots of unity for some s, and that $\log(1 + z)$ is not a bounded function in the disc $|z|_p < 1$.

III. Linear Forms in the Logarithms of Algebraic Numbers

This chapter is of an auxiliary nature, being mainly concerned with the relationship between bounds for linear forms in the logarithms of algebraic numbers in different (archimedean and non-archimedean) metrics. This material will later be used in the analysis of Thue and Thue-Mahler equations. Elliptic and hyperelliptic equations, and equations of hyperelliptic type, will be analysed using direct bounds for linear forms in the logarithms of algebraic numbers, and the necessary results are stated without proof at the end of the chapter.

1. Direct Bounds and Connections Between Bounds in Different Metrics

Throughout $\alpha_1, \alpha_2, \ldots, \alpha_n$ denote algebraic numbers different from 0 or 1 and $h_1, h_2, \ldots, h_{n-1}$ are integers. In subsequent chapters we will use nontrivial bounds for the difference $\alpha_1^{h_1} \ldots \alpha_{n-1}^{h_{n-1}} - \alpha_n$ in archimedean and non-archimedean metrics, that is, bounds for the ordinary absolute value and the p-adic norm of the difference. But first let us see what can be achieved with simpler considerations; for example by Liouville's inequality.

Suppose the following: the heights of $\alpha_1, \ldots, \alpha_{n-1}$ do not exceed B; the height of α_n does not exceed A; $\mathbb{G}$ is an algebraic number field including $\alpha_1, \alpha_2 \ldots, \alpha_n$, $[\mathbb{G} : \mathbb{Q}] = g$, $H = \max |h_i|$ $(i = 1, \ldots, n - 1)$; $b_i > 0$ are the minimal natural numbers for which $b_i \alpha_i$ and $b_i \alpha_i^{-1}$ are algebraic integers $(i = 1, 2, \ldots, n - 1)$; and a_n is the minimal natural number such that $a_n \alpha_n$ is an algebraic integer. Let us assume that $\beta = \alpha_1^{h_1} \ldots \alpha_{n-1}^{h_{n-1}} - \alpha_n \neq 0$. Then

$$\gamma = b_1^{|h_1|} \ldots b_{n-1}^{|h_{n-1}|} a_n \beta \neq 0$$

is an algebraic integer, and

$$1 \leq |\operatorname{Nm}(\gamma)| \leq |\gamma| \overline{|\gamma|}^{g-1} \leq$$

$$\leq |\beta| B^{(n-1)H} A \left(A \prod_{i=1}^{n-1} \overline{\left| b_i^{|h_i|} \alpha_i^{h_i} \right|} + B^{(n-1)H} \overline{\left| a_n \alpha_n \right|} \right)^{g-1} \leq$$

$$\leq |\beta| A B^{(n-1)H} \left(A (g B^2)^{(n-1)H} + B^{(n-1)H} g A \right)^{g-1},$$

since $b_i^{|h_i|}\alpha_i^{h_i}$ is of the form $(b_i\alpha_i)^{h_i}$ (respectively of the form $(b_i\alpha_i^{-1})^{|h_i|}$) for $h_i > 0$ (respectively < 0), and

$$\max\left(\overline{|(b_i\alpha_i)^{h_i}|},\ \overline{|(b_i\alpha_i^{-1})^{|h_i|}|}\right) < (gB^2)^H$$

on account of (1.4) of Chap. II. Consequently,

$$|\beta| > 2^{-g}A^{-g}B^{-ngH}. \tag{1.1}$$

Similarly, if $\mathfrak{p}$ is a prime ideal of $\mathbb{G}$, we have

$$1 \leq |\operatorname{Nm}(\gamma)||\operatorname{Nm}(\gamma)|_{\mathfrak{p}} < 2^g A^g B^{ngH}|\gamma|_{\mathfrak{p}}\ ,$$

and since $|\gamma|_{\mathfrak{p}} \leq |\beta|_{\mathfrak{p}}$, we obtain an analogue of (1.1):

$$|\beta|_{\mathfrak{p}} > 2^{-g}A^{-g}B^{-ngH}. \tag{1.2}$$

For our purposes we need lower bounds for $|\beta|$ and $|\beta|_{\mathfrak{p}}$ in terms of $e^{-\delta H}$ for some δ in the range $0 < \delta \leq 1$. Evidently, (1.1) and (1.2) do not imply such bounds, and complicated manipulations will be needed to obtain them. To date there have been two main approaches to securing 'non-trivial' bounds (that is, stronger than (1.1) and (1.2)). In one approach (which we call *direct*), the derivation assumes that $\beta \neq 0$; the other assumes that for some prime ideal $\mathfrak{q}$, $|\beta|_{\mathfrak{q}}$ is 'not small'. Both these approaches have been applied to the analysis of diophantine equations, and in this monograph we make use of both. In view of the copious computations which the bounds for $|\beta|$ and $|\beta|_{\mathfrak{p}}$ require, we will give the details of the second approach only, that is, of the connection between these bounds and a hypothetical bound for $|\beta|_{\mathfrak{q}}$. Surveys of results for direct bounds are given in Baker's [25] and van der Poorten's [158] works.

Suppose that our initial assumptions on $\alpha_1, \alpha_2, \ldots, \alpha_n$ are fulfilled, and that $A \geq 80g$, $B \geq 80g$, and δ is an arbitrary number satisfying $0 < \delta \leq 1$.

Lemma 1.1 *Put*

$$\sigma = 2^{10}\delta^{-1}(n^2g^2\ln B)^2, \tag{1.3}$$

let q be a prime number such that $\sqrt{2}q^g \leq \sigma$, and let $|\ \ |_q$ be the q-adic valuation of the field $\mathbb{G}$ induced by the prime ideal $\mathfrak{q}$. Let

$$S = \operatorname{N}(q^{2e})\left(1 - \frac{1}{\operatorname{N}(\mathfrak{q})}\right), \qquad e = \operatorname{ord}_{\mathfrak{q}}q, \qquad |\alpha_i|_q = 1 \quad (i = 1, 2, \ldots, n-1).$$

Then the inequality

$$\left|\alpha_1^{h_1}\ldots\alpha_{n-1}^{h_{n-1}} - \alpha_n\right| < e^{-\delta H} \tag{1.4}$$

implies

$$H < \sigma^{4(n+1)}\ln A \tag{1.5}$$

provided that either $|\alpha_n|_q \neq 1$, or $|\alpha_n|_q = 1$ and

$$\left|h_1\log(\alpha_1^S) + \ldots + h_{n-1}\log(\alpha_{n-1}^S) - \log(\alpha_n^S)\right|_q > q^{-\frac{1}{8}H\sigma^{-4n-2}}, \tag{1.6}$$

where log( ) *denotes the q-adic logarithm defined in the metric* $|\ |_q$.

The condition (1.6) is rather stronger than the analogous trivial estimate of the type (1.2) obtained from the assumption $\beta \neq 0$, and it carries more information than that assumption. As a result, the derivation of inequalities of the type (1.5) from the assumption (1.4) is easier given (1.6) rather than just $\beta \neq 0$. Application of (1.6) is easy in the context of Thue and Thue-Mahler equations, and in this way we get a 'quick' proof of an improvement of Liouville's inequality on approximation of algebraic numbers by rationals.

Clearly one can consider in place of (1.4) an analogous inequality for the linear form

$$h_1 \ln \alpha_1 + \ldots + h_{n-1} \ln \alpha_{n-1} - \ln \alpha_n \tag{1.7}$$

in logarithms of the algebraic numbers $\alpha_1, \ldots, \alpha_n$ (taking any fixed branch of the logarithm). Thus the subject of Lemma 1.1 is really a relationship between bounds for linear forms in logarithms of these algebraic numbers in archimedean and non-archimedean metrics. A similar relationship holds between bounds in different non-archimedean metrics:

Lemma 1.2 *Suppose that the conditions of Lemma 1.1 are satisfied, and that p is a prime number, with* $\sqrt{2}q^9 p^{2n+3} \leq \sigma$, *and with the p-adic valuation* $|\ |_p$ *defined on the field* $\mathbb{G}$, *and suppose*

$$|\alpha_i - 1|_p < p^{-1/(p-1)} \qquad (i = 1, 2, \ldots, n-1). \tag{1.8}$$

Then the inequality

$$\left| \alpha_1^{h_1} \ldots \alpha_{n-1}^{h_{n-1}} - \alpha_n \right|_p < p^{-\delta H} \tag{1.9}$$

implies (1.5) under the hypotheses of Lemma 1.1.

The next five sections are devoted to the proof of these lemmas.

2. Preliminary Facts

We shall first establish some general results, more or less independent and, on the face of it, far removed from the problem under consideration; but we will soon see the essential role they play in our arguments.

We begin with linear inequalities.

In the modern theory of diophantine approximation, the problem of constructing auxiliary polynomials or auxiliary transcendental functions (which is what lies in wait) demands that we find a non-zero vector of not too large a height, and yielding sufficiently small values from a system of linear forms, where the number of linear forms is less than the number of components of the vector. The following two lemmas give typical results.

Lemma 2.1 *Given M linear forms*

$$L_j = \sum_{k=1}^{N} \alpha_{jk} z_k \qquad (j = 1, 2, \ldots, M)$$

in $N > M$ variables with real coefficients α_{jk}, suppose

$$|\alpha_{jk}| \le U \qquad (j = 1, 2, \ldots, M; \; k = 1, 2, \ldots, N)$$

for some integer U. Then for any integer $T > 1$ there are integers $z_1, z_2, \ldots, z_N$ not all zero such that

$$\begin{aligned} |L_j| &< 2T^{-1} & (j &= 1, 2, \ldots, M), \\ |z_k| &\le V = (NUT)^{M/(N-M)} & (k &= 1, 2, \ldots, N). \end{aligned} \qquad (2.1)$$

Proof. Put $Z = \lfloor (NUT)^{M/(N-M)} \rfloor$ and consider the values of the linear forms $L_1, \ldots, L_M$ in integer vectors $z = (z_1, \ldots, z_N)$ subject to $0 \le z_k \le Z$ ($k = 1, 2, \ldots, N$). These values form a set $\mathbb{S}$ of $(Z+1)^N$ points in the M-fold topological product of the interval $[-NUZ, NUZ]$ with itself. Break up each of these intervals into $D = NUTZ$ equal parts, thereby obtaining D^M hypercubes with edge length $2T^{-1}$. Since $D/NUT = Z \le (NUT)^{M/(N-M)} < Z+1$, we have

$$D^M \le (NUT)^{MN(N-M)} < (Z+1)^N.$$

Therefore the number of hypercubes is less than the number of elements of the set $\mathbb{S}$, so at least one hypercube contains two points of $\mathbb{S}$, corresponding to integer vectors $z^{(1)}$ and $z^{(2)}$, say. Then the integer vector $z^{(0)} = z^{(1)} - z^{(2)} \ne 0$ has components which satisfy (2.1), and since the points

$$\left(L_1(z^{(1)}), \ldots, L_M(z^{(1)}) \right), \qquad \left(L_1(z^{(2)}), \ldots, L_M(z^{(2)}) \right),$$

lie in the same hypercube of edge length $2T^{-1}$, we have

$$\max_j \left| L_j(z^{(0)}) \right| = \max_j \left| L_j(z^{(1)}) - L_j(z^{(2)}) \right| \le 2T^{-1}.$$

Lemma 2.2 *If in the previous lemma the α_{jk} are complex numbers and if $N > 2M$, then there exists a non-trivial set of integers $z_1, z_2, \ldots, z_N$ satisfying*

$$|L_j| < 4T^{-1} (j = 1, 2, \ldots, M),$$
$$|z_k| \le V' = (NUT)^{2M/(N-2M)} (k = 1, 2, \ldots, N).$$

Proof. We replace each linear form L_j having at least one complex coefficient by the pair of forms

$$L_j' = \sum_{k=1}^{N} \operatorname{Re} \alpha_{jk} z_k, \qquad L_j'' = \sum_{k=1}^{N} \operatorname{Im} \alpha_{jk} z_k \qquad (j = 1, 2, \ldots, M).$$

Then the result follows on applying Lemma 2.1 to the new system.

The following two lemmas concern the binomial coefficients

$$\binom{\gamma}{m} = \gamma(\gamma - 1)\ldots(\gamma - m + 1)/m!,$$

where $\gamma, m \geq 0$ are integers.

Lemma 2.3 *For any integers* $\gamma, m, s \geq 0$, *there exist integers* $\sigma_{r,s}^{(m)}$ *satisfying* $0 \leq \sigma_{r,s}^{(m)} \leq 2^s(m + s)^s$ $(r = 0, 1, \ldots, s)$ *such that*

$$\binom{\gamma}{m}\gamma^s = \sum_{r=0}^{s} \sigma_{r,s}^{(m)}\binom{\gamma}{m + r}.$$

Proof. Let Δ be a linear operator acting on the linear space of infinite sequences $(\alpha_0, \alpha_1, \ldots)$ according to the rule

$$\Delta[\alpha_k] = (k + 1)\alpha_{k+1} + k\alpha_k \qquad (k = 0, 1, \ldots).$$

Then for any integer s,

$$\Delta^s[\alpha_k] = \sum_{r=0}^{s} \sigma_{r,s}^{(k)}\alpha_{k+r}, \tag{2.2}$$

where $\sigma_{r,s}^{(k)}$ are integers. Let

$$\tau_s^{(k)} = \max_{0 \leq r \leq s} \sigma_{r,s}^{(k)}. \tag{2.3}$$

In particular, $\tau_1^{(k)} = k + 1$. For $s \geq 1$ we have

$$\Delta^{s+1}[\alpha_k] = \sum_{r=0}^{s} \sigma_{r,s}^{(k)}\Delta[\alpha_{k+r}] =$$

$$= \sum_{r=0}^{s-1}(k + r + 1)(\sigma_{r,s}^{(k)} + \sigma_{r+1,s}^{(k)})\alpha_{k+r+1} + (k + s + 1)\sigma_{s,s}^{(k)}\alpha_{k+s+1}.$$

We see that $\tau_{s+1}^{(k)} \leq 2(k + s)\tau_s^{(k)}$, and so

$$\tau_{s+1}^{(k)} < 2^s(k + s)^s\tau_1^{(k)} < 2^{s+1}(k + s + 1)^{s+1},$$

that is,

$$\tau_s^{(k)} < 2^s(k + s)^s \qquad (s = 1, 2, \ldots). \tag{2.4}$$

Now consider the differential operator

$$D = (1 + x)\frac{d}{dx}$$

acting on the field of formal power series

$$g = \sum_{k=0}^{\infty} \alpha_k x^k.$$

Clearly

$$Dg = (1+x)\frac{dg}{dx} = \sum_{k=0}^{\infty} \Delta[\alpha_k] x^k, \tag{2.5}$$

where Δ is the operator defined above. Take the special series

$$g = (1+x)^{\gamma} = \sum_{m=0}^{\infty} \binom{\gamma}{m} x^m.$$

Then

$$D^s g = \gamma^s (1+x)^{\gamma} = \sum_{m=0}^{\infty} \binom{\gamma}{m} \gamma^s x^m.$$

On the other hand, (2.5) yields

$$D^s g = \sum_{k=0}^{\infty} \Delta^s \left[\binom{\gamma}{k} \right] x^k,$$

and so

$$\binom{\gamma}{m} \gamma^s = \Delta^s \left[\binom{\gamma}{k} \right] \qquad (m = 0, 1, \ldots).$$

The lemma then follows from (2.2), (2.3) and (2.4).

Lemma 2.4 *Let γ, m, a, b be integers, with $|\gamma| \le a$, $1 \le m \le b$, and $a/b = \lambda > 2$. Then*

$$\left| \binom{\gamma}{m} \right| < (e\lambda + 1)^b. \tag{2.6}$$

Proof. If $0 \le \gamma < m$, then $\binom{\gamma}{m} = 0$. If $\gamma \ge m$, then $\binom{\gamma}{m} = \dfrac{\gamma!}{m!(\gamma - m)!}$ is the number of combinations of γ elements taken m at a time, and we have $0 < \binom{\gamma}{m} \le \binom{a}{m} \le \binom{a}{b}$, since $b \le a/2$. To estimate $\binom{a}{b}$ we apply Stirling's formula

$$n! = \sqrt{2\pi n} \left(\frac{n}{e} \right)^{n+\theta/12n} \qquad (0 < \theta < 1).$$

We find

$$\binom{a}{b} \leq \frac{1}{\sqrt{2\pi}} \sqrt{\frac{a}{b(a-b)}} \frac{a^a e^{-a+b+a-b+\theta'}}{b^b (a-b)^{a-b}} <$$

$$< \frac{e^{1/6}}{\sqrt{2\pi}} \sqrt{\frac{\lambda}{b(\lambda-1)}} \frac{(\lambda b)^{\lambda b}}{b^b b^{\lambda b - b}(\lambda-1)^{\lambda b - b}} =$$

$$= \frac{e^{1/6}}{\sqrt{2\pi}} \sqrt{\frac{\lambda}{b(\lambda-1)}} \left(\frac{\lambda^\lambda}{(\lambda-1)^{\lambda-1}} \right)^b .$$

Since

$$\frac{\lambda^\lambda}{(\lambda-1)^{\lambda-1}} = \frac{\lambda}{(1-\frac{1}{\lambda})^{\lambda-1}} < \lambda e^{\frac{1}{\lambda}(\lambda-1)} \leq \lambda e,$$

we have

$$\binom{a}{b} < \frac{e^{1/6}}{\sqrt{2\pi}} \sqrt{2}(e\lambda)^b < (e\lambda)^b. \tag{2.7}$$

If $\gamma < 0$, then

$$\binom{\gamma}{m} = \frac{(-|\gamma|-1)\dots(-|\gamma|-m+1)}{m!} =$$

$$= (-1)^m \frac{|\gamma|(|\gamma|+1)\dots(|\gamma|+m-1)}{m!} = (-1)^m \binom{|\gamma|+m-1}{m}.$$

Therefore $\left| \binom{\gamma}{m} \right| < \binom{|\gamma|+m}{m} \leq \binom{a+b}{b}$, and on applying (2.7) to $\binom{a+b}{b}$, we obtain (2.6).

3. The Auxiliary System of Linear Equations

We turn now to the proof of Lemma 1.1. This proof decomposes into two stages: firstly, the derivation of an auxiliary system of linear equations (3.1) from the conditions of the lemma, its premise (1.4), and the contrary of its conclusion (1.5); and secondly, the extraction of a contradiction by analysis of this system. For the first stage we use constructions and extrapolation techniques from Baker ([8], [13]), with improvements taken from [215]. For the second stage we use the p-adic method described in [203], [209].

Lemma 3.1 *The contrary of (1.5) implies the existence of integers $p(\lambda) = p(\lambda_1, \dots \lambda_n)$, not all zero, such that*

$$\sum_{\lambda_1=0}^{L_1} \dots \sum_{\lambda_n=0}^{L_n} p(\lambda) \left(\alpha_1^{\lambda_1} \dots \alpha_n^{\lambda_n} \right)^\ell \binom{\gamma_1}{m_1} \dots \binom{\gamma_{n-1}}{m_{n-1}} = 0, \tag{3.1}$$

where

$$\gamma_i = \lambda_i + h_i \lambda_n,$$

$$\binom{\gamma_i}{m_i} = \gamma_i(\gamma_i - 1)\ldots(\gamma_i - m + 1)/m_i! \qquad (i = 1, 2, \ldots, n-1), \qquad (3.2)$$

and where the integer parameters $\ell, m_1, \ldots, m_{n-1}$ vary in the ranges

$$0 \le \ell \le \sigma^{2(n+1)},$$

$$0 \le m_i \le \frac{1}{2}\sigma^{-2(n+1)}H \qquad (i = 1, 2, \ldots, n-1),$$

and where

$$L_i = L = \lfloor \sigma^{-2(n+2)}H \rfloor \qquad (i = 1, 2, \ldots, n-1), \qquad L_n = \lfloor \sigma^{2n} \rfloor.$$

Proof. The proof decomposes, in turn, into three steps which we deal with over the next several sections.

Consider a system of linear forms with respect to unknowns $p(\lambda) = p(\lambda_1, \ldots \lambda_n)$:

$$\sum_\lambda p(\lambda)(\alpha_1^{\lambda_1} \ldots \alpha_n^{\lambda_n})^\ell \binom{\gamma_1}{m_1} \ldots \binom{\gamma_{n-1}}{m_{n-1}}, \qquad (3.3)$$

where $\ell, m_1, \ldots, m_{n-1}$ vary in the ranges $0 \le \ell \le \sigma$, $0 \le m_1, \ldots, m_{n-1} \le \sigma^{-2(n+1)}H$. We shall find a bound for the coefficients of this system, and of the conjugate systems.

The size

$$\left| (\alpha_1^{\lambda_1} \ldots \alpha_n^{\lambda_n})^\ell \binom{\gamma_1}{m_1} \ldots \binom{\gamma_{n-1}}{m_{n-1}} \right|$$

does not exceed

$$(gB)^{nL\ell}(gA)^{L_n\ell} \left(\max_i \left| \binom{\gamma_i}{m_i} \right| \right)^{n-1},$$

because $\overline{|\alpha_i|} \le gB$ $(1 \le i \le n-1)$, and $\overline{|\alpha_n|} \le gA$ by (1.4) of Chap. II. Multiplying (3.3) by $(a_1^{L_1} \ldots a_n^{L_n})^\ell$, where each a_i is the least natural number such that $a_i\alpha_i$ is integer, we have a system of linear forms with respect to $p(\lambda)$ with coefficients from I_G, the ring of integers of the field G. It is essential here that the numbers (3.2) be integral. The size of the coefficients is bounded by

$$(gB^2)^{nL\ell}(gA^2)^{L_n\ell} \left(\max_i \left| \binom{\gamma_i}{m_i} \right| \right)^{n-1}. \qquad (3.4)$$

Application of Lemma 2.4 shows that

$$\max_i \left| \binom{\gamma_i}{m_i} \right| < e^{2(n+1)\sigma^{-2(n+1)}\ln \sigma \cdot H}, \qquad (3.5)$$

in view of the restrictions on γ_i and m_i imposed by the conditions of Lemma 3.1. Therefore (3.4) does not exceed

$$\exp\{3n\ln B\cdot\sigma^{-2n-3}H+3\ln A\cdot\sigma^{2n+1}+2(n^2-1)\sigma^{-2(n+1)}H\ln\sigma\}<e^{\sigma^{-2n-1}H}$$

3.6

because of (1.3). The number of linear forms is bounded:

$$M\le 2\sigma(\sigma^{-2(n+1)}H+1)^{n-1}\le$$

$$\le 2\sigma\sigma^{-2(n^2-1)}H^{n-1}\left(1+\frac{\sigma^{2(n+1)}}{H}\right)^{n-1}<2e\sigma^{-2n^2+3}H^{n-1},$$

where we have assumed the opposite of (1.5), namely

$$H\ge\sigma^{4(n+1)}\ln A.\tag{3.7}$$

The number of independent variables $p(\lambda)$ is bounded above and below:

$$N\ge(\sigma^{-2(n+2)}H)^{n-1}\sigma^{2n}=\sigma^{-2n^2+4}H^{n-1},$$

$$N\le\sigma^{-2n^2+4}H^{n-1}\left(1+\frac{\sigma^{2(n+2)}}{H}\right)^{n-1}\left(1+\frac{1}{\sigma^{2n}}\right)<2e\sigma^{-2n^2+4}H^{n-1},$$

applying (3.7). Hence the quotient $2M/(N-2M)$ is bounded above by

$$\frac{4e\sigma^{-2n^2+3}H^{n-1}}{\sigma^{-2n^2+4}H^{n-1}-4e\sigma^{-2n^2+3}H^{n-1}}=\frac{4e}{\sigma-4e}<\frac{16}{\sigma}.$$

We now apply Lemma 2.2 to the system of linear forms (3.3) multiplied by $a_1^{L_1}\ldots a_n^{L_n}$. Give U the value (3.6), and give T the value $\exp(2g\sigma^{-2n-1}H)$. Then there exist integers $p(\lambda)$ satisfying

$$0\ne\max_\lambda|p(\lambda)|\le V',\tag{3.8}$$

with

$$V'=(NUT)^{2M/(N-2M)}<$$

$$<\left(2e\sigma^{-2n^2+4}H^{n-1}e^{(2g+1)\sigma^{-2n-1}}H\right)^{16/\sigma}<e^{32(g+1)\sigma^{-2n-2}H}.\tag{3.9}$$

The corresponding absolute values of the linear forms (3.3), after multiplication by $(a_1^{L_1}\ldots a_n^{L_n})^\ell$, are at most

$$4T^{-1}=4e^{-2g\sigma^{-2n-1}H}.\tag{3.10}$$

Let Λ be one of these values. Thus $\Lambda\in I_{\mathbb{G}}$, and if $\Lambda\ne 0$, then $|\mathrm{Nm}(\Lambda)|\ge 1$, hence $|\Lambda|\ge\lceil\Lambda\rceil^{-g+1}$. Because (3.4) is bounded by (3.6), and V' is at most (3.9), we obtain

$$\lceil\Lambda\rceil\le NV'e^{\sigma^{-2n-1}H}<$$

$$<2e\sigma^{-2n^2+4}H^{n-1}e^{(32(g+1)\sigma^{-2n-2}+\sigma^{-2n-1})H}<e^{2\sigma^{-2n-1}H}.$$

Consequently, $|\Lambda| > e^{-2(g-1)\sigma^{-2n-1}H}$, which is inconsistent with inequalities (3.10) and (3.7). We must conclude that all the numbers Λ are zero. Thus we have the following intermediate result:

(I) *There exist rational integers* $p(\lambda) = p(\lambda_1,\ldots,\lambda_n)$ *satisfying inequalities (3.8) and (3.9) such that all the sums (3.3) are zero, for every* $\ell, m_1,\ldots,m_{n-1}$ *in the range*

$$0 \leq \ell \leq \sigma, \qquad 0 \leq m_i \leq \sigma^{-2(n+1)}H \qquad (i=1,2,\ldots,n-1). \tag{3.11}$$

4. Transition to the Auxiliary Function

We now introduce an auxiliary function of complex variables $z_1,\ldots,z_{n-1}$:

$$\Phi = \Phi(z_1,\ldots,z_{n-1}) = \sum_\lambda p(\lambda)\alpha_1^{\gamma_1 z_1}\ldots\alpha_{n-1}^{\gamma_{n-1}z_{n-1}},$$

where $p(\lambda) = p(\lambda_1,\ldots,\lambda_n)$ are the integers whose existence is asserted by the conclusion **(I)** of the previous paragraph, and

$$\alpha_i^{\gamma_i z_i} = \exp\{\gamma_i z_i \ln \alpha_i\} \qquad (i=1,2,\ldots,n-1),$$

in which we take the principal branch of the logarithm. For brevity, we will write

$$\Phi^*_{m_1,\ldots,m_{n-1}}(z_1,\ldots,z_{n-1}) = \prod_{i=1}^{n-1} \alpha_i^{m_i z_i}\Phi_{m_1,\ldots,m_{n-1}}(z_1,\ldots,z_{n-1}),$$

$$\Phi_{m_1,\ldots,m_{n-1}} = \frac{D_1^{m_1}}{m_1!}\cdots\frac{D_{n-1}^{m_{n-1}}}{m_{n-1}!}\Phi,$$

where D_i is the differential operator $\alpha_i^{-z_i}\dfrac{d}{dz_i}$ $(1 \leq i \leq n-1)$.

For $\ell \leq \sigma$, we have by (1.4)

$$\left|\left(\alpha_1^{h_1}\ldots\alpha_{n-1}^{h_{n-1}}\right)^{\lambda_n\ell} - \alpha_n^{\lambda_n\ell}\right| \leq \lambda_n\ell(1+|\alpha_n|)^{\lambda_n\ell}e^{-\delta H} \leq$$

$$\leq L_n\ell(1+gA)^{L_n\ell}e^{-\delta H} \leq \sigma^{2n+1}A^{2\sigma^{2n+1}}e^{-\delta H} < e^{-\frac{1}{2}\delta H}, \tag{4.1}$$

given that (3.7) holds. The inequalities $(gB)^{-1} \leq |\alpha_i| \leq gB$ $(i=1,2,\ldots,n-1)$ imply

$$|\ln\alpha_i| \leq \big|\ln|\alpha_i|\big| + \pi \leq \ln(gB) + \pi < B \qquad (i=1,2,\ldots,n-1).$$

Thus we have, for all $m_1,\ldots,m_{n-1}$,

$$\left|\prod_{i=1}^{n-1}(\ln\alpha_i)^{m_i}\right| < B^{n\sigma^{-2(n+1)}H}.$$

In view of (3.5) we have

$$\left|\prod_{i=1}^{n-1} \alpha_i^{\lambda_i \ell}(\ln \alpha_i)^{m_i}\binom{\gamma_i}{m_i}\right| \leq$$

$$\leq (gB)^{nL\ell}B^{n\sigma^{-2(n+1)}}H e^{2(n+1)\sigma^{-2(n+1)}\ln \sigma \cdot H} < e^{\sigma^{-2n-1}H}. \quad (4.2)$$

We now estimate the difference between the sum (3.3) multiplied by

$$P = \prod_{i=1}^{n-1}(\ln \alpha_i)^{m_i} \quad (4.3)$$

and the expression

$$\Phi^*_{m_1,\dots,m_{n-1}}(z_1,\dots,z_{n-1})\big|_{z_i=\ell} = P\sum_{\lambda}p(\lambda)\prod_{i=1}^{n-1}\alpha_i^{\gamma_i\ell}\binom{\gamma_i}{m_i}.$$

Clearly the absolute value of this difference does not exceed

$$|P|\sum_{\lambda}|p(\lambda)|\prod_{i=1}^{n-1}\left|\binom{\gamma_i}{m_i}\right|\left|(\alpha_1^{\lambda_1}\dots\alpha_n^{\lambda_n})^\ell - (\alpha_1^{\gamma_1}\dots\alpha_{n-1}^{\gamma_{n-1}})^\ell\right|. \quad (4.4)$$

Since $\gamma_i = \lambda_i + h_i\lambda_n$,

$$\left|(\alpha_1^{\lambda_1}\dots\alpha_n^{\lambda_n})^\ell - (\alpha_1^{\gamma_1}\dots\alpha_{n-1}^{\gamma_{n-1}})^\ell\right| = \left|\alpha_1^{\lambda_1}\dots\alpha_{n-1}^{\lambda_{n-1}}\right|^\ell\left|\alpha_n^{\lambda_n\ell} - (\alpha_1^{h_1}\dots\alpha_{n-1}^{h_{n-1}})^{\lambda_n\ell}\right|.$$

The bounds (3.8), (3.9), (4.1), (4.2) imply an upper bound of $e^{-\delta H/3}$ for the sum (4.4). Thus we have the following intermediate result as a consequence of assertion (I) of the last section:

(II) *For any integers* $\ell, m_1,\dots,m_{n-1}$ *in the ranges (3.11),*

$$|\Phi^*_{m_1,\dots,m_{n-1}}(\ell,\dots,\ell)| < e^{-\delta H/3}. \quad (4.5)$$

Thus we have passed from equations (3.1) valid in the ranges (3.11) to inequalities (4.5) valid in the same ranges. The following arguments extend (4.5) onto a wider range for ℓ under slight restrictions on the ranges for $m_1,\dots,m_{n-1}$, and thence yield (3.1) on this wider range for ℓ.

We now bound the absolute value of $\Phi^*_{m_1,\dots,m_{n-1}}(z,\dots,z)$ for complex z. Since

$$\Phi^*_{m_1,\dots,m_{n-1}}(z,\dots,z) = P\sum_{\lambda}p(\lambda)\prod_{i=1}^{n-1}\alpha_i^{\gamma_i z}\binom{\gamma_i}{m_i},$$

we have by arguments similar to those above

$$\left|\Phi^*_{m_1,\dots,m_{n-1}}(z,\dots,z)\right| < e^{c_1\sigma^{-2n-2}\ln \sigma \cdot H}\max_{\lambda}\left|\alpha_1^{\gamma_1 z}\dots\alpha_{n-1}^{\gamma_{n-1}z}\right|, \quad (4.6)$$

where $c_1 = 32(g+1) + 2n^2 + n\ln B + 3$. Since

$$\left|\alpha_i^{\lambda_i z}\right| = \exp\{\mathrm{Re}(\lambda_i z \ln \alpha_i)\} \le \exp\{L|z||\ln \alpha_i|\} \le$$
$$\le \exp\{\sigma^{-2(n+2)} H|z|(\ln(gB) + \pi\} < \exp\{3 \ln B \cdot \sigma^{-2(n+2)} H|z|\},$$

we find

$$\left|\alpha_1^{\gamma_1 z} \ldots \alpha_{n-1}^{\gamma_{n-1} z}\right| = \left|\alpha_1^{\lambda_1 z} \ldots \alpha_{n-1}^{\lambda_{n-1} z} \left(\alpha_1^{h_1} \ldots \alpha_{n-1}^{h_{n-1}}\right)^{\lambda_n z}\right| <$$
$$< \exp\{3n \ln B \cdot \sigma^{-2(n+2)} H|z|\} \left|\left(\alpha_1^{h_1} \ldots \alpha_{n-1}^{h_{n-1}}\right)^{\lambda_n z}\right|. \quad (4.7)$$

Put $\alpha = \alpha_1^{h_1} \ldots \alpha_{n-1}^{h_{n-1}} \alpha_n^{-1}$. Then (1.4) implies

$$|\alpha - 1| \le |\alpha_n|^{-1} e^{-\delta H} \le gA e^{-\delta H} < 1/2,$$

and so

$$|\alpha^{\lambda_n z}| \le \exp\{L_n|z||\ln \alpha|\} \le \exp\{\sigma^{2n}|z|(\ln \tfrac{3}{2} + \pi)\}.$$

Now

$$\left|\left(\alpha_1^{h_1} \ldots \alpha_{n-1}^{h_{n-1}}\right)^{\lambda_n z}\right| = |(\alpha \alpha_n)^{\lambda_n z}| \le$$
$$\le e^{\sigma^{2n}|z|(|\ln \alpha_n|+5)} \le e^{\sigma^{2n}|z|(\ln(gA)+\pi+5)} \le e^{3\sigma^{2n}|z|\ln A}.$$

Finally, by (4.6), (4.7) and the last bound,

$$\left|\Phi^*_{m_1,\ldots,m_{n-1}}(z,\ldots,z)\right| < \sigma^{(c_1+1)\sigma^{-2n-2} H} A^{3\sigma^{2n}|z|}. \quad (4.8)$$

We now bound

$$\Phi^*_{m_1,\ldots,m_{n-1}}(\ell,\ldots,\ell) \qquad (0 \le \ell \le \sigma^{2(n+1)}) \quad (4.9)$$

for integer ℓ. Put

$$Q = P' \sum_\lambda p(\lambda) q(\lambda, \ell), \qquad P' = (a_1 \ldots a_{n-1})^{L\ell} a_n^{L_n \ell},$$
$$q(\lambda, l) = (\alpha_1^{\lambda_1} \ldots \alpha_n^{\lambda_n})^\ell \binom{\gamma_1}{m_1} \ldots \binom{\gamma_{n-1}}{m_{n-1}}. \quad (4.10)$$

By analogy with the previous bounds, we obtain

$$P' < B^{nL\ell} A^{L_n \ell},$$

and using (3.5) we find that

$$\overline{|q(\lambda, \ell)|} <$$
$$< \exp\left\{(n\sigma^{-2n-4} \ln gB \cdot H + \sigma^{2n} \ln gA)\ell + 2n^2 \sigma^{-2n-2} \ln \sigma \cdot H\right\} <$$
$$< \exp\left\{c_2 \sigma^{-2n-4} H\ell + 2n^2 \sigma^{-2n-2} \ln \sigma \cdot H\right\},$$

where $c_2 = 3n \ln B + 1$. Now applying (3.8) and (3.9), we find

$$\overline{|Q|} < \exp\left\{c_3\sigma^{-2n-2}\ln\sigma\cdot H + c_4\sigma^{-2n-4}H\ell\right\},$$

where $c_3 = 2(n-1)^2 + 32(g+1)$ and $c_4 = 4n\ln B + 2$. Since $Q \in I_{\mathbb{G}}$, provided $Q \neq 0$ we have

$$|Q| > \exp\left\{-g(c_3\sigma^{-2n-2}\ln\sigma\cdot H + c_4\sigma^{-2n-4}H\ell)\right\}. \tag{4.11}$$

Now (1.4) implies

$$\left|(\alpha_1^{\gamma_1}\ldots\alpha_{n-1}^{\gamma_{n-1}})^\ell - (\alpha_1^{\lambda_1}\ldots\alpha_{n-1}^{\lambda_{n-1}})^\ell\alpha_n^{\lambda_n\ell}\right| =$$

$$= \left|\alpha_1^{\lambda_1}\ldots\alpha_{n-1}^{\lambda_{n-1}}\right|^\ell\left|(\alpha_1^{h_1}\ldots\alpha_{n-1}^{h_{n-1}})^{\lambda_n\ell} - \alpha_n^{\lambda_n\ell}\right| <$$

$$< (gB)^{nL\ell}\lambda_n\ell(1+|\alpha_n|)^{\lambda_n\ell}e^{-\delta H} < \exp\left\{c_5\sigma^{-2n-4}H\ell - \delta H\right\},$$

where $c_5 = 2n\ln B + 2$. So we find

$$\left|P^{-1}\Phi^*_{m_1,\ldots,m_{n-1}}(\ell,\ldots,\ell) - Q/P'\right| <$$

$$< NV'\exp\left\{2(n^2-1)\sigma^{-2n-1}\ln\sigma\cdot H + c_5\sigma^{-2n-4}H\ell - \delta H\right\} <$$

$$< \exp\left\{c_5(\sigma^{-2n-1}H + \sigma^{-2n-4}H\ell) - \delta H\right\} < e^{-\delta H/2}.$$

Thus when $Q = 0$ we have

$$\left|\Phi^*_{m_1,\ldots,m_{n-1}}(\ell,\ldots,\ell)\right| < |P|e^{-\delta H/2} < e^{-\delta H/3}. \tag{4.12}$$

If $Q \neq 0$, we obtain from (4.11)

$$\left|P^{-1}\Phi^*_{m_1,\ldots,m_{n-1}}(\ell,\ldots,\ell)\right| > |Q|(P')^{-1} - e^{-\delta H/2} >$$

$$> \exp\left\{-g(c_3\sigma^{-2n-2}\ln\sigma\cdot H + c_4\sigma^{-2n-4}H\ell) -\right.$$

$$\left. - n\ln B\cdot\sigma^{-2n-4}H\ell - \sigma^{-2n-4}H\ell\right\} - e^{-\delta H/2} >$$

$$> \exp\left\{-c_6(\sigma^{-2n-2}\ln\sigma\cdot H + \sigma^{-2n-4}H\ell)\right\} - e^{-\delta H/2}$$

$$> \frac{1}{2}\exp\left\{-c_6(\sigma^{-2n-2}\ln\sigma\cdot H + \sigma^{-2n-4}H\ell)\right\},$$

where $c_6 = (4g+1)n\ln B + gc_3$. Therefore,

$$\left|\Phi^*_{m_1,\ldots,m_{n-1}}(\ell,\ldots,\ell)\right| > \frac{1}{2}|P|\exp\left\{-c_6(\sigma^{-2n-2}\ln\sigma\cdot H + \sigma^{-2n-4}H\ell)\right\}. \tag{4.13}$$

To obtain the final bound it is necessary to estimate $|P|$ from below. We note that for any complex z the inequality $|e^z - 1| \leq |z|e^{|z|}$ holds, as a consequence of the power series expansion for e^z. Therefore, for $i = 1, 2, \ldots, n-1$,

$$|\alpha_i - 1| = |e^{\ln\alpha_i} - 1| \leq |\ln\alpha_i|e^{|\ln\alpha_i|} < |\ln\alpha_i|e^{\ln(gB)+\pi},$$

$$|\ln\alpha_i| \geq |\alpha_i - 1|e^{-\pi}(gB)^{-1} > |\alpha_i - 1|B^{-2}.$$

Now bound $|\alpha_i - 1|$ from below: Since $a_i(\alpha_i - 1) \in I_{\mathbb{G}}$,

$$|a_i(\alpha_i - 1)| \geq \lceil a_i(\alpha_i - 1) \rceil^{-g+1} > B^{-3(g-1)},$$

and so

$$|\ln \alpha_i| > B^{-3(g-1)} a_i^{-1} B^{-2} \geq B^{-3g} \qquad (i = 1, 2, \ldots, n-1).$$

Hence $P > B^{-3gn\sigma^{-2n-2}H}$, and the right-hand side of (4.13) exceeds

$$\frac{1}{2} \exp\left\{-(c_6 + 3gn)(\sigma^{-2n-2} \ln \sigma \cdot H + \sigma^{-2n-4} H\ell)\right\}. \tag{4.14}$$

As a result we have the following assertion concerning the values (4.9):

(III) *For any integer ℓ in the range $0 \leq \ell \leq \sigma^{2(n+1)}$, either inequality (4.12) holds, or inequality (4.13) holds with (4.14) replacing the right-hand side. Here $m_1, \ldots, m_{n-1}$ take any values in the range $0 \leq m_i \leq \sigma^{-2n-2} H$ ($i = 1, 2, \ldots, n-1$).*

This will allow the extension of equations (3.1) from the initial range $0 \leq \ell \leq \sigma$ to a strictly larger range $0 \leq \ell \leq \sigma^{2(n+1)}$.

5. Analytic-arithmetical Extrapolation

Let $S = \lfloor \sigma^{-2n-5/2} H \rfloor$, $M_k = \sigma^{-2n-2} H - kS$, $\ell_k = \lfloor \sigma^k \rfloor$, and $R_k = \ell_k \sigma^{3/2}$ ($k = 1, 2, \ldots, 2(n+1)$).

It follows from assertion (II) that for $k = 1$ we have

$$\left|\Phi^*_{m_1,\ldots,m_{n-1}}(\ell, \ldots, \ell)\right| < e^{-\delta H/3}, \tag{5.1}$$

where $\ell, m_1, \ldots, m_{n-1}$ run through all integers in the ranges $0 \leq \ell \leq \ell_k$, $0 \leq m_i \leq M_k$ ($i = 1, 2, \ldots, n-1$). We suppose that (5.1) holds for a given $k < 2(n+1)$, and show that the corresponding assertion holds for $k+1$ instead of k. Thus we extend the range of ℓ step by step, while restricting the range of $m_1, \ldots, m_{n-1}$.

For fixed $m_1, \ldots, m_{n-1} \leq M_{k+1}$, set

$$f(z) = \Phi^*_{m_1,\ldots,m_{n-1}}(z, \ldots, z);$$

then

$$f^{(s)}(z) = \left(\frac{\partial}{\partial z_1} + \ldots + \frac{\partial}{\partial z_{n-1}}\right)^s \Phi^*_{m_1,\ldots,m_{n-1}}(z_1, \ldots, z_{n-1})\Big|_{z_i = z}$$

$$= \sum_{s_1+\ldots+s_{n-1}=s} \frac{s!}{s_1! \ldots s_{n-1}!} \prod_{i=1}^{n-1} (\ln \alpha_i)^{m_i+s_i} \sum_{\lambda} p(\lambda) \prod_{i=1}^{n-1} \binom{\gamma_i}{m_i} \gamma_i^{s_i} \alpha_i^{\gamma_i z}.$$

By Lemma 2.3, $\binom{\gamma_i}{m_i} \gamma_i^{s_i} = \sum_{r_i=0}^{s_i} \sigma_{r_i,s_i}^{(m_i)} \binom{\gamma_i}{m_i + r_i}$, where the $\sigma_{r_i,s_i}^{(m_i)}$ are integers,

with $0 \leq \sigma_{r_i,s_i}^{(m_i)} \leq 2^{s_i}(m_i + s_i)^{s_i}$ ($1 \leq i \leq n-1$). Denoting $\sigma_{r_1,s_1}^{(m_1)} \ldots \sigma_{r_{n-1},s_{n-1}}^{(m_{n-1})}$ by $\sigma_{\vec{r},\vec{s}}^{(\vec{m})}$, we have

$$f^{(s)}(z) =$$

$$\sum_{s_1+\ldots+s_{n-1}=s} \frac{s!}{s_1!\ldots s_{n-1}!} \sum_{r=0}^{\bar{s}} \sigma_{\bar{r},\bar{s}}^{(\bar{m})} \prod_{i=1}^{n-1} (\ln \alpha_i)^{s_i-r_i} \Phi^*_{m_1+r_1,\ldots,m_{n-1}+r_{n-1}}(z,\ldots,z).$$

If $s \leq S$ and $m_i \leq M_{k+1}$, then $m_i + s_i \leq M_k$ $(1 \leq i \leq n-1)$, and by the inductive hypothesis we have, for all integer ℓ in the range $0 \leq \ell \leq \ell_k$,

$$|f^{(s)}(\ell)| < s!s^n \max_{s_1+\ldots+s_{n-1}=s} \prod_{i=1}^{n-1} \frac{2^{s_i}(m_i + s_i)^{s_i}(s_i+1)}{s_i!} \cdot B^s e^{-\delta H/3},$$

noting that $|\ln \alpha_i| < B$ $(i = 1, 2, \ldots, n-1)$. Furthermore,

$$\prod_{i=1}^{n-1} 2^{s_i}(s_i+1) < 2^s \left(\frac{1}{n-1} \sum_{i=1}^{n-1} s_i + 1 \right)^{n-1} < 2^s(s+1)^n < e^{2\sigma^{-2n-5/2}H}.$$

Since $s_i! > (s_i/e)^{s_i}$ for $s_i \geq 1$, we find that if all the s_i are non-zero then

$$\prod_{i=1}^{n-1} \frac{(m_i + s_i)^{s_i}}{s_i!} < \prod_{i=1}^{n-1} e^{s_i} \left(\frac{m_i + s_i}{s_i} \right)^{s_i} = e^s \prod_{i=1}^{n-1} \underbrace{\left(\frac{m_i + s_i}{s_i} \right) \ldots \left(\frac{m_i + s_i}{s_i} \right)}_{s_i \text{times}}$$

$$\leq e^s \left(\frac{\sum_{i=1}^{n-1}(m_i + s_i)}{\sum_{i=1}^{n-1} s_i} \right)^{\sum_{i=1}^{n-1} s_i} \leq e^s \left(\frac{(n-1)M_k}{s} \right)^s.$$

If some of the s_i are zero, we get a similar estimate by considering non-zero s_i. Since the function $(M_k/s)^s$ increases in the interval $1 \leq s \leq S$,

$$\prod_{i=1}^{n-1} \frac{(m_i + s_i)^{s_i}}{s_i!} < (ne)^S \left(\frac{M_k}{S} \right)^S < (6n)^S \sigma^{S/2} < e^{\sigma^{-2n-5/2} \ln \sigma \cdot H}.$$

Thus

$$\frac{1}{s!}|f^{(s)}(\ell)| < e^{2\sigma^{-2n-5/2} \ln \sigma \cdot H} e^{-\delta H/3} < e^{-\delta H/4} \qquad (5.2)$$

for all ℓ and s in the ranges $0 \leq \ell \leq \ell_k$, $0 \leq s \leq S$. We wish to show that for all integers ℓ in the range $0 \leq \ell \leq \ell_{k+1}$ we have

$$|f(\ell)| < e^{-\delta H/3}, \qquad (5.3)$$

since this corresponds to (5.1). To this end we use (5.2) and the assertion (III) of the previous section.

Put $F(z) = [z(z-1)\ldots(z-\ell_k+1)]^{S+1}$. Computing the integral

$$\frac{1}{2\pi i} \int_{|\xi|=R} \frac{f(\xi)\, d\xi}{F(\xi)(\xi-z)}, \qquad |z| < R = R_k$$

by the Cauchy residue theorem, we have

$$f(z) = \frac{1}{2\pi i} \int_C \frac{F(z)f(\xi)}{F(\xi)(\xi - z)} d\xi - \frac{1}{2\pi i} \sum_{s=0}^{S} \sum_{t=0}^{\ell_k - 1} \frac{f^{(s)}(t)}{s!} \int_{C_t} \frac{F(z)(\xi - t)^s}{F(\xi)(\xi - z)} d\xi,$$

$$(5.4)$$

where C is the circle with centre 0 and radius R, C_t is the circle with centre t and radius $\frac{1}{2}$, and $|z| \leq \ell_{k+1}$.

To bound the integrand on C for $z = \ell$, $0 \leq \ell \leq \ell_{k+1}$, note that

$$|F(\ell)| < (\ell + \ell_k)^{\ell_k(S+1)} < (2\ell_{k+1})^{\ell_k(S+1)},$$

$$|F(\xi)| > (R - \ell_k)^{\ell_k(S+1)} > \left(\tfrac{1}{2}R\right)^{\ell_k(S+1)},$$

$$|\xi - \ell| > R - \ell_{k+1} > \tfrac{1}{2}R.$$

In addition, by (4.8) and (3.7),

$$|f(\xi)| < \sigma^{(c_1+1)\sigma^{-2n-2}H} A^{3\sigma^{2n}R} \leq \sigma^{(c_1+1)\sigma^{-2n-2}H} e^{3\sigma^{-2n-4}HR}.$$

So the integral over the circle C is bounded by the value

$$\frac{1}{2\pi}(2\ell_{k+1})^{\ell_k(S+1)} \left(\tfrac{1}{2}R\right)^{-\ell_k(S+1)} \left(\tfrac{1}{2}R\right)^{-1} 2\pi R \times$$

$$\times \exp\{(c_1 + 1)\sigma^{-2n-2} \ln \sigma \cdot H + 3\sigma^{-2n-4} HR\} <$$

$$< 2(4\ell_{k+1}\ell_k^{-1}\sigma^{-3/2})^{\ell_k(S+1)} \exp\{(c_1 + 1)\sigma^{-2n-2} \ln \sigma \cdot H + 3\sigma^{-2n-5/2} H\ell_k\}$$

$$< \exp\left\{-\frac{\ell_k}{4}\sigma^{-2n-5/2} \ln \sigma \cdot H + (c_1 + 1)\sigma^{-2n-2} \ln \sigma \cdot H + 3\sigma^{-2n-5/2} H\ell_k\right\}$$

$$< \exp\left\{-\frac{\ell_k}{8}\sigma^{-2n-5/2} \ln \sigma \cdot H\right\},$$

where it must be remembered that $\ell_k \geq \ell_1 = \sigma$, and that σ is determined by equation (1.3).

On the circles C_t we have

$$|\xi - t| = \frac{1}{2}, \qquad |\xi - \ell| \geq \frac{1}{2}, \qquad |F(\xi)| \geq 2^{-\ell_k(S+1)}.$$

Hence the integrals on the circles C_t are bounded by

$$\frac{1}{2\pi}(2\ell_{k+1})^{\ell_k(S+1)} 2^{\ell_k(S+1)} \frac{2\pi}{2} < (4\ell_{k+1})^{\ell_k(S+1)} < \exp\left\{3n\ell_k\sigma^{-2n-5/2} \ln H\right\}.$$

Now equation (5.4) and inequality (5.2) imply

$$|f(\ell)| < \exp\left\{-\frac{\ell_k}{8}\sigma^{-2n-5/2} \ln \sigma \cdot H\right\} +$$

$$+ \ell_k(S + 1) \exp\left\{-\frac{\delta}{4}H + 3n\ell_k\sigma^{-2n-5/2} \ln \sigma \cdot H\right\} <$$

$$< 2\exp\left\{-\frac{\ell_k}{8}\sigma^{-2n-5/2} \ln \sigma \cdot H\right\}.$$

By assertion (III), if (5.3) is false then

$$|f(\ell)| > \frac{1}{2}\exp\left\{-(c_6 + 3gn)(\sigma^{-2n-2}\ln\sigma \cdot H + \sigma^{-2n-4}H\ell)\right\}$$

must hold. But comparison with the previous inequality excludes this possibility, since (1.3) implies that

$$(c_6 + 3gn)(\sigma^{-2n-2}\ln\sigma \cdot H + \sigma^{-2n-4}H\ell_{k+1}) + \ln 2 < \frac{\ell_k}{8}\sigma^{-2n-5/2}\ln\sigma \cdot H - \ln 2.$$

Thus we obtain (5.1) for all $k \leq 2(n+1)$, and in particular for $k = 2(n+1)$. To complete the proof of Lemma 3.1, we make a transition from the inequalities (5.1) to equations (3.1) by arguments similar to those used in passing from equations (3.1) with $\ell \leq \sigma$ to the inequalities (4.5).

Since it was established previously that

$$|P^{-1}\Phi^*_{m_1,\dots,m_{n-1}}(\ell,\dots,\ell) - Q/P'| < e^{-\delta H/2},$$

we now find

$$|Q| < |P'P^{-1}||\Phi^*_{m_1,\dots,m_{n-1}}(\ell,\dots,\ell)| + P'e^{-\delta H/2} <$$

$$< \exp\left\{nL\ell\ln B + L_n\ell\ln A + 3gn\sigma^{-2n-2}H\ln B - \frac{\delta}{3}H\right\} +$$

$$+ \exp\left\{nL\ell\ln B + L_n\ell\ln A - \frac{\delta}{2}H\right\} < e^{-\delta H/4}.$$

If Q were not zero then (4.11) would hold, which is inconsistent with the latter inequality. Thus we have completed the proof of Lemma 3.1.

6. Completion of the Proof of Lemma 1.1

We now prove that, given $|\alpha_n|_q \neq 1$ and (1.6) the linear equations (3.1) have no non-trivial solutions. Then we must conclude, in view of Lemma 3.1, that inequality (1.5) holds, thus proving Lemma 1.1.

The following arguments rely on the theory of p-adic analytic functions described in §3 of Chap. II.

Lemma 6.1. *Let q be a prime, $L > 0$ an integer, and*

$$p(z) = \sum_{j=0}^{L}\sigma_j z^j,$$

with $|p(z)|_q < \mu$ for all z in the disc $|z - 1|_q \leq |\delta|_q \leq 1$, where $\sigma_j, z, \delta \in \Omega_q$ $(j = 0, 1, \dots, L)$. Then

$$\max_j|\sigma_j|_q < \mu|\delta|_q^{-L} \qquad (j = 0, 1, \dots, L). \tag{6.1}$$

Proof. Express the derivatives $p^{(k)}(1)$ using the Schnirelman integral (see Lemma 3.4, Chap. II):

$$p^{(k)}(1) = k! \int_{1,\delta} \frac{p(z)(z-1)}{(z-1)^{k+1}} dz.$$

To bound the integral we use inequality (3.3) of Chap. II:

$$\left| p^{(k)}(1)/k! \right|_q \leq \max_{|z-1|_q = |\delta|_q} \left(|p(z)|_q |z-1|_q^{-k} \right) < \mu |\delta|_q^{-k}.$$

Consequently,

$$|\sigma_0 + \sigma_1 + \ldots + \sigma_L|_q \leq \mu$$
$$|\sigma_1 + \ldots + L\sigma_L|_q \leq \mu |\delta|_q^{-1}$$

$$\vdots$$

$$|\sigma_L|_q \leq \mu |\delta|_q^{-L},$$

whence (6.1) follows.

Lemma 6.2 *Let* $L_1, L_2, \ldots, L_n, L, M$ *be natural numbers with* $L_i \leq L$ *for* $i = 1, 2, \ldots, n-1$, *and suppose that* $\theta_1, \ldots, \theta_n$, $\beta_1, \ldots, \beta_{n-1}$ *are elements of* Ω_q *satisfying*

$$|\theta_i - 1|_q \leq q^{-e_q} \qquad (i = 1, 2, \ldots, n), \tag{6.2}$$

where $e_2 = 2$, $e_q = 1$ $(q \geq 3)$, $|\beta_i|_q \leq 1$ $(i = 1, 2, \ldots, n-1)$, *and let*

$$\Delta = \log \theta_n - \beta_1 \log \theta_1 - \ldots - \beta_{n-1} \log \theta_{n-1} \neq 0$$

(q-adic logarithms). If

$$|\Delta|_q > q^{-\frac{1}{2}ML_n^{-1} + 2(n-1)LL_n^{-1} + 1}, \tag{6.3}$$

then the system of equations in $\rho(\lambda) = \rho(\lambda_1, \ldots, \lambda_n)$

$$\sum_{\lambda_1=0}^{L_1} \ldots \sum_{\lambda_n=0}^{L_n} \rho(\lambda)(\theta_1^{\lambda_1} \ldots \theta_n^{\lambda_n})^\ell \binom{\gamma_1}{m_1} \ldots \binom{\gamma_{n-1}}{m_{n-1}} = 0, \tag{6.4}$$

where $\gamma_i = \lambda_i + \beta_i \lambda_n$ $(i = 1, 2, \ldots, n-1)$ *and the integers* $\ell, m_1, \ldots, m_{n-1}$ *range over the intervals*

$$0 \leq \ell \leq L_n + 1, \qquad 0 \leq m_i \leq M \quad (i = 1, 2, \ldots, n-1),$$

has only the trivial solution $\rho(\lambda) = 0$ $(0 \leq \lambda_i \leq L_i, i = 1, 2, \ldots, n)$.

Proof. For the q-adic exponential function

$$(1 + \omega)^\gamma = \sum_{m=0}^{\infty} \binom{\gamma}{m} \omega^m, \qquad \omega, \gamma \in \Omega_q,$$

provided $|\omega|_q \le q^{-e_q}$ and $|\gamma|_q \le 1$ we have

$$\left|(1+\omega)^\gamma - \sum_{m=0}^M \binom{\gamma}{m}\omega^m\right|_q < q^{-M/2}. \tag{6.5}$$

Let $u_1, \ldots, u_{n-1}$ be numbers in Ω_q with $|u_i|_q \le q^{-e_q}$ $(i = 1, 2, \ldots, n-1)$, and let ℓ be an arbitrary integer in the range $1 \le \ell \le L_n + 1$. Setting $\omega_i' = u_i - \ell \log \theta_i$, we find from (6.2) that $|\omega_i'|_q \le q^{-e_q}$ $(1 \le i \le n-1)$. Hence for $\omega_i = \exp \omega_i' - 1$, we have

$$|\omega_i|_q = |\exp \omega_i' - 1|_q = |\omega_i'|_q \le q^{-e_q} \qquad (i = 0, 1, \ldots, n-1).$$

On multiplying (6.4) by $\omega_1^{m_1} \ldots \omega_{n-1}^{m_{n-1}}$ and summing the equations so obtained over $m_1, \ldots, m_{n-1}$ from 0 to M, we have from (6.5)

$$\left|\sum_\lambda \rho(\lambda)(\theta_1^{\lambda_1} \ldots \theta_n^{\lambda_n})^\ell \exp(\gamma_1\omega_1' + \ldots + \gamma_{n-1}\omega_{n-1}')\right|_q < q^{-M/2}, \tag{6.6}$$

assuming that $\max_\lambda |\rho(\lambda)|_q = 1$. The latter may be assumed, since the system of equations (6.4) is homogeneous with respect to $\rho(\lambda)$. Now by supposing that (6.4) has a non-zero solution, we shall arrive at a contradiction.

Since $\gamma_i = \lambda_i + \beta_i\lambda_n$ $(i = 1, 2, \ldots, n-1)$, (6.6) may be rewritten as

$$\left|\sum_\lambda \rho(\lambda) \exp(u_1\lambda_1 + \ldots + u_n\lambda_n)\right|_q < \frac{1}{2}q^{-M/2}, \tag{6.7}$$

where $u_n = \beta_1 u_1 + \ldots + \beta_{n-1}u_{n-1} + \ell\Delta$. Setting

$$\sigma(\lambda_n) = \sigma(\lambda_n; u_1, \ldots, u_{n-1}) =$$
$$= e^{(\beta_1 u_1 + \ldots + \beta_{n-1}u_{n-1})\lambda_n} \sum_{\lambda_1, \ldots, \lambda_{n-1}} \rho(\lambda) \exp\{u_1\lambda_1 + \ldots + u_{n-1}\lambda_{n-1}\}, \tag{6.8}$$

we rewrite (6.7) as

$$\left|\sum_{\lambda_n=0}^{L_n} \sigma(\lambda_n) \exp(\ell\Delta\lambda_n)\right|_q < q^{-M/2}. \tag{6.9}$$

We now consider (6.9) for fixed $u_1, \ldots, u_{n-1}$, with ℓ running through all integer values $1 \le \ell \le L_n + 1$. This system of inequalities shows that at points $z_\ell = \exp(\ell\Delta)$, the polynomial $P(z) = \sum_{\lambda_n=0}^{L_n} \sigma(\lambda_n)z^{\lambda_n}$ satisfies

$$|P(z_\ell)|_q < q^{-M/2} \qquad (\ell = 1, 2, \ldots, L_n + 1). \tag{6.10}$$

Applying the Lagrange interpolation formula, we find

$$P(z) = \sum_{\ell=1}^{L_n+1} \frac{A(z)}{(z - z_\ell)A'(z_\ell)}P(z_\ell), \quad \text{where } A(z) = \prod_{\ell=1}^{L_n+1}(z - z_\ell),$$

whence it follows that

$$\sigma(\lambda_n) = \sum_{\ell=1}^{L_n+1} \frac{a_{\lambda_n}^{(\ell)}}{A'(z_\ell)} P(z_\ell), \qquad (6.11)$$

where the $a_{\lambda_n}^{(\ell)}$ are determined by the equations

$$\frac{A(z)}{z - z_\ell} = \sum_{\ell_n=0}^{L_n} a_{\lambda_n}^{(\ell)} z^{\lambda_n}.$$

From (6.2) we have $|\Delta|_q \leq q^{-e_q}$, so $|z_i|_q \leq 1$, and so $|a_{\lambda_n}^{(\ell)}|_q \leq 1$ ($1 \leq \ell \leq L_n + 1$). Hence from (6.10) and (6.11), we conclude that

$$|\sigma(\lambda_n)|_q < q^{-M/2} \left(\min_\ell |A'(z_\ell)|_q \right)^{-1}. \qquad (6.12)$$

Since $A'(z_\ell) = \prod_{\substack{i=1 \\ i \neq \ell}}^{L_n+1} (z_\ell - z_i)$, and

$$|z_\ell - z_i|_q = |\exp\{(\ell - i)\Delta\} - 1|_q = |\ell - i|_q |\Delta|_q,$$

we find

$$|A'(z_\ell)|_q = |\Delta|_q^{L_n} \prod_{\substack{i=1 \\ i \neq \ell}}^{L_n+1} |\ell - i|_q = |\Delta|_q^{L_n} |(\ell-1)!(L_n+1-\ell)!|_q \geq |\Delta|_q^{L_n} q^{-L_n/(q-1)},$$

noting that the exponent of the power to which q divides $(\ell-1)!(L_n+1-\ell)!$ does not exceed

$$\left\lfloor \frac{\ell-1}{q} \right\rfloor + \left\lfloor \frac{L_n+1-\ell}{q} \right\rfloor + \left\lfloor \frac{\ell-1}{q^2} \right\rfloor + \left\lfloor \frac{L_n+1-\ell}{q^2} \right\rfloor + \ldots \leq \frac{L_n}{q-1}.$$

Therefore (6.12) implies that

$$|\sigma(\lambda_n)|_q < q^{-\frac{1}{2}M + L_n/(q-1)} |\Delta|_q^{-L_n} < q^{-2(n-1)L},$$

the latter step being valid because of (6.3).

Turning to equation (6.8), we now see that for any λ_n ($0 \leq \lambda_n \leq L_n$) and for any $u_1, \ldots, u_{n-1} \in \Omega_q$ with $\max_i |u_i|_q \leq q^{-e_q}$,

$$\left| \sum_{\lambda_1, \ldots, \lambda_{n-1}} \rho(\lambda) \exp\{u_1\lambda_1 + \ldots + u_{n-1}\lambda_{n-1}\} \right|_q < q^{-2(n-1)L}$$

holds. This means that for any $z_1, \ldots, z_{n-1} \in \Omega_q$ satisfying $|z_i - 1|_q \leq q^{-e_q}$ ($i = 1, 2, \ldots, n-1$), and for fixed λ_n, we have

$$\left| \sum_{\lambda_1=0}^{L_1} \cdots \sum_{\lambda_{n-1}=0}^{L_{n-1}} \rho(\lambda_1,\ldots,\lambda_n) z_1^{\lambda_1} \cdots z_{n-1}^{\lambda_{n-1}} \right|_q < q^{-2(n-1)L}.$$

Applying Lemma 6.1 to each variable $z_1,\ldots,z_{n-1}$ in succession, we find

$$|\rho(\lambda_1,\ldots,\lambda_n)|_q < q^{-2(n-1)L+e_q(n-1)L} \le 1$$

for any $\lambda_1,\ldots,\lambda_n$. This contradicts the assumption $\max_\lambda |\rho(\lambda)|_q = 1$; and so Lemma 6.2 is proved.

Lemma 6.3 *Let $L_1,\ldots,L_n,L,M$ be natural numbers and suppose that $L_i \le L$ $(i = 1,2,\ldots,n-1)$, and that $\theta_1,\ldots,\theta_n,\beta_1,\ldots,\beta_{n-1} \in \Omega_q$ with*

$$|\theta_i - 1|_q \le q^{-e_q} \qquad (i = 1,2,\ldots,n-1), \tag{6.13}$$

$$|\theta_n|_q = q^t \qquad (t \ne 0) \tag{6.14}$$

where $e_2 = 2, e_q = 1$ $(q \ge 3)$, $|\beta_i|_q \le 1$ $(i = 1,2,\ldots,n-1)$, and where if $t > 0$ then $M > 4(n-1)L+2t$, while if $t < 0$ then $M > 4(n-1)L+2|t|L_n^2$. Then the system of equations (6.4) has only the trivial solution $\rho(\lambda) = \rho(\lambda_1,\ldots,\lambda_n) = 0$ $(0 \le \lambda_i \le L_i, i = 1,2,\ldots,n)$.

Proof. If there is a non-trivial solution, then we may assume $\max_\lambda |\rho(\lambda)|_q = 1$, in view of the homogeneity of the system (6.4). As before, we shall see that this assumption leads to a contradiction.

At the start of the proof of the previous lemma, we established that (6.13) and (6.5) imply (6.6) for $\omega_i' = u_i - \ell \log \theta_i$, where the u_i are arbitrary numbers in Ω_q with $|u_i|_q \le q^{-e_q}$ $(i = 1,2,\ldots,n-1)$ and ℓ is an arbitrary integer in the range $1 \le \ell \le L_n + 1$. Noting that $\gamma_i = \lambda_i + \beta_i \lambda_n$ $(1 \le i \le n-1)$, we can reduce (6.6) to

$$\left| \sum_{\lambda_n=0}^{L_n} \tau(\lambda_n) \left(\theta_n \theta_1^{-\beta_1} \cdots \theta_{n-1}^{-\beta_{n-1}} \right)^{\lambda_n \ell} \right|_q < q^{-M/2}, \tag{6.15}$$

where

$$\tau(\lambda_n) = \exp\left(\lambda_n \sum_{i=1}^{n-1} u_i \beta_i \right) \sum_{\lambda_1,\ldots,\lambda_{n-1}} \rho(\lambda) \exp\left(u_1\lambda_1 + \ldots + u_{n-1}\lambda_{n-1} \right).$$

The latter inequality means that the polynomial $T(z) = \sum_{\lambda_n=0}^{L_n} \tau(\lambda_n) z^{\lambda_n}$ satisfies

$$|T(z_\ell)|_q < q^{-M/2} \tag{6.16}$$

at the points $z_\ell = (\theta_n \theta_1^{-\beta_1} \cdots \theta_{n-1}^{-\beta_{n-1}})^\ell$ $(1 \le \ell \le L_n + 1)$. As in the proof of the previous lemma, applying the Lagrange interpolation formula yields

$$\tau(\lambda_n) = \sum_{\ell=1}^{L_n+1} \frac{a_{\lambda_n}^{(\ell)}}{A'(z_\ell)} T(z_\ell). \tag{6.17}$$

In addition it must be remembered that $z_\ell \neq z_{\ell'}$ ($\ell \neq \ell'$), since (6.13) and (6.14) imply that

$$|z_\ell|_q = |\theta_n|_q^\ell = q^{t\ell}, \tag{6.18}$$

and so $|z_\ell|_q \neq |z_{\ell'}|_q$ ($\ell \neq \ell'$). We see that for $t > 0$

$$|a_{\lambda_n}^{(\ell)}|_q \leq q^{t(2+3+\ldots+L_n+1)} < q^{t(L_n+1)(L_n+2)/2}, \tag{6.19}$$

and for $t < 0$

$$|a_{\lambda_n}^{(\ell)}|_q \leq 1 \qquad (0 \leq \lambda_n \leq L_n;\ 1 \leq \ell \leq L_n + 1). \tag{6.20}$$

Hence from (6.16), (6.17), (6.19) and (6.20) we conclude that

$$|\tau(\lambda_n)|_q < q^{-M/2 + t(L_n+1)(L_n+2)/2} \left(\min_\ell |A'(z_\ell)|_q \right)^{-1} \tag{6.21}$$

if $t > 0$, and if $t < 0$ then

$$|\tau(\lambda_n)|_q < q^{-M/2} \left(\min_\ell |A'(z_\ell)|_q \right)^{-1}. \tag{6.22}$$

We now bound the q-adic norm of $A'(z_\ell) = \prod_{\substack{i=1 \\ i \neq \ell}}^{L_n+1} (z_\ell - z_i)$ from below.

From (6.18),

$$|z_i - z_\ell|_q = \max(|z_\ell|_q;\, |z_i|_q) = \begin{cases} q^{t\max(\ell,i)} & (t > 0), \\ q^{t\min(\ell,i)} & (t < 0). \end{cases}$$

Thus for $t > 0$

$$|A'(z_\ell)|_q = \prod_{\substack{i=1 \\ i \neq \ell}}^{L_n+1} q^{t\max(\ell,i)} \geq q^{\frac{1}{2}(L_n+1)(L_n+2)t - t}, \tag{6.23}$$

and for $t < 0$

$$|A'(z_\ell)|_q = \prod_{\substack{i=1 \\ i \neq \ell}}^{L_n+1} q^{t\min(\ell,i)} \geq q^{\frac{1}{2}L_n(L_n+1)t}. \tag{6.24}$$

From (6.21) and (6.23) we have, for $t > 0$,

$$|\tau(\lambda_n)|_q < q^{-\frac{1}{2}M + t},$$

and from (6.22) and (6.24) we have, for $t < 0$,

$$|\tau(\lambda_n)|_q < q^{-\frac{1}{2}M + \frac{1}{2}L_n(L_n+1)|t|};$$

that is, in either case it follows from the lemma's hypothesis that

$$|\tau(\lambda_n)|_q < q^{-2(n-1)L}.$$

Equation (6.15) shows that the proof of the lemma is now complete, exactly as for the previous lemma.

We shall use Lemmas 6.2 and 6.3 to disprove the existence of non-trivial solutions of the system (3.1) using the premises of Lemma 1.1.

Suppose that in Lemmas 6.2 and 6.3 the values $L_1, \ldots, L_n$ are the same as in Lemma 3.1, and that $M = \lfloor \frac{1}{2} \sigma^{-2n-2} H \rfloor$, $\theta_i = \alpha_i^S$ $1 \leq i \leq n$), and $\beta_i = h_i$ $(1 \leq i \leq n-1)$. Since $\sqrt{2}q^g \leq \sigma$, we have $S < N(q^{2e}) \leq q^{2g} \leq \frac{1}{2}\sigma^2$, whence

$$S^{-1}\sigma^{2(n+1)} > 2\sigma^{2n} > L_n + 1. \tag{6.25}$$

Because the parameter ℓ in equations (3.1) ranges over the interval $0 \leq \ell \leq \sigma^{2(n+1)}$, we may conclude from (6.25) that (3.1) contains a subsystem (6.4), with restrictions on $\ell, m_1, \ldots, m_{n-1}$ as given in the statement of Lemma 6.2. (In (3.1) we take ℓ divisible by S.) Since

$$S = N(q^{2e})\left(1 - \frac{1}{N(q)}\right) = \Phi(q^{2e})$$

is the Euler function, for all numbers $\alpha \in \mathbb{G}$ with $(\alpha, q) = 1$ we have

$$\alpha^{\Phi(q^{2e})} \equiv 1 \pmod{q^{2e}}.$$

Consequently

$$|\alpha^S - 1|_q \leq q^{-2} \leq q^{-e_q}. \tag{6.26}$$

Since by the conditions of Lemma 1.1 we have $(\alpha_i, q) = 1$ $(1 \leq i \leq n-1)$, it follows from (6.26) that (6.13) holds, and if $(\alpha_n, q) = 1$ then (6.2) holds.

Suppose first that $|\alpha_n|_q = q^r$ $(r \neq 0)$. This corresponds to Lemma 6.3, where the exponent of t defined by (6.14) is rS. We bound the exponent r. Let a be the least natural number such that $a\alpha$ is integer. Then

$$|\alpha_n|_q \geq |a\alpha_n|_q \geq |\mathrm{Nm}(a\alpha_n)|^{-1}, \quad |\mathrm{Nm}(a\alpha_n)| \leq \lceil \overline{a\alpha_n} \rceil^g \leq (gA^2)^g.$$

Therefore

$$|\alpha_n|_q \geq (gA^2)^{-g},$$

whence we obtain $r > -6g \ln A$. On the other hand,

$$|\alpha_n|_q \leq |a|_q^{-1} \leq A,$$

and so $r < 2\ln A$. Thus, from the negation of (1.5) we have

$$|r| < 6g \ln A < 6g\sigma^{-4n-4}H.$$

Now we see that

$$M = \left\lfloor \frac{1}{2}\sigma^{-2n-2}H \right\rfloor > 4(n-1)\sigma^{-2n-4}H + 2|r|S\sigma^{2n} \geq 4(n-1)L + 2|t|L_n^2$$

by the hypotheses of Lemma 6.3. So by applying this lemma when $|\alpha_n|_q \neq 1$ we complete the proof of the corresponding statement of Lemma 1.1.

Now suppose $|\alpha_n|_q = 1$, and apply Lemma 6.2. We only need to check the relation between (1.6) and (6.3), that is, the inequality

$$\tfrac{1}{5}\sigma^{-4n-2}H \leq \tfrac{1}{2}ML_n^{-1} + 2(n-1)LL_n^{-1} + 1.$$

From the expressions for M, L, L_n in terms of σ and H, we see that this inequality holds.

Thus Lemma 1.1 is proved in its entirety.

7. Connection Between Bounds in Different Non-archimedean Metrics

We turn now to the proof of Lemma 1.2 on the relationship between bounds for linear forms in the logarithms of algebraic numbers in different non-archimedean metrics. Our arguments are a translation into the language of analytic functions on Ω_p of the arguments of §§4-5, and yield an analogue of Lemma 3.1 from the negation of inequality (1.9). Lemma 1.2 then follows immediately from results of the previous section. A critical feature of the forthcoming arguments is that we now work with functions defined in a p-adic disc of small radius, while previously in the analytic-arithmetical extrapolation we dealt with entire functions. Nevertheless, we will see that the use of the Schnirelman integral in place of the complex contour integral allows the arguments to be carried through in more or less parallel fashion.

We shall prove Lemma 3.1 using inequality (1.9). Clearly assertion (**I**) of §3 will be the basis for the extension of the range of ℓ in equations (3.1).

We introduce an auxiliary function

$$\Psi = \Psi(z_1, \ldots, z_{n-1}) = \sum_\lambda p(\lambda)\alpha_1^{\gamma_1 z_1} \ldots \alpha_{n-1}^{\gamma_{n-1} z_{n-1}}$$

of variables $z_1, \ldots, z_{n-1} \in \Omega_p$, $|z_i|_p \leq 1$ $(i = 1, 2, \ldots, n-1)$, where the $p(\lambda)$ are the integers determined in assertion (**I**) of §3, and

$$\alpha_i^{\gamma_i z_i} = \exp\{\gamma_i z_i \log \alpha_i\} \qquad (i = 1, 2, \ldots, n-1),$$

where the exponentials and logarithms are p-adic. Such a definition is possible in view of (1.8). Next let

$$\Psi^*_{m_1, \ldots, m_{n-1}}(z_1, \ldots, z_{n-1}) = \prod_{i=1}^{n-1} \alpha_i^{m_i z_i} \Psi_{m_1, \ldots, m_{n-1}}(z_1, \ldots, z_{n-1}),$$

$$\Psi_{m_1, \ldots, m_{n-1}} = \frac{D_1^{m_1}}{m_1!} \cdots \frac{D_{n-1}^{m_{n-1}}}{m_{n-1}!} \Psi,$$

where D_i is the differential operator $\alpha_i \frac{d}{dz_i}$ $(1 \leq i \leq n-1)$.

Using (1.9) we find

$$\left|(\alpha_1^{\lambda_1}\ldots\alpha_n^{\lambda_n})^\ell - (\alpha_1^{\gamma_1}\ldots\alpha_{n-1}^{\gamma_{n-1}})^\ell\right|_p =$$

$$= \left|(\alpha_1^{\lambda_1}\ldots\alpha_{n-1}^{\lambda_{n-1}})^\ell\right|_p \left|\alpha_n^{\lambda_n\ell} - (\alpha_1^{h_1}\ldots\alpha_{n-1}^{h_{n-1}})^{\lambda_n\ell}\right|_p \leq$$

$$\leq \left|\alpha_n - \alpha_1^{h_1}\ldots\alpha_{n-1}^{h_{n-1}}\right|_p < p^{-\delta H}. \quad (7.1)$$

Furthermore, setting $P = \prod_{i=1}^{n-1}(\log\alpha_i)^{m_i}$, we have

$$|P|_p = \prod_{i=1}^{n-1}|\log\alpha_i|_p^{m_i} = \prod_{i=1}^{n-1}|\alpha_i - 1|_p^{m_i} < 1, \qquad \left|\binom{\gamma_i}{m_i}\right|_p \leq 1.$$

Hence the difference between the sum (3.3), multiplied by P, and the expression

$$\left.\Psi^*_{m_1,\ldots,m_{n-1}}(z_1,\ldots,z_{n-1})\right|_{z_i=\ell} = P\sum_\lambda p(\lambda)\prod_{i=1}^{n-1}\alpha_i^{\gamma_i\ell}\binom{\gamma_i}{m_i}$$

is less than $p^{-\delta H}$. We have an analogue of assertion (II) of §4:

(II) *For any integers* $\ell, m_1,\ldots,m_{n-1}$ *in the ranges (3.11),*

$$\left|\Psi^*_{m_1,\ldots,m_{n-1}}(\ell,\ldots,\ell)\right|_p < p^{-\delta H}. \quad (7.2)$$

We now observe that for any $z \in \Omega_p$ in the disc $|z|_p \leq 1$

$$|\alpha_1^{\gamma_1 z}\ldots\alpha_{n-1}^{\gamma_{n-1} z}|_p = \left|\prod_{i=1}^{n-1}\exp(z\gamma_i\log\alpha_i)\right|_p = 1.$$

Therefore

$$|\Psi^*_{m_1,\ldots,m_{n-1}}(z_1,\ldots,z_{n-1})|_p \leq 1 \quad (7.3)$$

for all $z_i \in \Omega_p$, $|z_i|_p \leq 1$ $(i = 1, 2, \ldots, n-1)$.

In §4 it was shown that, for the number Q defined there, either $Q = 0$ or

$$|Q|_p > \exp\left\{-g(c_3\sigma^{-2n-2}\ln\sigma\cdot H + c_4\sigma^{-2n-4}H\ell)\right\}.$$

Applying (7.1), we observe that if $Q \neq 0$ then for any natural number ℓ,

$$\left|P^{-1}\Psi^*_{m_1,\ldots,m_{n-1}}(\ell,\ldots,\ell)\right|_p > \exp\left\{-c_6(\sigma^{-2n-2}\ln\sigma\cdot H + \sigma^{-2n-4}H\ell)\right\}.$$

Since

$$1 > |P|_p = \prod_{i=1}^{n-1}|\alpha_i - 1|_p^{m_i} \geq \prod_{i=1}^{n-1}|a_i(\alpha_i - 1)|_p^{m_i} \geq$$

$$\geq \prod_{i=1}^{n-1}|\operatorname{Nm}[a_i(\alpha_i - 1)]|_p^{m_i} \geq (B(gB+1))^{-gn\sigma^{-2n-2}H},$$

we obtain the following analogue of assertion (III) of §4:

(III) *For any integer ℓ in the range $0 \leq \ell \leq \sigma^{2(n+1)}$, either (7.2) holds, or*

$$\left| \Psi^*_{m_1, \ldots, m_{n-1}}(\ell, \ldots, \ell) \right|_p \tag{7.4}$$
$$> (B(gB+1))^{-gn\sigma^{-2n-2}H} \exp\left\{ -c_6(\sigma^{-2n-2} \ln \sigma \cdot H + \sigma^{-2n-4} H\ell) \right\}.$$

Now put $S = \lfloor \sigma^{-2n-5/2} H \rfloor$, $M_k = \sigma^{-2n-2} H - kS$, $\ell_k = \lfloor (\sigma/p)^k \rfloor$ $(k = 1, 2, \ldots, 2(n+1))$. As before, we shall prove by induction that the inequality

$$\left| \Psi^*_{m_1, \ldots, m_{n-1}}(\ell p, \ldots, \ell p) \right|_p < p^{-\delta H} \tag{7.5}$$

holds for all integers ℓ, m_i in the ranges $0 \leq \ell \leq \ell_k$, and $0 \leq m_i \leq M_k$ $(i = 1, 2, \ldots, n-1)$, if $k \leq 2(n+1)$. For $k = 1$, this follows from assertion (II) above, and we suppose it true for $k < 2(n+1)$.

Fix integers $m_1, \ldots, m_{n-1} \leq M_{k+1}$, and let $f(z) = \Psi^*_{m_1, \ldots, m_{n-1}}(z, \ldots, z)$. Then

$$f^{(s)}(z) = \left(\frac{\partial}{\partial z_1} + \ldots + \frac{\partial}{\partial z_{n-1}} \right)^s \Psi^*_{m_1, \ldots, m_{n-1}}(z_1, \ldots, z_{n-1}) \Big|_{z_i = z}$$

$$= \sum_{s_1 + \ldots + s_{n-1} = s} \frac{s!}{s_1! \ldots s_{n-1}!} \prod_{i=1}^{n-1} (\log \alpha_i)^{m_i + s_i} \sum_{\lambda} p(\lambda) \prod_{i=1}^{n-1} \binom{\gamma_i}{m_i} \gamma_i^{s_i} \alpha_i^{\gamma_i z}.$$

By Lemma 2.3,

$$\binom{\gamma_i}{m_i} \gamma_i^{s_i} = \sum_{r_i=0}^{s_i} \sigma^{(m_i)}_{r_i, s_i} \binom{\gamma_i}{m_i + r_i},$$

where the $\sigma^{(m_i)}_{r_i, s_i}$ are integers. Denoting $\sigma^{(m_1)}_{r_1, s_1} \ldots \sigma^{(m_{n-1})}_{r_{n-1}, s_{n-1}}$ by $\sigma^{(\bar{m})}_{\bar{r}, \bar{s}}$, we have

$$f^{(s)}(z) =$$
$$\sum_{s_1 + \ldots + s_{n-1} = s} \frac{s!}{s_1! \ldots s_{n-1}!} \sum_{\bar{r}=0}^{\bar{s}} \sigma^{(\bar{m})}_{\bar{r}, \bar{s}} \prod_{i=1}^{n-1} (\log \alpha_i)^{s_i - r_i} \Psi^*_{m_1 + r_1, \ldots, m_{n-1} + r_{n-1}}(z, \ldots, z).$$

If $s \leq S$ and $m_i \leq M_{k+1}$, then $m_i + s_i \leq M_k$ $(1 \leq i \leq n-1)$, and by the inductive hypothesis we have, for $0 \leq \ell < \ell_k$,

$$|f^{(s)}(\ell p)|_p < \max_{r_1 + \ldots + r_{n-1} \leq s} |\Psi^*_{m_1 + r_1, \ldots, m_{n-1} + r_{n-1}}(\ell p, \ldots, \ell p)|_p < p^{-\delta H}, \tag{7.6}$$

noting that $s!/(s_1! \ldots s_{n-1}!)$ and $\sigma^{(\bar{m})}_{\bar{r}, \bar{s}}$ are integers. We now prove that this implies

$$|f(\ell p)| < p^{-\delta H} \qquad (\ell_k \leq \ell < \ell_{k+1}), \tag{7.7}$$

thereby obtaining (7.5) for all integers ℓ, m_i in the ranges $0 \leq \ell < \ell_{k+1}$, $0 \leq m_i \leq M_{k+1}$ $(i = 1, 2, \ldots, n-1)$.

First observe that $f(z)$ is regular in some disc $|z|_p \leq R$ with $R > 1$; for it follows from (1.8) and from $|\gamma_i|_p \leq 1$ $(i = 1, \ldots, n-1)$ that the functions

$$\alpha_i^{\gamma_i z} = \exp(\gamma_i z \log \alpha_i) \qquad (i = 1, 2, \ldots, n-1) \tag{7.8}$$

are regular in the discs $|z|_p \leq p^{-1/(p-1)}|\log \alpha_i|_p^{-1} = p^{-1/(p-1)}|\alpha_i - 1|_p^{-1} = R_i$, where $R_i > 1$. So we can take $R = \min_i R_i$.

Next put $F(z) = [z(z-p)\ldots(z-(\ell_k-1)p)]^{S+1}$, and take $t \in \Omega_p$ with $(2p\ell_{k+1})^{-1} < |t|_p < (p\ell_{k+1})^{-1}$. Applying Lemma 3.3 of Ch.III with $z = jp$ $(\ell_k \leq j < \ell_{k+1})$ we find

$$f(z) = \int_{0,1} \frac{F(z)}{F(\xi)} \frac{f(\xi)\xi}{\xi - z} d\xi - \sum_{s=0}^{S} \sum_{\ell=0}^{\ell_k-1} \frac{f^{(s)}(\ell p)}{s!} \int_{\ell p, t} \frac{F(z)}{F(\xi)} \frac{(\xi - \ell p)^{s+1}}{\xi - z} d\xi. \tag{7.9}$$

By (7.3), $|f(\xi)|_p \leq 1$ for $\xi \leq 1$, and so inequality (3.3) of Ch.III yields

$$\left| \int_{0,1} \frac{F(jp)}{F(\xi)} \frac{f(\xi)\xi}{\xi - jp} d\xi \right|_p \leq \max_{|\xi=1|} \left(\frac{|F(jp)|_p}{|F(\xi)|_p} \frac{|f(\xi)\xi|_p}{|\xi - jp|_p} \right) < |F(jp)|_p \leq p^{-(S+1)\ell_k}, \tag{7.9}$$

noting that for $|\xi|_p = 1$, $|\xi - jp|_p = |F(\xi)|_p = 1$. Similarly,

$$\left| \int_{\ell p, t} \frac{F(jp)}{F(\xi)} \frac{(\xi - \ell p)^{s+1}}{\xi - jp} d\xi \right|_p \leq \max_{|\xi - \ell p| = |t|_p} \left(\frac{|F(jp)|_p}{|F(\xi)|_p} \frac{|\xi - \ell p|_p^{s+1}}{|\xi - jp|_p} \right). \tag{7.10}$$

From our choice of t we have $|\xi - jp|_p = p^{-1}|\ell - j|_p$, since $|\ell p - jp|_p > (p\ell_{k+1})^{-1} > |t|_p$. Further, for $0 \leq i < \ell_k$ we have

$$|\xi - ip|_p = \begin{cases} |\xi - \ell p|_p = |t|_p > (2p\ell_{k+1})^{-1}, & i = \ell, \\ |\ell p - ip|_p = p^{-1}|\ell - i|_p, & i \neq \ell. \end{cases}$$

Hence

$$|F(\xi)|_p = |t|_p^{S+1} \left(\prod_{\substack{t=0 \\ i\neq\ell}}^{\ell_k-1} p^{-1}|\ell - i|_p \right)^{S+1} >$$

$$> (2p\ell_{k+1})^{-S-1} p^{-\ell_k(S+1)} |\ell!(\ell_k - \ell - 1)!|_p^{S+1} >$$

$$> (2p\ell_{k+1})^{-S-1} p^{-\ell_k(S+1)\left(1 + \frac{1}{p-1}\right)}$$

since $\ell!(\ell_k - \ell - 1)!$ divides $\ell_k!$, and $|\ell_k!|_p > p^{-\ell_k/(p-1)}$.

Thus the integral in (7.10) is bounded by

$$p^{-\ell_k(S+1)} (2p\ell_{k+1})^{S+1} p^{\ell_k(S+1)\left(1 + \frac{1}{p-1}\right)} |t|_p^s < (2p\ell_{k+1})^{S+1} p^{\ell_k(S+1)/(p-1)}.$$

Now for $f(jp)$ we have from (7.6), (7.9) that

$$|f(jp)|_p < \max\left(p^{-(S+1)\ell_k}, p^{-\delta H}(2\ell_{k+1})^{S+1}p^{(S+1)(\ell_k+1)}\right),$$

which equals $p^{-(S+1)\ell_k}$ when $k \le 2n+1$. That this condition holds follows by substituting the expressions for S, ℓ_k, ℓ_{k+1} in terms of σ, H into (1.3).

Turning to assertion **(III)**, if (7.7) is false for $\ell = j$, then $|f(jp)|_p$ is bounded below by the right-hand side of (7.4). The last inequality therefore yields

$$(B(gB+1))^{g n \sigma^{-2n-2}}e^{c_6(\sigma^{-2n-2}\ln\sigma + \sigma^{-2n-4}\ell p)} > p^{\sigma^{-2n-5/2}(\sigma/p)^k},$$

which implies $\ell > \ell_{k+1}$, while we chose $\ell < \ell_{k+1}$. Hence (7.7) holds, completing the proof of (7.5) for all integers ℓ, m_i in the ranges

$$0 \le \ell < (\sigma/p)^{2n+2}, \qquad 0 \le m_i \le \frac{1}{2}\sigma^{-2n-2}H.$$

Following the argument at the close of §5, we obtain a system of equations

$$\sum_\lambda p(\lambda)(\alpha_1^{\lambda_1}\ldots\alpha_n^{\lambda_n})^{\ell p}\binom{\gamma_1}{m_1}\ldots\binom{\gamma_{n-1}}{m_{n-1}} = 0,$$

where $\ell, m_1, \ldots, m_{n-1}$ run over the ranges given above. Since $\sqrt{2}p^g p^{2n+3} \le \sigma$, one can select a subsystem of the form (6.4), where $\theta_i = \alpha_i^{pS}$ $(i = 1, 2, \ldots, n-1)$. Now the q-adic logarithms of the θ_i are $p\log(\alpha_i^S)$, and in view of (1.6), Lemma 6.2 may be applied. The argument at the close of §6 completes the proof of Lemma 1.2.

In the statement of Lemma 1.2, condition (1.8) can be simplified to

$$|\alpha_i|_p = 1 \qquad (i = 1, 2, \ldots, n-1), \tag{7.11}$$

by appropriately modifying (1.5) and the definition of σ. In fact, let $T = N(\mathfrak{p}^{2e'})\left(1 - \frac{1}{N(\mathfrak{p})}\right)$, $e' = \mathrm{ord}_\mathfrak{p}\, p$. Then

$$\alpha_i^T \equiv 1 \pmod{\mathfrak{p}^{2e'}} \qquad (i = 1, 2, \ldots, n-1),$$

and (1.8) holds for the numbers $\alpha_i' = \alpha_i^T$ $(i = 1, 2, \ldots, n-1)$. From the inequalities (1.4) and (1.5) of Ch.II, it follows that for any $\alpha \in \mathbf{G}$ we have $h(\alpha^T) \le 2^g g^{gT} h(\alpha)^{2gT}$. Consequently, applying Lemma 1.2 to the numbers α_i' $(i = 1, 2, \ldots, n-1)$ we obtain:

Lemma 7.1 *The assertion of Lemma 1.2 remains true if we replace (1.8) by (7.11), σ by $\tilde\sigma = 2^{12}\delta^{-1}(n^2 g^3 T \ln B)^2$, and inequality (1.5) by $H < 2g\tilde\sigma^{4(n+1)}T\ln A$.*

While the condition (7.11) is not burdensome in applications, it can be removed at some cost in the precision of the bounds.

8. Lemmas on Direct Bounds

Given the quite detailed investigation above of the connection between bounds in different non-archimedean metrics, we are now prepared for the transition to direct bounds for linear forms in logarithms of algebraic numbers [211]. Here we discuss only one approach to this transition.

Let the field $\mathbb{G}$, containing numbers $\alpha_1, \ldots, \alpha_n$, be normal, and let Γ be the Galois group of $\mathbb{G}$, that is, the group of automorphisms τ of $\mathbb{G}$ over $\mathbb{Q}$. Denoting the image of $\alpha \in \mathbb{G}$ under τ by α^τ, we obtain from (3.1)

$$\sum_\lambda p(\lambda) \left((\alpha_1^\tau)^{\lambda_1} \ldots (\alpha_n^\tau)^{\lambda_n} \right)^\ell \binom{\gamma_1}{m_1} \ldots \binom{\gamma_{n-1}}{m_{n-1}} = 0, \qquad (8.1)$$

where $\ell, m_1, \ldots, m_{n-1}$ have the same range of values as in (3.1). Applying the method of §6 to the system (8.1) (taking $|\alpha_i^\tau|_q = 1$, $1 \leq i \leq n-1$), we may conclude that it has only trivial solutions, if we impose the condition

$$|h_1 \log(\alpha_1^{\tau S}) + \ldots + h_{n-1} \log(\alpha_{n-1}^{\tau S}) - \log(\alpha_n^{\tau S})|_q > q^{-\frac{1}{8} \sigma^{-4n-2} H}, \qquad (8.2)$$

generalising (1.6). Hence:

Lemma 8.1 *Suppose, under the conditions of Lemma 1.1, that the field $\mathbb{G}$ is normal with group of automorphisms Γ, and that for all $\tau \in \Gamma$ we have $|\alpha_i^\tau|_q = 1$ ($i = 1, 2, \ldots, n-1$). If (8.2) holds for any $\tau \in \Gamma$ then the assertion of Lemma 1.1 holds.*

Suppose that α_n is not contained in $\mathbb{K} = \mathbb{Q}(\alpha_1 \ldots \alpha_{n-1})$, and that τ is an automorphism of $\mathbb{G}$ over $\mathbb{K}$, with $\alpha_n^\tau \neq \zeta \alpha_n$ for any root of unity ζ. If neither (1.6) nor (8.2) holds, then

$$0 \neq |\log(\alpha_n^S) - \log(\alpha_n^{\tau S})|_q \leq q^{-\frac{1}{8} \sigma^{-4n-2} H}.$$

Since

$$\left| \log \left(\frac{\alpha_n}{\alpha_n^\tau} \right)^S \right|_q = |\alpha_n^S - \alpha_n^{\tau S}|_q \geq |\operatorname{Nm} a^S (\alpha_n^S - \alpha_n^{\tau S})|^{-1} \geq$$

$$\geq A^{-gS} (2\overline{|\alpha|})^{-gS} > A^{-3gS},$$

we find

$$\frac{1}{5} \sigma^{-4n-2} H < 3gS \ln A < \frac{3}{2} g\sigma^2 \ln A,$$

and consequently $H < \frac{15}{2} g\sigma^{4(n+1)} \ln A$, which differs only slightly from (1.5). Thus if α_n is not contained in $\mathbb{Q}(\alpha_1, \ldots, \alpha_{n-1})$, and if none of the quotients of conjugates $\alpha_n^{(i)} / \alpha_n^{(j)}$ ($i \neq j; i, j = 1, 2, \ldots, \deg \alpha_n$) is a root of unity, then (1.4) implies that $H < 8g\sigma^{4(n+1)} \ln A$.

If we apply similar considerations to α_k ($1 \leq k \leq n-1$) instead of α_n, the conclusions are also similar, by virtue of the relations

$$0 \neq |h_k(\log(\alpha_k^S) - \log(\alpha_k^{TS}))|_q \leq q^{-\frac{1}{8}\alpha^{-4n-2}H},$$

following from the negation of (1.6) and (8.2).

With these and other ideas, one can completely remove all conditions resembling (1.6) and (8.2), but this considerably complicates the arguments, and we eschew the details. It is clear, nonetheless, that no insurmountable barrier separates assertions on direct estimates and assertions on the connections between bounds in different metrics.

Through the work of Baker, Stark, van der Poorten and Loxton, and of many other investigators, considerable precision and generality has now been achieved in direct estimates in both archimedean and non-archimedean metrics. We shall later use the following result due to Baker [25]:

Lemma 8.2 *Let* $\alpha_1, \alpha_2, \ldots, \alpha_n$ *be algebraic numbers distinct from 0 and 1, of degree at most* d, *and of height at most* $A_1, A_2, \ldots, A_n$ *respectively, with* $4 \leq A_1 \leq \ldots \leq A_n$, *let* $h_1, h_2, \ldots, h_n$ *be rational integers, and let* $H \geq \max |h_i|$ $(i = 1, 2, \ldots, n)$, $H \geq 4$. *If we set*

$$V = \ln A_1 \ln A_2 \ldots \ln A_n, \qquad V' = V/\ln A_n, \qquad c_1 = (16nd)^{200n},$$

then

$$|h_1 \ln \alpha_1 + \ldots + h_n \ln \alpha_n| > H^{-c_1 V \ln V'} \tag{8.3}$$

holds, provided that the left-hand side of (8.3) is non-zero. (Principal values of the logarithms are taken.)

Van der Poorten [158] obtained the non-archimedean analogue of this theorem, and also proved the following result:

Lemma 8.3 *Suppose that the conditions of Lemma 8.2 hold, and that* $G = \mathbb{Q}(\alpha_1, \ldots, \alpha_n)$, $[G : \mathbb{Q}] = g$, *and that* p *is a prime number, with* $|\ |_p$ *the* p-adic *valuation in* G. *Then*

$$|\alpha_1^{h_1} \ldots \alpha_n^{h_n} - 1|_p > e^{-c_2 V (\ln H)^2}, \tag{8.4}$$

where $c_2 = (16(n+1)g)^{12(n+1)}p^g$, *provided that the left-hand side of (8.4) is non-zero.*

In Chapter 5, we use yet another result of Baker in §5 (see Lemma 5.1), and its p-adic version due to van der Poorten in §§1,4. To prove these theorems by the methods of §§3-5, an auxiliary system of linear equations in $p(\lambda)$ is constructed, more complicated than (3.1), and an auxiliary function more complicated than $\Phi(z_1, \ldots, z_{n-1})$ of §4 is used. The proof that the resulting system of equations is inconsistent flows not from p-adic methods, as in §6, but from the theory of Kummer fields, where the following well-known lemma is pivotal:

Suppose that $\mathbb{K}$ is a finite extension of the field of rational numbers and that p is a prime number. Then either $\mathbb{K}(\alpha_1^{1/p}, \ldots, \alpha_n^{1/p})$ is of degree p over $\mathbb{K}(\alpha_1^{1/p}, \ldots, \alpha_{n-1}^{1/p})$, or $\alpha_n = \alpha_1^{j_1} \ldots \alpha_{n-1}^{j_{n-1}} \gamma^p$, where $\gamma \in \mathbb{K}$ and $0 \le j_1, \ldots, j_{n-1} < p$.

For instance, in the fundamental work of Baker and Stark [28], which stimulated extensive applications of this lemma, it was used to prove that the inequality

$$0 \ne |\alpha_1 \ldots \alpha_{n-1}^{h_{n-1}} - \alpha_n| < e^{-\delta H}$$

implies that $H < c_3 (\ln A)^{1+\epsilon}$, with c_3 depending only on $\alpha_1, \ldots, \alpha_{n-1}, g, \delta$,ϵ, where $\epsilon > 0$ is arbitrary. A system of auxiliary equations of the form

$$\sum p(\lambda)(\alpha_1^{\lambda_1} \ldots \alpha_n^{\lambda_n})^{1/p} \gamma_1^{m_1} \ldots \gamma_{n-1}^{m_{n-1}} = 0$$

$$0 \le m_i \le M \qquad (i = 1, 2, \ldots, n-1)$$

is constructed, and its analysis is reduced via the lemma to another problem on bounds for linear forms in logarithms of algebraic numbers, with α_n related to $\alpha_1, \ldots, \alpha_{n-1}$ via $\alpha_n = \alpha_1^{j_1} \cdots \alpha_{n-1}^{j_{n-1}} \gamma^p$, where $0 \le j_1, \ldots, j_{n-1} < p$.

Work by Stark [226] and Baker [22] preceded and underlay the proofs of Lemmas 8.2 and 8.3. Due to the bulk and intricacy of these proofs, we refer the reader to the original papers cited above, and to the work of Waldschmidt [243]. Many other bounds are known under stronger restrictions, or highlighting the influence of some parameter (such as the A in Lemmas 1.2 and 1.3).

IV. The Thue Equation

At last, we pass to the analysis of diophantine equations, starting with the central problem of the representation of numbers by binary forms. We return, as in Chapter I, to the connection between the magnitude of solutions of Thue's equation and rational approximation of algebraic numbers; but now our approach is the opposite of Thue's: we obtain bounds for the approximation as a corollary to bounds for the solutions. We arrive at an effective improvement of Liouville's inequality and its generalisations; and we will see how fundamental parameters of the equation, in particular the height of the form and of the number represented by the form, influence the magnitude of the solutions.

1. Existence of a Computable Bound for Solutions

Let $f(x, y)$ be an integral binary form of degree $n \geq 3$, and $A \neq 0$ an integer. In this chapter we consider solutions x, y of the equation

$$f(x, y) = A, \tag{1.1}$$

which is often called *Thue's equation*, in honour of Axel Thue, who proved in 1909 that the equation had only finitely many solutions (cf. Ch.I§2). The investigation of Thue's equation and its generalisations was central to the development of the theory of diophantine equations in the early 20th century, when it was discovered that many diophantine equations in two unknowns could be reduced to it. Ultimately, this led Mordell to his fundamental theorem on the finite rank of the groups of rational points on an elliptic curve, and led Siegel to his theorem on the finitude of the number of integral points on an algebraic curve of genus greater than zero.

Until long after Thue's work, no method was known for the construction of bounds for the number of solutions of (1.1) in terms of the parameters of the equation (such as $|A|$, or the height or degree of the form). Only in 1968 was such a method introduced by Baker [13], based on his theory of bounds for linear forms in the logarithms of algebraic numbers. Modifications and refinements of this method are still the only way to explicitly bound the number of solutions of (1.1) and to determine the influence of the parameters on that number.

To appreciate the basic principle of the analysis of (1.1), let us consider one of many ways to reduce the equation to a bound from below for the absolute value of numbers β of the form

$$\beta = \alpha_1^{h_1} \ldots \alpha_r^{h_r} - \alpha, \tag{1.2}$$

where $\alpha_1, \ldots, \alpha_r, \alpha$ are algebraic numbers, and $h_1, \ldots, h_r$ are rational integers. It will suffice to know that $\beta \neq 0$ implies that

$$|\beta| > c_1 e^{-\delta H}, \qquad H = \max(|h_1|, \ldots, |h_r|) \tag{1.3}$$

with any $\delta > 0$, where the positive quantity c_1 is determined effectively by δ, $\alpha_1, \ldots, \alpha_r, \alpha$ (cf. Ch.II §8).

Let $\theta^{(1)}, \ldots, \theta^{(n)}$ be the roots of the polynomial $f(x, 1)$. If integers x, y satisfy (1.1), we have for $y \neq 0$ that

$$\left| \frac{x}{y} - \theta^{(1)} \right| \leq \frac{c_2}{y}, \qquad c_2 = \left(\frac{|A|}{|a_0|} \right)^{\frac{1}{n}},$$

where we agree that

$$\left| \frac{x}{y} - \theta^{(1)} \right| = \min_i \left| \frac{x}{y} - \theta^{(i)} \right|,$$

and where a_0 is the leading coefficient of $f(x, 1)$. Then for $i \neq 1$ and

$$|y| > 2(|A|/|a_0|)^{\frac{1}{n}} |\theta^{(1)} - \theta^{(i)}|^{-1} \tag{1.4}$$

we find

$$\left| \frac{x}{y} - \theta^{(i)} \right| \geq |\theta^{(1)} - \theta^{(i)}| - \left| \frac{x}{y} - \theta^{(1)} \right| > \frac{1}{2} |\theta^{(1)} - \theta^{(i)}|. \tag{1.5}$$

Supposing (1.4) holds for all $i \neq 1$ we obtain

$$\frac{|A|}{|a_0|} = |y|^n \prod_{i=1}^n \left| \frac{x}{y} - \theta^{(i)} \right| > |y|^n 2^{-n+1} \left| \frac{x}{y} - \theta^{(1)} \right| \prod_{\substack{i=1 \\ i \neq 1}}^n |\theta^{(1)} - \theta^{(i)}|,$$

and then

$$\left| \frac{x}{y} - \theta^{(1)} \right| < \frac{c_3}{|y|^n}, \qquad c_3 = \frac{2^{n-1}|A|}{|f'(\theta^{(1)})|}. \tag{1.6}$$

It is obvious that if (1.4) does not hold for some $i \neq 1$, we will have an upper bound for $|y|$, and then also for $|x|$. Hence we may assume (1.4) for all $i \neq 1$, and consequently also (1.6).

The bounds thus obtained have a simple meaning: if x, y satisfy (1.1) then only one value $|x/y - \theta^{(i)}|$ can be "small" provided that y is "large" enough.

Consider now the quotient $\dfrac{x - \theta^{(2)} y}{x - \theta^{(3)} y}$. Writing it in the form

$$\frac{x - \theta^{(1)} y - (\theta^{(2)} - \theta^{(1)}) y}{x - \theta^{(1)} y - (\theta^{(3)} - \theta^{(1)}) y}$$

we see from (1.5) and (1.6) that it is "close" to $(\theta^{(2)} - \theta^{(1)})/(\theta^{(3)} - \theta^{(1)})$. More precisely,

$$\left| \frac{x - \theta^{(2)}y}{x - \theta^{(3)}y} - \frac{\theta^{(2)} - \theta^{(1)}}{\theta^{(3)} - \theta^{(1)}} \right| < c_4 |y|^{-n}. \tag{1.7}$$

Let $\mathbb{K}$ be a finite extension of $\mathbb{Q}$ containing $\theta = \theta^{(1)}$. Denote by $\varepsilon_1, \ldots, \varepsilon_r$ a system of fundamental units of $\mathbb{K}$. Writing (1.1) in the form $\mathrm{Nm}(x - \theta y) = Aa_0^{-1}$, we note that for all solutions x, y of (1.1), the elements $\mu = x - \theta y$ of $\mathbb{K}$ have a fixed norm. Therefore, multiplying μ by a suitable unit of $\mathbb{K}$ yields a number λ whose height is bounded in terms of A, a_0, n, and the chosen system $\varepsilon_1, \ldots, \varepsilon_r$ of fundamental units (cf. the proof of Lemma 2.2 of Ch.II). Thus $\mu = \lambda \eta$, where $h(\lambda) < c_5$, for some unit η of $\mathbb{K}$. Since $\overline{|\mu|} \leq |x| + \overline{|\theta|}|y|$, putting $\eta = \varepsilon_1^{h_1} \ldots \varepsilon_r^{h_r}$, we find $\overline{|\eta|} < c_6|y|$, and

$$\left| \sum_{j=1}^{r} h_j \ln |\varepsilon_j^{(i)}| \right| < \ln |y| + c_7 \qquad (i = 1, 2, \ldots, n).$$

From this system of inequalities one selects a subsystem with determinant differing from the regulator of the field by a power of 2, obtaining

$$\max_j |h_j| < c_8 (\ln |y| + 1) \qquad (j = 1, 2, \ldots, r). \tag{1.8}$$

Reverting now to inequality (1.7), setting

$$x - \theta^{(i)} y = \lambda^{(i)} \eta^{(i)} = \lambda^{(i)} \varepsilon_1^{(i)^{h_1}} \ldots \varepsilon_r^{(i)^{h_r}} \qquad (i = 2, 3)$$

yields

$$\left| \left(\frac{\varepsilon_1^{(2)}}{\varepsilon_1^{(3)}} \right)^{h_1} \ldots \left(\frac{\varepsilon_r^{(2)}}{\varepsilon_r^{(3)}} \right)^{h_r} - \frac{\lambda^{(3)}(\theta^{(2)} - \theta^{(1)})}{\lambda^{(2)}(\theta^{(3)} - \theta^{(1)})} \right| < c_9 |y|^{-n}, \tag{1.9}$$

where $c_9 = c_4 |\lambda^{(3)}||\lambda^{(2)}|^{-1}$. The left-hand side of this inequality represents a number of the form (1.2) which is not zero, and so, applying (1.3), is bounded from below by $c_1 e^{-\delta H}$. Using (1.8) to bound H, comparison with (1.9) gives

$$c_9 |y|^n < c_1 (e|y|)^{\delta c_8}.$$

Taking $\delta = (n-1)/c_8$, we get $|y| \leq e^{n-1} c_1 c_9^{-1}$. Since all the values $c_1, c_2, \ldots, c_9$ are effectively determined by the parameters of the equation, the bounds obtained for $|y|$, and hence for $|x|$, are also effective.

There are other ways based on (1.3) to analyse equation (1.1), but they all involve similar considerations, differing only in details. One such variant will be used in §2 below.

It is clear that great interest resides not so much in an effective bound for the solutions of (1.1), but in determining the best possible bound so as to exhibit the influence of the parameters of the equation on the magnitude of its solutions. But it is not obvious a priori what those parameters should be taken to be. Clearly, the bound for the solutions of (1.1) can be expressed in terms of $|A|$, the height H_f of the form $f(x,y)$, and its degree n; but our arguments above suggest the use of deeper parameters, notably the regulator of $\mathbb{K}$. Such an approach has proved productive, and has influenced investigations of other types of diophantine equations.

2. Dependence on the Number Represented by the Form

Historically, the first problem studied in connection with equation (1.1) was the estimation of the magnitude of its solutions as a function of A, with the form f fixed. This is closely related to the problem of rational approximation of a real root of the polynomial $f(x,1)$. Indeed, the arguments leading to (1.6) can be adapted to yield an opposite inequality

$$\left|\frac{x}{y} - \theta^{(1)}\right| > c_{10}|A|/|y|^n, \tag{2.1}$$

where $c_{10} > 0$ depends only on the form f. If one knows that the inequality

$$\left|\frac{x}{y} - \theta^{(1)}\right| < |y|^{-\nu}, \tag{2.2}$$

for some $\nu < n$, has only finitely many solutions in integers x, y ($y \neq 0$), then for all y with $|y|$ greater than some y_0, the opposite inequality holds. On comparison with (1.6) that yields

$$|A| = |f(x,y)| > c_{11}|y|^{n-\nu} \qquad (|y| > y_0). \tag{2.3}$$

Thus, it follows from Thue's theorem on rational approximation of algebraic numbers (Ch.I§2) that all solutions x, y of (1.1) satisfy

$$\max(|x|, |y|) < c_{12}|A|^{2/(n-2)+\epsilon}, \tag{2.4}$$

while Roth's theorem gives

$$\max(|x|, |y|) < c_{13}|A|^{1/(n-2)+\epsilon}, \tag{2.5}$$

where c_{12}, c_{13} depend only on the form f and the arbitrary number $\epsilon > 0$. However, the dependence cannot be made explicit in terms of the coefficients of f, that is, it is ineffective.

Our first goal is the proof of an effective bound for the solutions of (1.1):

Theorem 2.1 *Let $f = f(x,y)$ be an integral irreducible binary form of degree $n \geq 3$, and let $A > 0$ be an integer. All solutions of (1.1) in integers x, y satisfy*

$$X = \max(|x|, |y|) < c_{14}(|A|H_f)^{c_{15}} \tag{2.6}$$

where H_f is the height of the form f, and c_{14} and c_{15} are functions of n and of the regulator of the field $\mathbb{K}$ containing the zeros of $f(x,1)$.

The dependence on $|A|$ in (2.6) is similar to that in inequalities (2.4) and (2.5), though the exponent c_{15} is less explicit; and it depends on the regulator of $\mathbb{K}$ as well as on n. However, (2.6) contains information lacking in (2.4) and (2.5) on the dependence on the height H_f when the degree n and the field $\mathbb{K}$ are fixed.

It follows in particular from (2.6) that one has $|A| > c_{16} X^{1/c_{15}}$, which with (2.1) gives

$$\left| \frac{x}{y} - \theta^{(1)} \right| > c_{17} |y|^{-n+(1/c_{15})} \qquad (|y| \neq 0).$$

This inequality is a general effective improvement, long unattainable, of Liouville's theorem. It was obtained by successive improvements of Baker's first result

$$\left| \frac{x}{y} - \theta^{(1)} \right| > c_{18} |y|^{-n} \exp\left((\ln |y|)^{1/(n+1+\epsilon)} \right),$$

where c_{18} depends on n, f and ϵ only, with $\epsilon > 0$ arbitrary.

We will see two main routes to Theorem 2.1. The first is based on direct bounds for linear forms in logarithms of algebraic numbers; the basic idea was given in the previous section. This was the path followed by Baker [13], [14], Stark [227], and Feldman [69]. The second route is via the connection between bounds in different metrics for the form, and was pursued by the author [203], [206], [209]. Neither approach, however, yields inequality (2.6) with an absolute constant c_{15}, nor even with its value depending on n alone. None of the recent progress in bounds for linear forms in logarithms of algebraic numbers is germane here, and the need for radically new ideas in the main arguments is evident.

The first method is applied in two different situations in §§4, 5 below, but first we examine the second method. We rely on the lemmas of Ch.II and on Lemma 1.1 of Chapter III. In addition, we need some simple lemmas introduced as required below. First of all, we introduce the notion of an "exceptional" number.

Definition. An algebraic number θ of degree ≥ 4 is said to be exceptional if there exists an enumeration $\theta^{(1)}, \theta^{(2)}, \ldots, \theta^{(n)}$ of its conjugates such that for all $i, j (i \neq j; 3 \leq i, j \leq n)$

$$\frac{\theta^{(1)} - \theta^{(i)}}{\theta^{(2)} - \theta^{(i)}} \cdot \frac{\theta^{(2)} - \theta^{(j)}}{\theta^{(1)} - \theta^{(j)}} = \frac{1 - \zeta_j}{1 - \zeta_i} \qquad (2.7)$$

for roots of unity $\zeta_k \neq 1$ $(k = 3, 4, \ldots, n)$.

Accordingly, a binary form $f(x, y) = a_0 \operatorname{Nm}(x - \theta y)$ is called exceptional if θ is an exceptional number. In this paragraph we consider equation (1.1) for non-exceptional forms, postponing consideration of exceptional forms to the next paragraph.

Lemma 2.1 *Let* $\mathbb{K}$ *be an algebraic number field, and let* q *be a rational prime. Denote by* $\mid \;\mid_q$ *the* q-*adic valuation on the field* $\mathbb{K}(\zeta_0)$, *where* ζ_0 *is a primitive* S-*th root of unity. Suppose* θ *is in* $\mathbb{K}$. *Then there is some* S-*th root* ζ *of* 1 *such that*

$$|\theta - \zeta|_q \leq |\theta^S - 1|_q |S|_q^{-1} .$$

Proof. Let $|\theta - \zeta|_q = \min_{(\zeta')} |\theta - \zeta'|_q$, where the minimum is taken over all S-th roots of unity ζ'. Then

$$|\zeta - \zeta'|_q \leq \max\big(|\zeta - \theta|_q, |\theta - \zeta'|_q\big) = |\theta - \zeta'|_q .$$

Hence,

$$|S\zeta^{S-1}|_q = \prod_{\zeta' \neq \zeta} |\zeta - \zeta'|_q \leq \prod_{\zeta' \neq \zeta} |\theta - \zeta'|_q = \frac{|\theta^S - 1|_q}{|\theta - \zeta|_q},$$

and the lemma follows.

Lemma 2.2 *Suppose that* θ *is not an exceptional algebraic number and let* S *be a natural number,* q *a rational prime. Denote by* $\mid \;\mid_q$ *the* q-*adic valuation on the field* $\mathbb{Q}(\theta^{(1)}, \ldots, \theta^{(n)}, \zeta)$, *where the* $\theta^{(i)}$ *are the conjugates of* θ, *and* ζ *is a primitive* S-*th root of* 1. *Let* ζ_i $(i = 3, 4, \ldots, n)$ *be arbitrary* S-*th roots of unity distinct from* 1, *and set*

$$\xi_{ij} = (\zeta_i - 1)\frac{\theta^{(i)} - \theta^{(2)}}{\theta^{(1)} - \theta^{(i)}} - (\zeta_j - 1)\frac{\theta^{(j)} - \theta^{(2)}}{\theta^{(1)} - \theta^{(j)}},$$

where $i \neq j$, $3 \leq i, j \leq n$. *Then*

$$|\xi_{ij}|_q > (16n^2 h^4)^{-Sn^4},$$

where h *is the height of the number* θ.

Proof. Since θ is not exceptional, the numbers ξ_{ij} are non-zero. If a is a natural number for which $a\theta$ is an integer, then

$$\xi'_{ij} = a^2 \xi_{ij}(\theta^{(1)} - \theta^{(i)})(\theta^{(1)} - \theta^{(j)}) \neq 0$$

is an algebraic integer. We find

$$\left|\xi'_{ij}\right| \leq a^2 \left|\zeta_i - 1\right| \left|\theta^{(2)} - \theta^{(i)}\right| \left|\theta^{(1)} - \theta^{(j)}\right| +$$
$$+ a^2 \left|\zeta_j - 1\right| \left|\theta^{(j)} - \theta^{(2)}\right| \left|\theta^{(1)} - \theta^{(i)}\right| \leq 16n^2 h^4 .$$

The degree $[\xi'_{ij} : \mathbb{Q}]$ is less than Sn^4 since ξ'_{ij} is contained in the field $\mathbb{Q}(\theta^{(1)}, \theta^{(2)}, \theta^{(i)}, \theta^{(j)}\zeta)$, of degree no greater than $n(n-1)(n-3)(n-4)S$. Hence

$$|\mathrm{Nm}(\xi'_{ij})| \leq \left|\overline{\xi'_{ij}}\right|^{[\xi'_{ij}:\mathbb{Q}]} < (16n^2 h^4)^{Sn^4},$$

and then

$$|\xi_{ij}|_q \geq |\xi'_{ij}|_q \geq |\mathrm{Nm}(\xi'_{ij})|^{-1} > (16n^2 h^4)^{-Sn^4} .$$

We now proceed directly to the proof of Theorem 2.1.

Set $f = f(x,y) = a\,\mathrm{Nm}(x - \theta y)$, where $a > 0$ is an integer, and write $\mathbb{K} = \mathbb{Q}(\theta)$. Let $\mathbb{G} = \mathbb{Q}(\theta^{(1)}, \theta^{(2)}, \dots \theta^{(n)})$ denote the splitting field of the form f, and denote by $R = R_{\mathbb{K}}$ the regulator of the field $\mathbb{K}$.

Setting $\mu = x - \theta y$, we have the identity

$$(\theta^{(i)} - \theta^{(j)})\mu^{(l)} - (\theta^{(l)} - \theta^{(j)})\mu^{(i)} - (\theta^{(i)} - \theta^{(l)})\mu^{(j)} = 0 \qquad (2.8)$$

for any distinct i, j, l $(1 \le i, j, l \le n)$. We suppose that x and y are integers satisfying (1.1) and the enumeration of the conjugates of the field $\mathbb{K}$ is such that

$$|\mu^{(1)}| = \max_{(i)} |\mu^{(i)}|, \qquad |\mu^{(2)}| = \min_{(i)} |\mu^{(i)}| \qquad (i = 1, 2, \dots, n). \qquad (2.9)$$

Take out from (2.8) a subsystem

$$(\theta^{(1)} - \theta^{(2)})\mu^{(i)} - (\theta^{(i)} - \theta^{(2)})\mu^{(1)} - (\theta^{(1)} - \theta^{(i)})\mu^{(2)} = 0, \qquad (2.10)$$

$i = 3, 4, \dots, n$.

Let U be a group of units of the field $\mathbb{K}$ constructed by Lemma 2.1 of Ch.II. Then by Lemma 2.2, Ch.II there exists a unit $\eta \in U$ such that for $\nu = \mu\eta^{-1}$ we have

$$|\nu^{(i)}| = |A/a|^{1/n} e^{\theta c_{19} R} \qquad (i = 1, 2, \dots, n), \qquad |\theta| < 1, \qquad (2.11)$$

where $c_{19} = c_{19}(n)$, that is, it depends only on n.

It follows from (2.8), (2.9) that

$$1 \ge \frac{|\mu^{(j)}|}{|\mu^{(1)}|} \ge \frac{|\mu^{(2)}|}{|\mu^{(1)}|} = \left| \frac{(\theta^{(1)} - \theta^{(2)})\mu^{(i)}}{(\theta^{(1)} - \theta^{(i)})\mu^{(1)}} - \frac{\theta^{(i)} - \theta^{(2)}}{\theta^{(1)} - \theta^{(i)}} \right| =$$

$$= \frac{|\theta^{(1)} - \theta^{(2)}|}{|\theta^{(1)} - \theta^{(i)}|} \frac{|\nu^{(i)}|}{|\nu^{(1)}|} \left| \frac{\nu^{(1)}(\theta^{(i)} - \theta^{(2)})}{\nu^{(i)}(\theta^{(1)} - \theta^{(2)})} - \frac{\eta^{(i)}}{\eta^{(1)}} \right|.$$

But in view of (2.11) and the estimates

$$|\theta^{(1)} - \theta^{(i)}| \le 2nH_f,$$

$$|\theta^{(1)} - \theta^{(2)}| \ge a^{-n+1} \prod_{\substack{1 \le i < j \le n \\ (i,j) \ne (1,2)}} |\theta^{(i)} - \theta^{(j)}|^{-1} > (2nH_f)^{-\frac{(n+2)(n-1)}{2}+2},$$

we find

$$e^{2c_{19}R} > \left| \frac{\eta^{(j)}}{\eta^{(1)}} \right| > (2nH_f)^{-n^2} e^{-2c_{19}R} \left| \frac{\nu^{(1)}(\theta^{(i)} - \theta^{(2)})}{\nu^{(i)}(\theta^{(1)} - \theta^{(2)})} - \frac{\eta^{(i)}}{\eta^{(1)}} \right|, \qquad (2.12)$$

where $j = 2, 3, \dots, n$; $i = 3, 4, \dots, n$. To estimate the difference on the right in (2.12) from below, we apply Lemma 1.1 of Ch.III, noting that $\eta = \eta_1^{h_1} \cdots \eta_r^{h_r}$ for some integers $h_1, \dots, h_r$. Then

$$\frac{\eta^{(i)}}{\eta^{(1)}} = \left(\frac{\eta_1^{(i)}}{\eta_1^{(1)}}\right)^{h_1} \cdots \left(\frac{\eta_r^{(i)}}{\eta_r^{(1)}}\right)^{h_r}.$$

Using the inequalities (1.6), (1.7) of Ch.II, we find the estimates for the heights of algebraic numbers

$$h\left(\frac{\nu^{(1)}(\theta^{(i)} - \theta^{(2)})}{\nu^{(i)}(\theta^{(1)} - \theta^{(2)})}\right) < (|A| H_f e^R)^{c_{20}} ; \tag{2.13}$$

and, similarly,

$$h\left(\frac{\eta_j^{(i)}}{\eta_j^{(1)}}\right) < e^{c_{21}R} , \qquad (j = 1, 2, \ldots, r). \tag{2.14}$$

Let q be any prime (for example $q = 2$), and let $|\ |_q$ be the valuation on the field $\mathbb{G}$ induced by a prime ideal of $I_{\mathbb{G}}$ over q. Since the $\eta_j^{(i)}/\eta_j^{(1)}$ are units, they are relatively prime to q, and in accordance with the statement 1.1 of Ch.III we consider two cases:

$$\left|\frac{\nu^{(1)}(\theta^{(i)} - \theta^{(2)})}{\nu^{(i)}(\theta^{(1)} - \theta^{(2)})}\right|_q \neq 1 \tag{2.15}$$

or else this norm equals 1.

We first suppose that (2.15) holds for some i ($3 \le i \le n$). Then, if some δ ($0 \le \delta \le 1$) satisfies the inequality

$$\left|\frac{\nu^{(1)}(\theta^{(i)} - \theta^{(2)})}{\nu^{(i)}(\theta^{(1)} - \theta^{(2)})} - \left(\frac{\eta_1^{(i)}}{\eta_1^{(1)}}\right)^{h_1} \cdots \left(\frac{\eta_r^{(i)}}{\eta_r^{(1)}}\right)^{h_r}\right| < e^{-\delta H}, \tag{2.16}$$

where $H = \max_{(j)} |h_j|$, it follows from Ch.III, Lemma 1.1 and from the inequalities (2.13),(2.14) that

$$H < c_{22}(\delta^{-1} R^2)^{4(n+1)}(\ln|A| + \ln H_f + R), \tag{2.17}$$

where $c_{22} = c_{22}(n)$. If we suppose that the inequality (2.16) does not hold for the given δ, then it follows from (2.12) that

$$e^{2c_{19}R} > |\eta^{(j)}/\eta^{(1)}| > (2H_f)^{-c_{23}} e^{-2c_{19}R - \delta H} \qquad (j = 2, 3, \ldots, n). \tag{2.18}$$

Hence

$$\left|\sum_{i=1}^{r} h_i \ln\left|\frac{\eta_i^{(j)}}{\eta_i^{(1)}}\right|\right| < H_1 \qquad (j = 2, 3, \ldots, n), \tag{2.19}$$

where $H_1 = \delta H + c_{23} \ln(2H_f) + 2c_{19}R$. Adding all the inequalities (2.19) and recalling that

$$\sum_{j=2}^{n} \ln|\eta_i^{(j)}/\eta_i^{(1)}| = \ln \prod_{j=2}^{n} |\eta_i^{(j)}| - (n-1)\ln|\eta_i^{(1)}| = -n\ln|\eta_i^{(1)}|,$$

we obtain $\left|\sum_{i=1}^{r} h_i \ln |\eta_i^{(1)}|\right| < H_1$, and then it follows from (2.19) that

$$\left|\sum_{i=1}^{r} h_i \ln |\eta_i^{(j)}|\right| < 2H_1 \qquad (j = 1, 2, \ldots, n). \qquad (2.20)$$

Since the units $\eta_1, \ldots, \eta_r$ are independent, we can take a subsystem with determinant not less than $2^{-n}R$ in absolute value from the system of inequalities (2.20). Applying Lemma 2.1 of Ch.II, we obtain

$$H = \max_{(i)} |h_i| < c_{24}H_1 \qquad (i = 1, 2, \ldots, n),$$

where $c_{24} = c_{24}(n)$. If one takes $\delta = (2c_{24})^{-1}$, a stronger inequality than (2.17) for H follows. Hence, certainly

$$H \le c_{25}R^{8(n+1)}(\ln |A| + \ln H_f + R), \qquad (2.21)$$

where $c_{25} = c_{25}(n)$.

Suppose now that (2.15) does not hold for any $i = 3, 4, \ldots, n$. To apply Lemma 1.1 of Ch.III we must have the inequality

$$\left|\sum_{j=1}^{r} h_j \log \left(\frac{\eta_j^{(i)}}{\eta_j^{(1)}}\right)^S - \log \left(\frac{\nu^{(1)}(\theta^{(i)} - \theta^{(2)})}{\nu^{(i)}(\theta^{(1)} - \theta^{(2)})}\right)^S\right|_q > q^{-c_{26}H}, \qquad (2.22)$$

for at least one i $(3 \le i \le n)$. Here the logarithms are q-adic, the number S is determined in the assertion of the lemma, $c_{26} < c_{27}(\delta^{-1}R^2)^{-4r-6}$, and $c_{27} = c_{27}(n)$. That means we must have

$$\left|\left(\frac{(\theta^{(1)} - \theta^{(2)})\mu^{(i)}}{(\theta^{(i)} - \theta^{(2)})\mu^{(1)}}\right)^S - 1\right|_q > q^{-c_{26}H}. \qquad (2.23)$$

Suppose that for all $i = 3, 4, \ldots, n$ the opposite inequalities hold. Then by Lemma 2.1

$$\left|\frac{(\theta^{(1)} - \theta^{(2)})\mu^{(i)}}{(\theta^{(i)} - \theta^{(2)})\mu^{(1)}} - \zeta_i\right|_q \le |S|_q^{-1}q^{-c_{26}H} = q^{-b}, \qquad (2.24)$$

where ζ_i is some S-th root of 1. Turning to (2.10), we find

$$\left|\frac{(\theta^{(1)} - \theta^{(i)})\mu^{(2)}}{(\theta^{(i)} - \theta^{(2)})\mu^{(1)}} - (\zeta_i - 1)\right|_q \le q^{-b}$$

from which it follows that

$$\left|\frac{\mu^{(2)}}{\mu^{(1)}} - (\zeta_i - 1)\frac{\theta^{(i)} - \theta^{(2)}}{\theta^{(1)} - \theta^{(i)}}\right|_q \le q^{-b_1} \qquad (i = 3, 4, \ldots, n), \qquad (2.25)$$

where $b_1 = c_{26}H - c_{27}(\ln(2H_f)/\ln q) - 2g$. Hence, for any i, j such that $i \ne j$; $3 \le i, j \le n$, we find that

$$\left| (\zeta_i - 1)\frac{\theta^{(i)} - \theta^{(2)}}{\theta^{(1)} - \theta^{(i)}} - (\zeta_j - 1)\frac{\theta^{(j)} - \theta^{(2)}}{\theta^{(1)} - \theta^{(j)}} \right|_q \leq q^{-b_1}. \qquad (2.26)$$

We may suppose that $\zeta_i \neq 1$ for all $i = 3, 4, \ldots, n$. Indeed, if $\zeta_i = 1$ for some i, then it follows from (2.25) that

$$|a\mu^{(2)}|_q \leq |a\mu^{(1)}|_q q^{-b_1} \leq q^{-b_1}.$$

On the other hand

$$|a\mu^{(2)}|_q \geq |\,\mathrm{Nm}(a\mu^{(2)})|^{-1} = |a^{n-1}A|^{-1} \geq H_f^{-n}|A|^{-1},$$

and then $b_1 \leq (\ln|A| + n\ln H_f)/\ln q$, which gives a stronger estimate for H than (2.21).

Now applying Lemma 2.2 we obtain

$$b_1 \ln q < Sn^4(4\ln H_f + \ln(16n^2)),$$

from which, once again, a stronger estimate than (2.21) follows for H. Thus, we may suppose that (2.22) holds for some i, $(3 \leq i \leq n)$, and that the arguments described above lead to an estimate for H not worse than (2.21).

To complete the proof of Theorem 2.1 in the case of non-exceptional forms it only remains to estimate $|x|$ and $|y|$ for the solutions x, y of equation (1.1). Using (2.21) we find

$$\overline{|\mu|} = \overline{|\nu\eta|} \leq \overline{|\nu|}\,\overline{|\eta|} < |A|^{1/n}e^{c_{19}R}\overline{|\eta_1|}^{n|h_1|} \cdots \overline{|\eta_r|}^{n|h_r|} <$$

$$< \exp\{c_{28}R^{8(n+1)}(\ln|A| + \ln H_f + R)\ln\max_{(j)} \overline{|\eta_j|}\}.$$

where $c_{28} = c_{28}(n)$. Lemmas 1.1 and 2.1 of Ch.II show that $\ln\max_{(j)}\overline{|\eta_j|} < c_{29}R$. Since $\mu = x - \theta y$ we obtain

$$\max(|x|, |y|) < \exp\{c_{30}R^{8(n+1)+1}(\ln|A| + \ln H_f + R)\}$$

which corresponds to (2.6) with the values

$$c_{14} = e^{c_{30}R^{8(n+1)+2}}, \quad c_{15} = c_{30}R^{8(n+1)+1}, \quad c_{30} = c_{30}(n). \qquad (2.27)$$

It is possible to give explicit presentation of c_{30} as a function of n, and, as we shall see below, to replace the exponents $8(n+1)+2$ and $8(n+1)+1$ in the expressions for c_{14} and c_{15} by absolute constants. However it does not seem to be possible to exclude the dependence of c_{15} on R.

3. Exceptional Forms

First of all we prove that there is no exceptional form of degree $n \geq 5$.

Lemma 3.1 *Let z_1, z_2 be distinct complex numbers different from 1 but with absolute value 1, and set $w = (1 - z_1)/(1 - z_2)$. Then $z_1 = (1 - w)/(1 - \bar{w})$ and $z_2 = z_1\bar{w}/w$, where the bar denotes complex conjugation.*

Proof. Set $z_1 = e^{2\pi i\lambda}$, $(0 < \lambda < 1)$ and $z_2 = e^{2\pi i\mu}$, $(0 < \mu < 1)$. Then

$$\frac{1 - z_1}{1 - z_2} = \frac{e^{\pi i\lambda}(e^{-\pi i\lambda} - e^{\pi i\lambda})}{e^{\pi i\mu}(e^{-\pi i\mu} - e^{\pi i\mu})} = \frac{\sin\pi\lambda}{\sin\pi\mu}e^{\pi i(\lambda - \mu)}.$$

Hence $\arg w = \pi(\lambda - \mu)$ and we see that $z_1 = z_2e^{2\pi i(\lambda-\mu)} = z_2w/\bar{w}$, from which it follows that $z_1 = (1 - w)/(1 - \bar{w})$.

Now let $n \geq 5$. Denote the left-hand side of (2.7) by α_{ij}. By Lemma 3.1 we see that $\zeta_j = (1 - \alpha_{ij})/(1 - \bar{\alpha}_{ij})$. Now take a pair of distinct values, say k and l, $(k \neq l; k, l \neq j; k, l, j \geq 3)$ of the index i. Then

$$\zeta_j = (1 - \alpha_{kj})/(1 - \bar{\alpha}_{kj}) = (1 - \alpha_{lj})/(1 - \bar{\alpha}_{lj}),$$

and it follows that $(1 - \alpha_{kj})/(1 - \alpha_{lj})$ is a real number. But

$$\frac{1 - \alpha_{kj}}{1 - \alpha_{lj}} = \left(1 - \frac{\theta^{(1)} - \theta^{(k)}}{\theta^{(2)} - \theta^{(k)}} \cdot \frac{\theta^{(2)} - \theta^{(j)}}{\theta^{(1)} - \theta^{(j)}}\right) : \left(1 - \frac{\theta^{(1)} - \theta^{(l)}}{\theta^{(2)} - \theta^{(l)}} \cdot \frac{\theta^{(2)} - \theta^{(j)}}{\theta^{(1)} - \theta^{(j)}}\right) =$$

$$= \left(\frac{\theta^{(1)} - \theta^{(j)}}{\theta^{(2)} - \theta^{(j)}} - \frac{\theta^{(1)} - \theta^{(k)}}{\theta^{(2)} - \theta^{(k)}}\right) : \left(\frac{\theta^{(1)} - \theta^{(j)}}{\theta^{(2)} - \theta^{(j)}} - \frac{\theta^{(1)} - \theta^{(l)}}{\theta^{(2)} - \theta^{(l)}}\right)$$

and, after some additional transformations, we find that this quotient is

$$\frac{\theta^{(j)} - \theta^{(k)}}{\theta^{(2)} - \theta^{(k)}} \cdot \frac{\theta^{(2)} - \theta^{(l)}}{\theta^{(j)} - \theta^{(l)}}. \tag{3.1}$$

Thus the expressions (3.1) are real for any distinct $j, k, l \geq 3$.

Of course the field $\mathbb{G} = \mathbb{Q}(\theta^{(1)}, \ldots, \theta^{(n)})$ contains the complex conjugate $\bar{\theta}^{(i)}$ of each $\theta^{(i)}$ (they are zeros of the same minimal polynomial over $\mathbb{Q}$). It follows from (2.7) and Lemma 3.1 that $\zeta_i, \zeta_j \in \mathbb{G}$.

Denote by Γ the Galois group of the minimal polynomial of θ; i.e. the galois group of the field $\mathbb{G}$. Because Γ is transitive, it contains an element which we may view as a permutation τ taking 1 into 3. If $\tau(2) = \tau_2 > 2$, then there is a symbol i, say, mapping into 2, and another symbol j, say, mapping into a symbol τ_j distinct from 1. Applying the permutation τ to (2.7), we obtain

$$\frac{\theta^{(3)} - \theta^{(2)}}{\theta^{(\tau_2)} - \theta^{(2)}} \cdot \frac{\theta^{(\tau_2)} - \theta^{(\tau_j)}}{\theta^{(3)} - \theta^{(\tau_j)}} = \frac{1 - \zeta_j^\tau}{1 - \zeta_i^\tau}, \tag{3.2}$$

where ζ^τ denotes the image of ζ under the automorphism τ. Because the left-hand side of this relation is of the form (3.1) it is a real number. Similarly, if $\tau_2 = 2$ we can find two symbols $i, j \geq 3$ not mapping into 1, and if $\tau_2 = 1$ such symbols occur for the inverse permutation τ^{-1}. In all these cases an application of τ, or τ^{-1} as appropriate, to (2.7) results in a real number. However, the right-hand side of (3.2) cannot be real, since the numbers ζ_j^τ, ζ_i^τ are distinct from 1; and if they do not coincide, then by Lemma 3.1 the number $w = (1 - \zeta_j^\tau)/(1 - \zeta_i^\tau)$ must be complex. Moreover, if $\zeta_j^\tau = \zeta_i^\tau$ then $\zeta_j = \zeta_i$. Then (2.7) entails $\theta^{(i)} = \theta^{(j)}$, which cannot occur due to the assumed irreducibility of the form.

Therefore we see that the relations (2.7) are impossible if $n \geq 5$. In the case $n = 4$, they reduce to just one equality

$$\frac{\theta^{(1)} - \theta^{(3)}}{\theta^{(2)} - \theta^{(3)}} \cdot \frac{\theta^{(2)} - \theta^{(4)}}{\theta^{(1)} - \theta^{(4)}} = \frac{1 - \zeta_4}{1 - \zeta_3}. \tag{3.3}$$

Obviously not all the $\theta^{(i)}$ here can be real; there is at least one pair of complex conjugates among them. It turns out not to be difficult to show that exactly one pair of complex conjugates occurs, and that the right-hand side of (3.3) is a root of unity. Furthermore, one can check that the zeros of the polynomial $x^4 - 2x^2 - 1$ can be numbered in such a way that (3.3) holds with $\zeta_4 = i$, $\zeta_3 = i^3$. Hence, exceptional numbers and exceptional forms of the 4-th degree exist. The detailed classification of such numbers and forms is given in an article by Avanesov [5].

Lemma 8.1 of Ch.III may be used for a further reduction of the set of exceptional numbers to show that on the right-hand side of (3.3) one can always put either $e^{\pi i/3}$ or $e^{-\pi i/3}$. An application of the Frobenius theorem on primes in the automorphism sections of a Galois group makes it possible to analyse equation (1.1) in the case of exceptional forms of degree 4 as well, but then one must have an estimate for the least prime q for which a certain integral irreducible polynomial is reduced mod q into non-linear irreducible factors. Similar arguments may be applied to cubic forms not equivalent to forms for which

$$\frac{\theta^{(1)} - \theta^{(2)}}{\theta^{(3)} - \theta^{(2)}} = e^{\pi i/3} \text{ or } e^{-\pi i/3}.$$

The latter constitute the set of the forms considered by Gelfond, since they satisfy the relation (4.4) of Ch.I. We will not describe the details of the necessary arguments since those take us far beyond our main topic. However it may turn out that these arguments could prove useful in the solution of other problems (cf. [211], [118]).

In the next chapter we again deal with exceptional numbers and forms, but then in a more general sense: Let $\mathbb{K}$ be an algebraic number field of finite degree over $\mathbb{Q}$, and let θ be an algebraic number of degree n over $\mathbb{K}$, with $\theta^{(1)}, \ldots, \theta^{(n)}$ denoting the zeros of its minimal polynomial over $\mathbb{K}$. We say that θ is exceptional relative to $\mathbb{K}$ if some permutation of $\theta^{(1)}, \ldots, \theta^{(n)}$ satisfies the equalities (2.7). The arguments we used above to prove the absence of exceptional numbers if $n \geq 5$ work in this case as well, provided we replace the field $\mathbb{G}$ by the field $\mathbb{G}^* = \mathbb{K}(\theta^{(1)}, \ldots, \theta^{(n)}, \zeta)$, where ζ is a primitive root of unity such that $\mathbb{K}(\zeta)$ contains all the roots of unity involved in (2.7). One can again construct permutations τ leading from the equalities (2.7) and obtain a contradiction if $n \geq 5$.

Thus if $[\theta : \mathbb{K}] \geq 5$, then θ is not an exceptional number relative to $\mathbb{K}$.

Thus (4.4), (4.5) and (4.6) yields

$$X < \exp\left\{c_{41}\left[H_f^{\frac{3}{2}(n-1)}\left(\ln|A| + H_f^{\frac{3}{2}(n-1)}\right)\right]^{1+\varepsilon}\right\},$$

where $c_{41} = c_{41}(n,\varepsilon)$, and $\varepsilon > 0$ is arbitrary.

5. Norm Forms with Two Dominating Variables

We now show that the arguments developed above apply to a wide class of diophantine equations in several unknowns which are essential generalisations of the Thue equation, but with sufficiently much of its rough features for our methods to be applicable [216].

Let $\mathbb{K}$ be an algebraic number field of degree $n \geq 3$ over $\mathbb{Q}$, and let θ be an algebraic integer in $\mathbb{K}$ of degree n. Further, let $A \neq 0$ be an integer and δ a real number in the interval $0 < \delta < 1$. We consider the equation

$$\mathrm{Nm}(x + \theta y + \lambda) = A,\qquad (5.1)$$

in unknown rational integers x, y with λ an integer in $\mathbb{K}$ satisfying

$$|\lambda| \leq X^{1-\delta}\ \text{where}\ X = \max(|x|,|y|).\qquad (5.2)$$

If $\lambda = 0$ the equation (5.1) is Thue's equation (the general equation (1.1) is reduced to (5.1) with $\lambda = 0$ by multiplying by the n-th power of the leading coefficient of $f(x,1)$). But if $\lambda \neq 0$ (5.1) has an essentially different structure and the possibility of analyzing it relying, say, on rational approximations to θ is dubious. Nonetheless we shall show that arguments close to those described above using the theory of linear forms in the logarithms of algebraic numbers lead to the following result:

Theorem 5.1 *There are effectively computable numbers $c_{42} > 0$ and $c_{43} > 0$, depending only on n and on the regulator of $\mathbb{K}$, such that for any δ in the interval $0 < \delta \leq c_{42}$, $(c_{42} < 1)$ expressions (5.1) and (5.2) imply that*

$$X < (2|A|H)^{c_{43}(1/\delta)\ln(1/\delta)},\qquad (5.3)$$

where H denotes the height of θ.

As a corollary to this theorem one can see that on completing $1, \theta = \theta_1$ until one obtains an integral basis $1, \theta_1, \ldots, \theta_{n-1}$ of the field $\mathbb{K}$, and considering the equation

$$\mathrm{Nm}(x_0 + \theta_1 x_1 + \ldots + \theta_{n-1}x_{n-1}) = A\qquad (5.4)$$

in which $x_0, x_1, \ldots, x_{n-1}$ are unknown rational integers subject to the conditions

$$\max_{2 \leq i \leq n-1} |x_i| \leq (\max(|x_0|,|x_1|))^{1-\delta},$$

all solutions will have bounds of the form (5.3). In particular, if $n = 3$ and $A = 1$, the solutions of (5.4) define units of the cubic field, all components of which (with respect to their representation in terms of the integral basis of the field $\mathbb{K}$) have "almost the same" order of magnitude. It also is easy to see that diophantine inequalities

$$|x_0 + \theta_1 x_1 + \ldots + \theta_{n-1} x_{n-1}| < X^{-n+1+c_{44}/((1/\delta)\ln(1/\delta))}, \qquad (5.5)$$

$$\max(|x_0|, |x_1|) \leq X, \quad \max_{2 \leq i \leq n-1} |x_i| \leq X^{1-\delta}, \qquad (5.6)$$

with $X > X_0$ have no rational integral solutions $(x_0, x_1, \ldots, x_{n-1}) \neq 0$. Since δ is not subject to any condition other than $0 < \delta \leq c_{42}$, one may suppose that δ changes with X. Taking $C > e$ and

$$\delta \sim \ln C \ln \ln X / (c_{44} \ln X)$$

we find from (5.5), (5.6) that the system of inequalities

$$|x_0 + \theta_1 x_1 + \ldots + \theta_{n-1} x_{n-1}| < C X^{-n+1}$$

$$\max(|x_0|, |x_1|) \leq X, \max_{2 \leq i \leq n-1} |x_i| \leq X(\ln X)^{-c_{45}}$$

has no integral solutions $(x_0, x_1, \ldots, x_{n-1}) \neq (0)$. On the other hand, if we take $\delta = c_{42}$, we find that for any fixed $\theta, \xi \in \mathbb{K}$, with $\deg \theta = n$,

$$|x_0 + \theta x_1 + \xi| > X^{-n+1+c_{46}},$$

where $c_{46} > 0$. Thus, we obtain an improvement of the inhomogeneous 'Liouville inequality'.

The proof of Theorem 5.1 relies on the following lemma due to Baker [20]:

Lemma 5.1 *Let* $\alpha_1, \alpha_2, \ldots, \alpha_n$ *be algebraic numbers of degree not greater than* d *and different from 0 or 1. Suppose that* α_n *has height not greater than* A *and that* $u_1, \ldots, u_{n-1}$ *are rational integers. Then for any* δ_1 *in the interval* $0 < \delta_1 < 1/2$, *we have*

$$|\alpha_1^{u_1} \cdots \alpha_{n-1}^{u_{n-1}} - \alpha_n| > \delta_1^{c_{47} \ln A} e^{-\delta_1 U}, \qquad U = \max_{1 \leq i \leq n-1} |u_i|$$

where c_{47} *is effectively determined in terms of* n, d *and the numbers* $\alpha_1, \alpha_2,$ $\ldots, \alpha_{n-1}$; *provided only that the left-hand side of the inequality is not zero.*

Proof of Theorem 5.1. Suppose that x, y and λ satisfy (5.1) and (5.2) and set $\mu = x + \theta y + \lambda$. Suppose that the conjugates $\mu^{(i)}$ are indexed in such a way that

$$|\mu^{(1)}| = \max_{(i)} |\mu^{(i)}|, \qquad |\mu^{(2)}| = \min_{(i)} |\mu^{(i)}| \qquad (i = 1, 2, \ldots, n).$$

Clearly it follows from $\mu^{(i)} = x + \theta^{(i)} y + \lambda^{(i)}$ $(i = 1, 2, 3)$ on eliminating x and y that

$$(\theta^{(3)} - \theta^{(2)})(\mu^{(1)} - \lambda^{(1)}) + (\theta^{(1)} - \theta^{(3)})(\mu^{(2)} - \lambda^{(2)}) - $$
$$-(\theta^{(1)} - \theta^{(2)})(\mu^{(3)} - \lambda^{(3)}) = 0. \quad (5.7)$$

Because the height of θ is H, it follows from (1.4), Ch.II that $\overline{|\theta|} \leq nH$, whence

$$|\mu^{(1)}| \leq |x| + \overline{|\theta|}|y| + \overline{|\lambda|} < 2nHX.$$

We may suppose that $|\mu^{(2)}| \leq 1$, for if not, we easily obtain an estimate for X which is better than (5.3). Similarly, we may suppose that

$$X = \max(|x|, |y|) > 3^{1/\delta}(2nH)^{n(n-1)/2\delta}. \quad (5.8)$$

Then we obtain

$$|\mu^{(1)}| \geq |\mu^{(1)} - \mu^{(2)}| - |\mu^{(2)}| \geq |\theta^{(1)} - \theta^{(2)}||y| - 2\overline{|\lambda|} - 1 >$$
$$> (2nH)^{-(n-1)n/2+1}|y| - 2X^{1-\delta} - 1,$$

since $|\theta^{(1)} - \theta^{(2)}| > (2nH)^{-(n-1)n/2+1}$. In the same way we obtain

$$|\theta^{(2)}||\mu^{(1)}| > (2nH)^{-(n-1)n/2+1}|x| - (2X^{1-\delta} + 1)nH.$$

Hence from (5.8) we find

$$|\mu^{(1)}| > X(2(2nH)^{-(n-1)n/2} - 3X^{-\delta}) > (2nH)^{-(n-1)n/2}X.$$

Therefore

$$(2nH)^{-(n-1)n/2}X < |\mu^{(1)}| < 2nHX. \quad (5.9)$$

Now from (5.7) and (5.2) we obtain

$$|\theta^{(3)} - \theta^{(2)} + (\theta^{(1)} - \theta^{(3)})\mu^{(2)}/\mu^{(1)} - (\theta^{(1)} - \theta^{(2)})\mu^{(3)}/\mu^{(1)}| \leq$$
$$\leq 3\overline{|\lambda|}(2nH)|\mu^{(1)}|^{-1} < 6nHX^{1-\delta}(2nH)^{n(n-1)/2}X^{-1} =$$
$$= 3(2nH)^{n(n-1)/2+1}X^{-\delta}.$$

Hence

$$\frac{|\theta^{(1)} - \theta^{(3)}||\mu^{(2)}|}{|\theta^{(3)} - \theta^{(2)}||\mu^{(1)}|} + 3(2nH)^{n(n-1)}X^{-\delta} \geq \left|1 - \frac{\theta^{(1)} - \theta^{(2)}}{\theta^{(3)} - \theta^{(2)}} \cdot \frac{\mu^{(3)}}{\mu^{(1)}}\right|. \quad (5.10)$$

Next we need an estimate from below for the difference

$$\left|1 - \frac{\theta^{(1)} - \theta^{(2)}}{\theta^{(3)} - \theta^{(2)}} \cdot \frac{\mu^{(3)}}{\mu^{(1)}}\right|. \quad (5.11)$$

First of all we note that we may assume that the difference is not zero. Indeed, suppose that

$$(\theta^{(3)} - \theta^{(2)})\mu^{(1)} - (\theta^{(1)} - \theta^{(2)})\mu^{(3)} = 0. \quad (5.12)$$

Much as in the previous § let Γ be the galois group of the field $\mathbb{Q}(\theta^{(1)}, \ldots, \theta^{(n)})$ and consider its elements as a group of permutations τ. Because the group Γ is transitive, it contains a substitution τ taking 2 into 1. On applying τ to (5.12) we obtain the relation

$$(\theta^{(k)} - \theta^{(1)})\mu^{(l)} - (\theta^{(l)} - \theta^{(1)})\mu^{(k)} = 0\,,$$

where $k \neq 1$, $l \neq 1$ and $k \neq l$. It follows that

$$|(\theta^{(k)} - \theta^{(1)})(x + \theta^{(l)}y) - (\theta^{(l)} - \theta^{(1)})(x + \theta^{(k)}y)| \leq 4nH\lceil\lambda\rceil,$$
$$|\theta^{(k)} - \theta^{(l)}||x + \theta^{(1)}y| \leq 4nH\lceil\lambda\rceil,$$

whence

$$|\mu^{(1)}| \leq \frac{4nH\lceil\lambda\rceil}{|\theta^{(k)} - \theta^{(l)}|} + \lceil\lambda\rceil < 3(2nH)^{n(n-1)/2}X^{1-\delta}\,.$$

Comparing this with the left-hand side of (5.9) we see that

$$X < 3^{1/\delta}(2nH)^{n(n-1)/\delta}$$

which is stronger than the assertion of the theorem. Thus, we may suppose that (5.12) does not hold.

Recalling that $\mathrm{Nm}(\mu) = A$ we apply Lemma 2.2, Ch.II to fix a unit η with $\lceil\mu/\eta\rceil = e^{\theta c'_{48}}|A|^{1/n}$. Here $|\theta| < 1$ and $c'_{48} > 0$ is a function of n and the regulator of the field $\mathbb{K}$. Set $\mu = \eta\nu$; thence $h(\nu) < c_{49}|A|$. If $\eta = \eta_1^{u_1}\ldots\eta_r^{u_r}$, where $u_1, \ldots, u_r$ are unknown integers, then

$$\frac{\theta^{(1)} - \theta^{(2)}}{\theta^{(3)} - \theta^{(2)}} \cdot \frac{\mu^{(2)}}{\mu^{(1)}} = \frac{\theta^{(1)} - \theta^{(2)}}{\theta^{(3)} - \theta^{(2)}} \cdot \frac{\nu^{(3)}}{\nu^{(1)}} \cdot \left(\frac{\eta_1^{(3)}}{\eta_1^{(1)}}\right)^{u_1} \cdots \left(\frac{\eta_r^{(3)}}{\eta_r^{(1)}}\right)^{u_r}\,.$$

We can bound the logarithm of the height of the number

$$\frac{(\theta^{(1)} - \theta^{(2)})\nu^{(3)}}{(\theta^{(3)} - \theta^{(2)})\nu^{(1)}}$$

from above by $c_{50}\ln(2|A|H)$, and then on applying Lemma 5.1 we obtain for (5.11) an estimate from below in the form

$$\delta_1^{c_{51}\ln(2|A|H)}e^{-\delta_1 U}, \text{ with } U = \max|u_i| \qquad (i = 1, 2, \ldots, r)\,, \tag{5.13}$$

where δ_1 is any number in the interval $0 < \delta_1 < 1/2$.

Since $\lceil\mu\rceil \leq 2nHX$ and $\mathrm{Nm}(\mu) = A$, we have

$$|\mu^{(i)}| \geq (2nHX)^{-n+1}|A| \qquad (i = 1, 2, \ldots, n)\,,$$

and since

$$c_{48}|A|^{1/n} \geq \lceil\mu/\eta\rceil \geq (2nHX)^{-n+1}|A||\eta^{(i)}|^{-1}\,,$$

we find from (5.8)

$$|\eta^{(i)}|^{-1} \leq c_{48}|A|^{\frac{1}{n}-1}(2nHX)^{n-1} < c_{48}X^n \qquad (i = 1, 2, \ldots, n)\,. \tag{5.14}$$

Hence

$$1 = \prod_{i=1}^{n}|\eta^{(i)}|^{-1} < |\eta^{(j)}|^{-1}(c_{48}X^n)^{n-1}\,,$$

which gives $|\eta^{(j)}| < (c_{48}X^n)^{n-1}$. Applying (5.14), we obtain

$$\max_{(i)} |\ln|\eta^{(i)}|| < c_{52}\ln X \qquad (i = 1, 2, \ldots, n).$$

Substituting $\eta^{(i)} = \eta_1^{(i)u_1}\ldots\eta_r^{(i)u_r}$ we obtain the system of linear inequalities in $u_1,\ldots,u_r$ from which $U < c_{53}\ln X$ follows.

Thus, the lower bound (5.13) for (5.11) may be replaced by the value

$$\delta_1^{c_{51}\ln(2|A|H)}X^{-c_{53}\delta_1}.$$

Suppose now that $\delta_1 = \delta/2c_{53} < 1/2$ and that

$$X > [6(2nH)^{n(n-1)}(2|A|H)^{c_{51}\ln(1/\delta_1)}]^{2/\delta}. \tag{5.15}$$

Then

$$3(2nH)^{n(n-1)}X^{-\delta} < \frac{1}{2}(2|A|H)^{-c_{51}\ln(1/\delta_1)}X^{-c_{53}\delta_1},$$

and so from (5.10) we have

$$\frac{|\theta^{(1)} - \theta^{(3)}||\mu^{(2)}|}{|\theta^{(3)} - \theta^{(2)}||\mu^{(1)}|} > \frac{1}{2}\delta_1^{c_{51}\ln(2|A|H)}X^{-c_{53}\delta_1}.$$

Hence, for $i = 1, 2, \ldots, n$

$$1 \geq \frac{|\mu^{(i)}|}{|\mu^{(1)}|} \geq \frac{|\mu^{(2)}|}{|\mu^{(1)}|} > \frac{1}{2}(2nH)^{-n(n-1)/2}(2|A|H)^{-c_{51}\ln(1/\delta_1)}X^{-\delta/2},$$

from which it follows that

$$\left|\ln\left|\frac{\eta^{(i)}}{\eta^{(1)}}\right|\right| < \frac{\delta}{2}\ln X + c\ln(2|A|H)\ln\frac{1}{\delta}, \tag{5.16}$$

if one takes into account that $\mu^{(i)} = \nu^{(i)}\eta^{(i)}$ and

$$\exp\{-2c'_{48}\} < |\nu^{(i)}|/|\nu^{(1)}| < \exp\{2c'_{48}\}.$$

Treating (5.16) as a system of linear inequalities in $u_1,\ldots,u_r$ yields

$$U < c_{54}\delta\ln X + c_{55}\ln(2|A|H)\ln(1/\delta),$$

which gives

$$\overline{|\eta|} < X^{c_{56}\delta}(2|A|H)^{c_{57}\ln(1/\delta)}.$$

Since $\overline{|\mu|} \leq \overline{|\nu|}\,\overline{|\eta|}$, it follows from (5.9) that

$$X < e^{c_{48}}|A|^{1/n}(2nH)^{n(n-1)/2}X^{c_{56}\delta}(2|A|H)^{c_{57}\ln(1/\delta)}.$$

But we may suppose that $c_{56}\delta \leq 1/2$, and then we find that

$$X < (2|A|H)^{c_{58}\ln(1/\delta)}. \tag{5.17}$$

This last inequality was obtained under the supposition (5.15). Hence, X does not exceed the maximum of the values (5.15), (5.17) and that yields the asserted inequality (5.3).

6. Equations in Relative Fields

To investigate a wider class of diophantine equations (including the problem of integral points on algebraic curves) it is useful to generalise the Thue equation to the case of binary forms with algebraic coefficients and with the unknowns in a subring of some given algebraic number field ([193],[26]). Relying on the ideas of the previous paragraphs, we consider the most general problem of this type and give the generalisations of the results obtained above on the equation of Thue, from which we derive as a corollary the bounds for the solutions of the norm-form equations in several unknowns (cf. [90]).

Let $\mathbb{K}$ be an algebraic number field of degree k over $\mathbb{Q}$ and $\mathbb{L}$ a finite extension of $\mathbb{K}$. Denote by $f = f(x,y)$ a binary form of degree n which has coefficients integers of $\mathbb{K}$ and which splits into linear factors over $\mathbb{L}$. We suppose that the polynomial $f(x,1)$ has at least three distinct zeros. Finally, $\alpha \neq 0$ is an integer of $\mathbb{K}$. The equation

$$f(x,y) = \alpha; \quad x,y \in I_{\mathbb{K}}, \tag{6.1}$$

is the obvious and natural generalisation of Thue's equation (1.1). We shall find bounds for $X = \max(\overline{|x|}, \overline{|y|})$ in terms of the parameters of the equation. In particular, we express the bounds in terms of the size $\overline{|\alpha|}$ of the number α and the height H_f (the maximum of the sizes of its coefficients) of the form f.

Denote by α_0 the leading coefficient of the polynomial $f(x,1)$, and let σ run through the isomorphisms of the field $\mathbb{L}$ into the field of complex numbers. In particular take τ to be the isomorphism for which $|y^\tau| = \overline{|y|}$ (given x, y, satisfying (6.1)). By the arguments detailed at the beginning of the §1, one easily finds that if X exceeds some particular power of $\overline{|\alpha|}H_f$, then there is a zero θ_1 of the polynomial $f(x,1)$ which satisfies

$$|x^\tau - \theta_1^\tau y^\tau| < X^{-1}. \tag{6.2}$$

Let θ_2, θ_3 be two other zeros of $f(x,1)$, distinct from θ_1 and different. Put $\mu_i = x - \theta_i y$, $(i = 1,2,3)$ and observe that by (6.2) we may suppose that

$$|\mu_2^\tau| = |x^\tau - \theta_2^\tau y^\tau| > \frac{1}{2}|\theta_1^\tau - \theta_2^\tau||y^\tau| \geq 1. \tag{6.3}$$

It is clear that

$$\overline{|\theta_i|} \leq nH_f\overline{|\alpha_0^{-1}|} \leq nH_f\overline{|\alpha_0|}^{k-1} \leq nH_f^k \quad (i = 1,2,3), \tag{6.4}$$

so

$$\lceil \overline{\mu_i} \rceil \leq 2nH_f^k X \qquad (i = 1, 2, 3).$$ (6.5)

By Lemma 2.1, Ch.II we can choose a system $\eta_1, \ldots, \eta_r$, of independent units in $\mathbb{L}$ so that

$$\prod_{j=1}^{r} \ln \lceil \overline{\eta_j} \rceil < c_{59} R,$$ (6.6)

where R is the regulator of the field $\mathbb{L}$, and c_{59} and the following similar quantities depend only on $l = [\mathbb{L} : \mathbb{Q}]$ if nothing else is indicated. By Lemma 2.2, Ch.II we find the units ξ_i, generated by the powers of $\eta_1, \ldots, \eta_r$ for which

$$|(\mu_i / \xi_i)^\sigma| = e^{\gamma_i c_{60} R} |\operatorname{Nm}(\mu_i)|^{1/l} \qquad (i = 1, 2, 3; \forall \sigma),$$

where $|\gamma_i| < 1$. Since $\lambda_i = \alpha_0 \mu_i$ is an integer dividing $\alpha_0^{n-1} \alpha$, we find that $\operatorname{Nm}(\lambda_i) = \operatorname{Nm}(\alpha_0) \operatorname{Nm}(\mu_i)$ divides $\operatorname{Nm}(\alpha_0^{n-1} \alpha)$. Hence

$$|\operatorname{Nm}(\mu_i)| \leq |\operatorname{Nm}(\alpha_0)|^{n-2} |\operatorname{Nm}(\alpha)| \leq (\lceil \overline{\alpha_0} \rceil^n \lceil \overline{\alpha} \rceil)^l.$$

Writing $\mu_i = \xi_i \nu_i$, $(i = 1, 2, 3)$, we notice that

$$\lceil \overline{\nu_i} \rceil \leq e^{c_{60} R} H_f^n \lceil \overline{\alpha} \rceil \qquad (i = 1, 2, 3),$$ (6.7)

and from (6.5) we find that

$$\lceil \overline{\xi_i} \rceil \leq \lceil \overline{\mu_i} \rceil \lceil \overline{\nu_i^{-1}} \rceil < 2nX e^{l c_{60} R} H_f^{nl} \lceil \overline{\alpha} \rceil^l \qquad (i = 1, 2, 3).$$ (6.8)

Hence, recalling that $|\xi_i^\sigma| \geq \lceil \overline{\xi_i} \rceil^{-l+1}$, we find from (6.8) that

$$|\ln |\xi_i^\sigma|| < c_{61} [\ln X + \ln(\lceil \overline{\alpha} \rceil H_f) + R] \qquad (i = 1, 2, 3; \forall \sigma).$$

Setting $\xi_i = \eta_1^{u_{1i}} \cdots \eta_r^{u_{ri}}$, we find from (6.6) that

$$U_i = \max_{1 \leq j \leq r} |u_{ji}| < c_{62} [\ln X + \ln(\lceil \overline{\alpha} \rceil H_f) + R] \qquad (i = 1, 2, 3).$$ (6.9)

We now use the equation

$$(\theta_3 - \theta_2) \mu_1 + (\theta_1 - \theta_3) \mu_2 - (\theta_1 - \theta_2) \mu_3 = 0,$$

from which we find

$$\frac{|\theta_3^\tau - \theta_2^\tau| |\mu_1^\tau|}{|\theta_1^\tau - \theta_3^\tau| |\mu_2^\tau|} = \left| \frac{\theta_1^\tau - \theta_2^\tau}{\theta_1^\tau - \theta_3^\tau} \cdot \frac{\mu_3^\tau}{\mu_2^\tau} - 1 \right|.$$ (6.10)

By (6.2), (6.3) and (6.4) we see that the left-hand side of this equality is bounded above by a value of the form

$$X^{-1} (2H_f)^{c_{63}}.$$ (6.11)

Since the left-hand side of (6.10) is not zero and

$$\frac{\mu_3^\tau}{\mu_2^\tau} = \frac{\nu_3^\tau \xi_3^\tau}{\nu_2^\tau \xi_2^\tau} = \frac{\nu_3^\tau}{\nu_2^\tau} (\eta_1^\tau)^{v_1} \ldots (\eta_r^\tau)^{v_r}, \qquad v_j = u_{j3} - u_{j2},$$

we can bound the right-hand side of (6.10) by applying a suitable estimate for the corresponding linear form in logarithms of algebraic numbers. Given (6.4) and (6.7), the logarithm of the height of number

$$\frac{(\theta_1^\tau - \theta_2^\tau)\nu_3^\tau}{(\theta_1^\tau - \theta_3^\tau)\nu_2^\tau}$$

does not exceed a value of order $c_{64}(\ln\lceil\alpha\rceil + \ln H_f + R)$. By Lemma 8.2 of Ch.III and (6.6), (6.9) the right-hand side of (6.10) is not less than

$$\exp\{-c_{65}R\ln R \cdot (\ln\lceil\alpha\rceil + \ln H_f + R)\ln[X + \ln(\lceil\alpha\rceil H_f) + R]\}.$$

Comparing this with the upper bound (6.11), we obtain

$$X < \exp\{c_{66}R\ln R \cdot (\ln(\lceil\alpha\rceil H_f) + R)\ln(\ln(\lceil\alpha\rceil H_f + R))\}. \qquad (6.12)$$

If one applies Lemma 5.1, then the right-hand side of (6.10) can be seen to have a lower bound

$$(c_{67}\lceil\alpha\rceil H_f)^{c_{68}\ln(1/\delta_1)} X^{-\delta_1},$$

where $0 < \delta_1 < 1/2$, and c_{67}, c_{68} depend only on l and R. Comparing again with (6.11) we obtain

$$X < c_{69}(\lceil\alpha\rceil H_f)^{c_{70}}, \qquad (6.13)$$

where c_{69} and c_{70} depend on l and R.

Thus, we have a complete analogue of previous results on the Thue equations:

Theorem 6.1 *All solutions x, y of equation (6.1) satisfy the inequalities (6.12) and (6.13), where R is the regulator of the field $\mathbb{L}$, c_{66} is effectively determined in terms of the degree of $\mathbb{L}$, and c_{69}, c_{70} may be expressed in terms of the degree and the regulator of $\mathbb{L}$.*

Though the bounds (6.12), (6.13) resemble the corresponding bounds (4.1) and (2.6) they do not reduce to those bounds if $\mathbb{K} = \mathbb{Q}$. This is because the estimates (6.12), (6.13) involve the regulator of the splitting field of the form f, while the estimates (4.1), (2.6) involve just the regulator of the minimal field containing a zero of the polynomial $f(x, 1)$.

Lemma 2.3 of Ch.II shows that the estimates cannot really be improved. The difference arises from the initial conditions under which the corresponding results are obtained: in Theorem 6.1 there is no supposition that the form f is irreducible over $\mathbb{K}$. Lemma 2.4 of Ch.II allows one to replace the regulator R by the value $c_{71}|D|^{1/2}(\ln|D|)^{l-1}$, where D is the discriminant of $\mathbb{L}$ and $c_{71} = c_{71}(l)$. If the form f is irreducible over $\mathbb{K}$, the discriminant D may be bounded in terms of the discriminant D_f of the form f, and then in terms of H_f, similar to what we did in §4. In the general case a rough bound in terms of H_f may be obtained as follows. It is clear from the previous arguments that instead of the splitting field $\mathbb{L}$ one can take the field $\mathbb{L}'$ containing the zeros $\theta_1, \theta_2, \theta_3$. Next choose a generating element of the field $\mathbb{L}'$ of the form of

$\theta = \theta_1 + t\theta_2 + t^2\theta_3$, where t is a rational integer, which because of (6.4) may be taken in the interval $|t| < 3(2n)^m H_f^{m(k+1)}$, $m = [\mathbb{L}' : \mathbb{Q}]$. Setting $\theta^* = \alpha_0\theta$ we find that

$$\overline{|\theta^*|} < 27(2n)^{2m+1} H_f^{(2m+1)(k+1)},$$

and for the discriminant of the number θ^* we obtain

$$0 \neq |D(\theta^*)| < (2\overline{|\theta^*|})^{m(m-1)} < c_{72} H_f^{m(m-1)(2m+1)(k+1)},$$

where $c_{72} = c_{72}(n)$. If n_1, n_2, n_3 are the degrees of $\theta_1, \theta_2, \theta_3$ respectively, then $m \leq n_1 n_2 n_3 k < n^3 k$. Since the discriminant $D_{\mathbb{L}'}$ of the field $\mathbb{L}'$ divides the discriminant $D(\theta^*)$ of the number $\theta^* \in I_{\mathbb{L}'}$, we have a rough bound

$$|D_{\mathbb{L}'}| < c_{72} H_f^{2n^9 k^3 (k+1)}.$$

Hence, from (6.12) we obtain

$$X < \exp\{c_{73} H_f^{n^9 k^3 (k+1)} (\ln \overline{|\alpha|} + H_f^{n^9 k^3 (k+1)})^{1+\varepsilon}\},$$

where $c_{73} = c_{73}(n, \varepsilon)$, and $\varepsilon > 0$ is arbitrary.

One can obtain more precise estimates in terms of H_f, if they should be needed, from the theory of the differents of algebraic extensions.

Let us now show that Theorem 6.1 allows one to obtain bounds for the solutions in rational integers $x_0, x_1, \ldots, x_s$ of diophantine equations of the form

$$\mathrm{Nm}(x_0 + \alpha x_1 + \ldots + \alpha_s x_s) = A, \tag{6.14}$$

where $\alpha_1, \ldots, \alpha_s$ are algebraic integers, and $A \neq 0$ is a rational integer.

Theorem 6.2 *Set* $\mathbb{K}_i = \mathbb{Q}(\alpha_1, \ldots, \alpha_{i-1}, \alpha_{i+1}, \ldots, \alpha_s)$, $(i = 1, 2, \ldots, s)$, *and let* $\mathbb{L}$ *be the normal extension of* $\mathbb{K} = \mathbb{Q}(\alpha_1, \ldots, \alpha_s)$. *Suppose that*

$$[\mathbb{K} : \mathbb{K}_i] \geq 3 \qquad (i = 1, 2, \ldots, s). \tag{6.15}$$

Then all rational integral solutions $x_0, x_1, \ldots, x_s$ *of equation (6.14) satisfy*

$$X = \max_{0 \leq i \leq s} |x_i| < \exp\{c_{74} R \ln R \cdot T \ln T\}, \tag{6.16}$$

where $T = \ln |A| + \ln H + R$, $H = \max(\overline{|\alpha_1|}, \ldots, \overline{|\alpha_s|})$, *and* R *is the regulator of the field* $\mathbb{L}$. *The quantity* c_{74} *has an effective presentation in terms of the degree of* $\mathbb{L}$. *In addition,*

$$X < c_{75}(|A|H)^{c_{76}}, \tag{6.17}$$

where c_{75}, c_{76} *can be expressed in terms of the regulator and the degree of* $\mathbb{L}$.

Proof. The proof of the theorem is a simple consequence of Theorem 6.1.

Write equation (6.14) in the form

$$\mathrm{Nm}_{\mathbb{K}_1/\mathbb{Q}}(\mathrm{Nm}_{\mathbb{K}/\mathbb{K}_1}(x + \alpha_1 x_1)) = A,$$

where $x = x_0 + \alpha_2 x_2 + \ldots + \alpha_s x_s$. It follows that

$$\mathrm{Nm}_{\mathbb{K}/\mathbb{K}_1}(x + \alpha_1 x_1) = \alpha, \tag{6.18}$$

where α is an integer of the field $\mathbb{K}_1$ dividing A. By Lemma 2.2 of Ch.II we find the unit η^{-1} of the field $\mathbb{K}_1$ for which

$$\overline{|\alpha/\eta|} < e^{c_{77} R_1} |\mathrm{Nm}(\alpha)|^{1/m_1}, \qquad m_1 = [\mathbb{K}_1 : \mathbb{Q}], \tag{6.19}$$

where R_1 is the regulator of $\mathbb{K}_1$, and $c_{77} = c_{77}(m_1)$. Let $n_1 = [\mathbb{K} : \mathbb{K}_1]$. Set $\eta^{-1} = \xi^{n_1} \xi_1$, where ξ, ξ_1 are the units of the field $\mathbb{K}_1$ which one obtains on representing η^{-1} by the product of the generating units and reducing the exponents modulo n_1. Then we have

$$\overline{|\xi_1|} < e^{c_{78} R_1}, \qquad c_{78} = c_{78}(n). \tag{6.20}$$

Multiplying the equality (6.18) by ξ^{n_1}, putting $x' = \xi x$, $x_1' = \xi x_1$, $\alpha' = \xi^{n_1} \alpha$ and recalling that $\mathrm{Nm}(\alpha)$ divides A^{m_1}, we obtain from (6.19) and (6.20) that

$$\mathrm{Nm}_{\mathbb{K}/\mathbb{K}_1}(x' + \alpha_1 x_1') = \alpha', \qquad \overline{|\alpha'|} < e^{c_{79} R_1} |A|,$$

where $x', x_1' \in I_{\mathbb{K}_i}$. The equation we have obtained is of the type (6.1), and in view of condition (6.15), Theorem 6.1 may be applied. That leads to bounds for $\overline{|x_1'|}$ of the form (6.12), (6.13). Since $\overline{|x'|} = |x_1| \overline{|\xi|} \geq |x_1|$, we have similar bounds for $|x_1|$. Here Lemma 2.3 of Ch.II allows one to replace R_1 by $R = R_{\mathbb{L}}$.

By symmetry of the unknowns $x_1, \ldots, x_s$ in equation (6.14) similar bounds hold for the other unknowns. With this, a bound for x_0 follows directly from equation (6.14). Thus we obtain (6.16) and (6.17).

We should remark that inequality (6.17) yields a sharpening of the generalised Liouville inequality. We have

$$|x_0 + \alpha_1 x_1 + \ldots + \alpha_s x_s| > c_{80} X^{-d+1+1/c_{76}},$$

where $d = \deg \mathbb{K}$, and $c_{80} > 0$ and $c_{76} > 0$ may be effectively determined in terms of $\alpha_1, \ldots, \alpha_s$ and $X = \max_{(i)} |x_i| \neq 0$. For special numbers, inequalities with better exponents (close to $-s$) are known. The non-effective version of the inequality corresponds to the best possible exponent: $-s - \varepsilon$, where $\varepsilon > 0$ is arbitrary [185].

V. The Thue-Mahler Equation

We develop and deepen the arguments of the previous chapter mainly by further applications of p-adic analysis. This analysis allows one to observe qualitatively new facts, for example, that the speed of growth of the maximal prime divisor of a binary form can be bounded from below. And we can deepen the bounds for rational approximations of algebraic numbers by including the p-adic metrics. We begin by investigating the solution of the Thue equation in rational numbers with denominators comprised from a fixed set of prime numbers with unknown exponents.

1. Solution of the Thue Equation in S-integers

As at the beginning of the previous chapter, we suppose that $f = f(x, y)$ is an integral irreducible binary form of degree $n \geq 3$, and that $A \neq 0$ is an integer. Furthermore, $p_1, \ldots, p_s$ denotes a fixed set of prime numbers with $P = \max(p_1, \ldots, p_s)$.

In 1933 Mahler [129] published a paper on the investigation of the equation

$$f(x, y) = p_1^{z_1} \cdots p_s^{z_s}, \qquad (x, y) = 1, \tag{1.1}$$

in which $x, y, z_1 \geq 0, \ldots, z_s \geq 0$ are unknown integers. Mahler proved that this equation has only finitely many solutions. From this it follows that Thue's equation has only a finite number of solutions in S-integers, where S is the set of primes $p_1, \ldots, p_s$. S-integers are of course all rational numbers whose denominators contain only primes from S. It follows that the greatest prime divisor $P[f(x, y)]$ of the numerical value $f(x, y)$ of the binary form tends to infinity as $X = \max(|x|, |y|)$, $(x, y \in \mathbb{Z})$ goes to infinity. Mahler's arguments are based on a p-adic generalisation of Thue's method for investigating rational approximations to algebraic numbers. Thus, as are other results obtained by Thue's method, they are ineffective.

We consider the equation

$$f(x, y) = A p_1^{z_1} \cdots p_s^{z_s}, \qquad (x, y) = 1, \tag{1.2}$$

which combines Thue's equation and that studied by Mahler (1.1). To develop the arguments of the previous chapter means including estimates in p-adic metrics. We obtain bounds for the solutions of the equation (1.2) and discuss the principal consequences implied by an analysis of those bounds.

Because the left-hand side of (1.2) is homogeneous in x and y, it is obvious that a solution of the Thue equation

$$f(x', y') = A' \tag{1.3}$$

in S-integers x', y' reduces to a solution of equations of the form (1.2); conversely, solutions of the latter reduce to solutions of (1.3) in S-integers. Thus, analysing equations (1.2) is a proper step towards consideration of rational points on curves of type (1.3). In our further investigations of other types of equations we shall always be able to deal not only with integer solutions, but with S-integer solutions as well. Moreover, a significant difference between the ineffective and effective approaches to the problem should be borne in mind. The ineffective approach deals specifically with corresponding rational solutions that are subject to the strong restriction that their denominators be composed of just the primes of the given set. One can then declare that there are just finitely many such solutions. In principle this approach is silent on solutions not satisfying the given restrictions. On the other hand, the effective approach considers any rational solution and gives specific information, bounds of various kinds and the like, about that solution. Any solution can be considered, because any rational solution can be viewed as being S-integral simply by constructing the system S from the prime divisors of the denominators of the solution under consideration.

As in the previous chapter we start by considering the influence of the main parameter A of (1.2) on the magnitude of its solutions. Once again we rely on the results of Chapter III.

Theorem 1.1 *Let $f = f(x, y)$ be an integral irreducible binary form of degree $n \geq 3$, and let $A \neq 0$ be an integer; $p_1, \ldots, p_s$ denote distinct prime numbers. Then all integral solutions $x, y, z_1 \geq 0, \ldots, z_s \geq 0$ of (1.2) satisfy*

$$\max(|x|, |y|, p_1^{z_1}, \ldots, p_s^{z_s}) < c_1(|A|H_f)^{c_2}, \tag{1.4}$$

where H_f is the height of f, and where c_1 and c_2 can be expressed in terms of n, the regulator and class-number of the field $\mathbb{K}$ containing a root of the form f, and the primes $p_1, \ldots, p_s$.

Proof. Much as in our consideration of the Thue equation in the previous chapter, we prove Theorem 1.1 using lemmas connecting estimates for linear forms in various metrics; that allows us to obtain the theorem for all forms of degree $n \geq 5$ and for non-exceptional forms of degree 4. It will also be apparent from our arguments that application of the corresponding direct estimates for linear forms in the logarithms of algebraic numbers allows one to obtain the theorem for all forms of degree $n \geq 3$.

Let $f(x, y) = a \operatorname{Nm}(x - \theta y)$, with $a > 0$. Set $\mathbb{K} = \mathbb{Q}(\theta)$ and $\mathbb{G} = \mathbb{Q}(\theta^{(1)}, \ldots, \theta^{(n)})$. Then (1.2) assumes the shape

$$a \operatorname{Nm}(x - \theta y) = A p_1^{z_1} \cdots p_s^{z_s}, \qquad (x, y) = 1. \tag{1.5}$$

We may suppose that $(A, p_1 p_2 \cdots p_s) = 1$, since any p_i-component of A may be joined to $p_i^{z_i}$, $(1 \le i \le s)$, thus replacing the number A by a smaller number.
Equation (1.5) yields a factorization into ideals in $I_{\mathbb{K}}$:

$$(a(x - \theta y)) = \mathfrak{a} \mathfrak{p}_1^{U_1} \cdots \mathfrak{p}_t^{U_t}$$

where $\mathfrak{a}$ divides $a^{n-1} A$ and the prime ideals $\mathfrak{p}_1, \ldots, \mathfrak{p}_t$ divide the prime numbers $p_1, \ldots, p_s$. On setting $\theta_1 = a\theta$, and reducing by (ax, y) we obtain

$$(X - \theta_1 Y) = \mathfrak{a}' \mathfrak{p}_1^{U_1'} \ldots \mathfrak{p}_t^{U_t'}, \qquad \mathfrak{a}' | a^{n-1} A, \qquad (x, y) = 1. \tag{1.6}$$

Note that we may suppose that the prime ideals divide the corresponding primes: $\mathfrak{p}_i \mid p_i$; possibly after some extension of the ideal $\mathfrak{a}'$. Indeed, if, for example, $\mathfrak{p}_1$ and $\mathfrak{p}_2$ divide p_1, then $\mathfrak{p}_1 \mathfrak{p}_2$ divides the norm $\mathrm{N}(\mathfrak{p}_1) = \mathfrak{p}_1^{(1)} \cdots \mathfrak{p}_1^{(n)}$, and then $\mathfrak{p}_2$ divides $\mathfrak{p}_1^{(2)} \cdots \mathfrak{p}_1^{(n)}$. Assuming that $U_1' \ge U_2'$ (if $U_1' \ge U_2'$ we exchange the roles of $\mathfrak{p}_1$ and $\mathfrak{p}_2$), we find from (1.6) that

$$\mathfrak{p}_2^{U_2'} \mid \prod_{i=2}^n (X - \theta_1^{(i)} Y),$$

and if $\mathfrak{P}$ is a prime divisor of $\mathfrak{p}_2$ in $\mathbb{G}$, then for some $i \ne 1$ we have

$$\mathfrak{P}^{U_2''} \mid (X - \theta_1^{(i)} Y), \quad U_2'' \ge U_2'/(n-1).$$

We also find from (1.6) that $\mathfrak{P}^{U_2''} \mid (X - \theta_1^{(1)} Y)$, so that $\mathfrak{P}^{U_2''} \mid (\theta_1^{(1)} - \theta_1^{(i)})$, from which it follows that $\mathrm{N}(\mathfrak{P})^{U_2''} \mid D(\theta_1)^{n!}$, and then $n!(n-1)\operatorname{ord}_{\mathfrak{p}_2} D(\theta_1) \ge U_2'$. Thus, (1.6) may be replaced by

$$(X - \theta_1 Y) = \mathfrak{a}'' \mathfrak{p}_1^{U_1'} \cdots \mathfrak{p}_s^{U_s'}, \qquad \mathfrak{a}'' \mid A', \tag{1.7}$$

where $A' = a^{n-1} A D(\theta_1)^{(n-1)n!}$ and $\mathfrak{p}_i$ are prime ideals of $I_{\mathbb{K}}$ dividing the corresponding p_i, $(i = 1, 2, \ldots, s)$.

Let h be the class-number of the field $\mathbb{K}$. Then $\mathfrak{p}_i^h = (\rho_i)$ is a principal ideal and ρ_i is an integer in $\mathbb{K}$. Setting $U_i' = h u_i + r_i$, $0 \le r_i < h$, $(i = 1, 2, \ldots, s)$, we find from (1.7) that

$$(X - \theta_1 Y) = (\rho \rho_1^{u_1} \ldots \rho_s^{u_s}), \tag{1.8}$$

where (ρ) is a principal ideal:

$$(\rho) = \mathfrak{a}'' \mathfrak{p}_1^{r_1} \cdots \mathfrak{p}_s^{r_s}. \tag{1.9}$$

We may assume by applying Lemma 2.2 of Ch.II to (1.9) that

$$\overline{|\rho|} < e^{c_3 R} |A'| (p_1 \cdots p_s)^h, \qquad c_3 = c_3(n), \tag{1.10}$$

where R is the regulator of $\mathbb{K}$. Similarly,

$$|\overline{\rho_i}| < e^{c_3 R} P^h \qquad (i = 1, 2, \dots, s),\qquad (1.11)$$

where P is the maximum of the primes $p_1, \dots, p_s$. It follows from (1.8) that

$$X - \theta_1 Y = \rho \rho_1^{u_1} \dots \rho_s^{u_s} \xi, \qquad (1.12)$$

where ξ is a unit in $I_{\mathbb{K}}$. Now denote by $U_{\mathbb{K}}$ the free group generated by the system of independent units in $\mathbb{K}$ defined by Lemma 2.1 of Ch.II. Since the regulator of this system of units is bounded by $c_4 R$, $c_4 = c_4(n)$, it follows that its index $U_{\mathbb{K}}$ in the group $E_{\mathbb{K}}$ of all the units of infinite order in $\mathbb{K}$ does not exceed c_4. Hence, ξ may be expressed in the form $\xi = \xi_0 \eta$, where

$$\eta \in U_{\mathbb{K}}, \qquad \xi_0 \in E_{\mathbb{K}}, \qquad |\overline{\xi_0}| < e^{c_5 R}.$$

Setting $\eta = \eta_1^{v_1} \cdots \eta_r^{v_r}$, where $\eta_1, \dots, \eta_r$ are generating elements of $U_{\mathbb{K}}$, we find from (1.12)

$$\mu = X - \theta_1 Y = \rho_0 \rho^{u_1} \cdots \rho_s^{u_s} \eta_1^{v_1} \cdots \eta_r^{v_r}, \qquad (1.13)$$

where $\rho_0 = \rho \xi_0$. Hence, from (1.10) we have

$$|\overline{\rho_0}| < e^{(c_3 + c_5) R} a^{n-1} |A| |D(\theta_1)|^{(n-1)n!} (p_1 \cdots p_s)^h. \qquad (1.14)$$

Passing to conjugates in (1.13) and eliminating X, Y, we obtain the system of equalities

$$(\theta_1^{(1)} - \theta_1^{(2)})\mu^{(i)} - (\theta_1^{(i)} - \theta_1^{(2)})\mu^{(1)} - (\theta_1^{(1)} - \theta_1^{(i)})\mu^{(2)} = 0, \qquad (1.15)$$

where we assume that the enumeration of the values $\mu^{(i)}$ corresponds to the conditions

$$|\mu^{(1)}| = \max_{(i)} |\mu^{(i)}|, \quad |\mu^{(2)}| = \min_{(i)} |\mu^{(i)}| \qquad (i = 1, 2, \dots, n). \qquad (1.16)$$

It is convenient to give some preliminary consideration to the exponents u_r, $(l \leq r \leq s)$ in (1.13), before applying the results of Ch.II on the connection between estimates for linear forms in different non-archimedean metrics to the system of equalities (1.15). Let k be fixed and suppose that for some $i \neq 2$ we have

$$(\rho_1^{(i)u_1} \cdots \rho_s^{(i)u_s}, \rho_k^{(2)}) \neq 1.$$

If, say, $(\rho_l^{(i)u_l}, \rho_k^{(2)}) \neq 1$, then $l = k$ since $(\rho_j) = \mathfrak{p}_j^h$ and from $(\mathfrak{p}_l^{(i)}, \mathfrak{p}_k^{(2)}) \neq 1$ it follows that $(p_l, p_k) \neq 1$, $p_l = p_k$. Let $\mathfrak{P}$ be a prime ideal in $I_{\mathbb{G}}$ dividing $(\rho_k^{(i)}, \rho_k^{(2)})$. We find from (1.13) that $\mathfrak{P}^{u_k}$ divides $X - \theta_1^{(2)} Y$ and $X - \theta_1^{(i)} Y$; consequently, it divides $\theta_1^{(2)} - \theta_1^{(i)}$, since $(X, Y) = 1$. Hence, $\mathfrak{P}^{u_k}$ divides $D(\theta_1)$. So $u_r \leq n! \ln |D(\theta_1)|$.

We now assume that

$$(\rho_1^{(i)u_1} \cdots \rho_s^{(i)u_s}, \rho_k^{(2)}) = 1, \qquad ((\theta_1^{(2)} - \theta_1^{(i)})\rho_0^{(2)}, \rho_k^{(2)}) \neq 1.$$

Again let $\mathfrak{P}$ be a prime ideal in $I_{\mathbb{G}}$ such that

$$\mathfrak{P}^u \parallel ((\theta_1^{(i)} - \theta_1^{(2)})\rho_0^{(1)}, \rho_k^{(2)u_k}). \tag{1.17}$$

Then $\mathfrak{P}^u$ divides $D(\theta_1)\,\mathrm{Nm}(\rho_0)$, so $u \le n!(\ln|D(\theta_1)| + \ln|\mathrm{Nm}(\rho_0)|)$. If u_k exceeds this bound for u, then recalling (1.15) and (1.17), we obtain

$$\mathfrak{P}^u \parallel (\theta_1^{(i)} - \theta_1^{(2)})\rho_0^{(1)} \qquad \mathfrak{P}^u \parallel (\theta_1^{(1)} - \theta_1^{(2)})\rho_0^{(i)}$$

and then $\mathfrak{P}$ is not contained in the quotient

$$(\theta_1^{(i)} - \theta_1^{(2)})\rho_0^{(1)} \,/\, (\theta_1^{(1)} - \theta_1^{(2)})\rho_0^{(i)} \qquad (i = 3, 4, \ldots, n)$$

Consequently, in the equality

$$\left(\frac{\theta_1^{(i)} - \theta_1^{(2)})\mu^{(1)}}{(\theta_1^{(1)} - \theta_1^{(2)})\mu^{(i)}}\right) = \left[\frac{(\theta_1^{(i)} - \theta_1^{(2)})\rho_0^{(1)}}{(\theta_1^{(1)} - \theta_1^{(2)})\rho_0^{(i)}}\right] \prod_{j=1}^{s} \left[\frac{\rho_j^{(1)}}{\rho_j^{(i)}}\right]^{u_j} \prod_{j=1}^{r} \left[\frac{\eta_j^{(1)}}{\eta_j^{(i)}}\right]^{v_j} \tag{1.18}$$

for any prime ideal $\mathfrak{P}_k$ in I_G and dividing $\rho_k^{(2)}$, one of the two possibilities holds: either the number in the first set of square brackets is a $\mathfrak{P}_k$-adic unit, or in view of (1.14)

$$u_k \le n!(\ln|D(\theta_1)| + \ln|\mathrm{Nm}(\rho_0)|) <$$
$$< c_6(\ln|A| + \ln H_f + R + hs\ln P), \qquad c_6 = c_6(n). \tag{1.19}$$

We obtain from (1.15), (1.16)

$$1 \ge \frac{|\mu^{(j)}|}{|\mu^{(1)}|} \ge \frac{|\mu^{(2)}|}{|\mu^{(1)}|} = \left|\frac{(\theta^{(1)} - \theta^{(2)})\mu^{(i)}}{(\theta^{(1)} - \theta^{(i)})\mu^{(1)}} \cdot \frac{\theta^{(i)} - \theta^{(2)}}{\theta^{(1)} - \theta^{(i)}}\right| =$$

$$= \frac{|\theta^{(1)} - \theta^{(2)}|}{|\theta^{(1)} - \theta^{(i)}|} \frac{|\rho_0^{(i)}|}{|\rho_0^{(1)}|} \left|\frac{(\theta^{(i)} - \theta^{(2)})\rho_0^{(1)}}{(\theta^{(1)} - \theta^{(2)})\rho_0^{(i)}} - \prod_{j=1}^{s}\left[\frac{\rho_j^{(i)}}{\rho_j^{(1)}}\right]^{u_j} \prod_{j=1}^{r}\left[\frac{\eta_j^{(i)}}{\eta_j^{(1)}}\right]^{v_j}\right|.$$

Because of the estimate (1.14), the logarithm of the height of the number $(\theta^{(i)} - \theta^{(2)})\rho_0^{(1)}/(\theta^{(1)} - \theta^{(2)})\rho_0^{(i)}$ is bounded by a quantity similar to (1.19), and by (1.11) the logarithm of the heights of the numbers $\rho_j^{(i)}/\rho_j^{(1)}$ is bounded by $c_7(R + h\ln P)$, whilst the logarithm of the heights of the numbers $\eta_j^{(i)}/\eta_j^{(1)}$ is bounded by $c_8 R$. Without repeating the arguments described in §2 Ch.IV (see also §4 below) we note at once that for non-exceptional forms f we obtain from Lemma 1.1 of Ch.III:

$$\left|\frac{(\theta^{(i)} - \theta^{(2)})\rho_0^{(1)}}{(\theta^{(1)} - \theta^{(2)})\rho_0^{(i)}} - \prod_{j=1}^{s}\left[\frac{\rho_j^{(i)}}{\rho_j^{(1)}}\right]^{u_j} \prod_{j=1}^{r}\left[\frac{\eta_j^{(i)}}{\eta_j^{(1)}}\right]^{v_j}\right| > e^{-\delta U} \tag{1.20}$$

$$U = \max_{(i,j)}(u_i, |v_j|) \qquad (i = 1, 2, \ldots, s; \quad j = 1, 2, \ldots, r),$$

provided that for given δ, $(0 < \delta \le \sqrt{2}P^{n!})^{-1})$, the inequality

$$U > \left(c_9\delta^{-1}(R + h\ln P)^2 s^4\right)^{4(n+s+1)}(\ln|A| + \ln H_f + R + hs\ln P) \tag{1.21}$$

holds, where $c_9 = c_9(n)$; the same goes for $c_{10}, \ldots, c_{18}$ below. To estimate the exponents u_k exceeding (1.19), take the prime ideal $\mathfrak{P}$ in I_G dividing $\rho_k^{(2)}$ and note that from the equality (1.15) it follows that

$$\left| \left[\frac{(\theta_1^{(i)} - \theta_1^{(2)})\rho_0^{(1)}}{(\theta_1^{(1)} - \theta_1^{(2)})\rho_0^{(i)}} \right] - \prod_{j=1}^{s} \left[\frac{\rho_j^{(i)}}{\rho_j^{(1)}} \right]^{u_j} \prod_{j=1}^{r} \left[\frac{\eta_j^{(i)}}{\eta_j^{(1)}} \right]^{v_j} \right|_{\mathfrak{P}_k} \leq p_k^{-u_k/e_k}, \qquad (1.22)$$

where e_k is the ramification index of $\mathfrak{P}_k$ in p_k. Comparing this with (1.18) we see that each number in square brackets is a $\mathfrak{P}_k$-adic unit. To obtain a lower bound for the left-hand side of (1.22), we apply Lemma 1.1 of Ch.III arguing as in obtaining the estimate (1.20), whilst recalling that

$$T = T_k = \mathrm{N}(\mathfrak{P}_k^{2l_k})\left(1 - \frac{1}{\mathrm{N}(\mathfrak{P}_k)}\right) < P^{2n!}.$$

The left-hand side of (1.22) is bounded from below by the value $p_k^{-\delta U}$ if

$$U > (c_{10}\delta^{-1}s^4(R + h\ln P)^2)^{4(s+n+1)} \times$$
$$\times P^{16(n+s+1)n!+1}(\ln|A| + \ln H_f + R + hs\ln P). \qquad (1.23)$$

Having assumed this inequality, which is stronger than (1.21), we find from (1.22) that $u_k \leq n!\delta U$, $(k = 1, 2, \ldots, s)$. Therefore the inequalities

$$1 \geq \frac{|\mu^{(j)}|}{|\mu^{(1)}|} \geq \frac{|\mu^{(2)}|}{|\mu^{(1)}|} > e^{-\delta U}, \qquad (j = 3, 4, \ldots, n),$$

yield for $(j = 2, 3, \ldots, n)$:

$$\left| \sum_{i=1}^{r} v_i \ln \left| \frac{\eta_j^{(i)}}{\eta_j^{(1)}} \right| \right| < c_{11}\delta U(R + h\ln P) + c_{12}(\ln|A| + \ln H_f + R + hs\ln P).$$

It follows from this system of inequalities that

$$\max_{1 \leq j \leq r} |v_j| < c_{13}\delta U(R + h\ln P) + c_{14}(\ln|A| + \ln H_f + R + hs\ln P)s.$$

Taking $\delta = \min((2n!c_{13}(R + h\ln P))^{-1}, (\sqrt{2}P^{n!})^{-1})$, we find that

$$U < 2c_{14}(\ln|A| + \ln H_f + R + hs\ln P)s.$$

This inequality is inconsistent with (1.23), so we have to suppose that

$$U < (c_{15}(R + h\ln P)^3 s^4)^{4(s+n+1)} \times$$
$$\times P^{20(n+s+1)n!+1}(\ln|A| + \ln H_f + R + hs\ln P) <$$
$$< (c_{17}P^{5(n!+1)}(R + h\ln P)^3)^{4(s+n+1)}(\ln|A| + \ln H_f + R + hP),$$

where we used the inequalities $s < P$ and $s < c_{16}P/\ln P$; c_{16} is an absolute constant. Turning to (1.13), we have

$$|\mu| \leq \overline{|\rho_0|}\, \overline{|\rho_1|}^{u_1} \cdots \overline{|\rho_s|}^{u_s}\, \overline{|\eta_1|}^{(n!-1)|v_1|} \cdots \overline{|\eta_r|}^{(n!-1)|v_r|}. \qquad (1.24)$$

From this, and using the obtained estimate for U, we come to (1.4), where

$$c_1 = e^{c_2(R+hP)} \qquad (1.25)$$

$$c_2 = (c_{18}P^{5(n!+1)}(R+h\ln P)^3)^{4(s+n+1)}P(R+h\ln P).$$

That completes the proof of Theorem 1.1 for non-exceptional forms; hence, for all forms of degree $n \geq 5$, and for non-exceptional forms of degree 4. An application of the direct estimates for linear forms in the logarithms of algebraic numbers as before gives the proof of the theorem for any $n \geq 3$ (see Lemma 5.1 Ch.IV and its p-adic versions in [158]).

Suppose we apply, for example, Lemmas 8.2 and 8.3 of Ch.III. We obtain a lower bound for the difference in (1.20) by Lemma 8.2 Ch.III:

$$\exp\{-(c_{19}s)^{200(n+s)+1}(R+h\ln P)^s R\ln(R+h\ln P)(\ln|A|+$$
$$+\ln H_f + R + hs\ln P)\ln U\},$$

while the left-hand side of (1.22), estimated from below by Lemma 8.3 Ch.III, gives

$$u_k \leq (c_{20}s)^{12(s+n)+1}P^{n!}(R+h\ln P)^s R(\ln|A| + \ln H_f + R + hs\ln P)(\ln U)^2.$$

This leads to the estimate

$$U < (c_{21}s)^{200(n+s)+3}P^{n!}(R+h\ln P)^s R(\ln|A|+$$
$$+\ln H_f + R + hP)\ln^2(\ln|A| + \ln H_f + R + hP).$$

Hence, appealing to the inequality (1.24), we find

$$\ln\max(|x|,|y|) < (c_{22}s)^{200(n+s)+4}P^{n!}(R+h\ln P)^{s+1}RB(\ln B)^2, \qquad (1.26)$$

where $B = \ln|A| + \ln H_f + R + hP$. In all these inequalities the quantities c_{19}, c_{20}, c_{21}, and c_{22} are functions of n only. We now see directly from (1.2) that the powers $p_i^{z_i}$, $(i = 1, 2, \ldots, s)$ are bounded by expression of the type (1.26), say, with some other value c_{22}.

One may replace the regulator R, and the class-number h, in the estimates above by their bound in terms of the discriminant of the field $\mathbb{K}$ (see Lemma 2.4 Ch.II). After that one may estimate the discriminant in terms of the height H_f of the form f, as was done in §4 Ch.IV in the case of the Thue equation. In this way we obtain:

Theorem 1.2 *Under the conditions of Theorem 1.1 all the solutions of (1.2) satisfy the inequality (1.26), as well as the inequality*

$$\ln\max(|x|,|y|) < (c_{23}s)^{200(n+s)+4}P^{n!}(H_f^{3n/2}\ln P)^{s+1}H_f^{3n/2}B'(\ln B')^2,$$

where $B' = \ln|A| + PH_f^{3n/2}$, and c_{23} is effectively determined in terms of n.

2. Rational Approximation to Algebraic Numbers in Several Metrics

One consequence of Theorem 1.1 is a new assertion on rational approximation to algebraic numbers. We have already seen that the bounds for the solutions of the Thue equation allow one to improve Liouville's inequality (§2 Ch.IV), as was shown in proving Theorem 2.1 Ch.IV. Since Theorem 1.1 implies Theorem 2.1 Ch.IV, it is natural to expect that some new strengthening or deepening of Liouville's inequality follows from it. This is indeed so ([205], [206]).

Theorem 2.1 *Let θ be an algebraic number of degree $n \geq 3$ and S some finite set of distinct prime ideals $\mathfrak{p}$ of the field $\mathbb{Q}(\theta)$. Then for rational integers x, y with $(x, y) = 1$ we have the inequality*

$$|x - \theta y| \prod_{\mathfrak{p} \in S} |x - \theta y|_{\mathfrak{p}}^{n_{\mathfrak{p}}} > c_{24} X^{-n+1+1/c_2}, \qquad X = \max(|x|, |y|), \qquad (2.1)$$

where $\mathfrak{p}$ runs through all the ideals in S and the $n_{\mathfrak{p}}$ denote the local degrees $[\mathbb{Q}_{\mathfrak{p}}(\theta) : \mathbb{Q}_{\mathfrak{p}}]$. Moreover c_2 is a positive quantity effectively determined in terms of n, the regulator R and the class-number of the field $\mathbb{Q}(\theta)$, and the number and the maximal norm of the ideals $\mathfrak{p} \in S$; $c_{24} > 0$ is determined by those same parameters and the height of θ.

Proof. Let $\mathbb{K} = \mathbb{Q}(\theta)$. We apply the 'product formula' in the following form: for $\alpha \in \mathbb{K}$, $\alpha \neq 0$,

$$\prod_{\mathfrak{p}} |\alpha|_{\mathfrak{p}}^{n_{\mathfrak{p}}} = |\mathrm{Nm}(\alpha)|^{-1}. \qquad (2.2)$$

Here $\mathfrak{p}$ runs over all (finite) prime ideals of the ring $I_{\mathbb{K}}$. If α is an integer and $\mathrm{Nm}(\alpha) = uv$ with rational coprime integers u, v then one has

$$\prod_{\mathfrak{p}|u} |\alpha|_{\mathfrak{p}}^{n_{\mathfrak{p}}} = |u|^{-1}, \qquad \prod_{\mathfrak{p}|v} |\alpha|_{\mathfrak{p}}^{n_{\mathfrak{p}}} = |v|^{-1}. \qquad (2.3)$$

Indeed, each of the products

$$\prod_{\mathfrak{p}|u} |\alpha|_{\mathfrak{p}}^{-n_{\mathfrak{p}}}, \qquad \prod_{\mathfrak{p}|v} |\alpha|_{\mathfrak{p}}^{-n_{\mathfrak{p}}} \qquad (2.4)$$

is a natural number since we have

$$|\alpha|_{\mathfrak{p}} = p^{-(1/e_{\mathfrak{p}})\mathrm{ord}_{\mathfrak{p}}\alpha}, \qquad n_{\mathfrak{p}} = e_{\mathfrak{p}} f_{\mathfrak{p}},$$

where p and $f_{\mathfrak{p}}$ are defined by the equality $N(\mathfrak{p}) = p^{f_{\mathfrak{p}}}$ and $e_{\mathfrak{p}}$ is the ramification index of $\mathfrak{p}$ in p. The numbers (2.4) are coprime, the first containing only prime divisors of u, and the second only prime divisors of v. By (2.2) their product is $|uv|$; hence (2.3) holds.

Let $p_1, \ldots, p_s$ be those prime numbers which are divided by at least one of the prime ideals in S. First, we shall suppose that θ is an integer, and we consider $\mathrm{Nm}(x - \theta y)$ for rational integers x, y with $(x, y) = 1$. Separating powers of the prime numbers $p_1, \ldots, p_s$ from $\mathrm{Nm}(x - \theta y)$, we obtain

$$\mathrm{Nm}(x - \theta y) = A p_1^{z_1} \cdots p_s^{z_s}, \qquad (A, p_1 \ldots p_s) = 1. \tag{2.5}$$

We now apply the above remarks by setting $\alpha = x - \theta y$, $u = A$, and $v = p_1^{z_1} \cdots p_s^{z_s}$. Then we have

$$\prod_{\mathfrak{p} \mid A} |x - \theta y|_{\mathfrak{p}}^{n_{\mathfrak{p}}} = |A|^{-1}.$$

Since none of the prime ideals from S divides A, it follows that

$$\prod_{\mathfrak{p} \notin S} |x - \theta y|_{\mathfrak{p}}^{n_{\mathfrak{p}}} \le \prod_{\mathfrak{p} \mid A} |x - \theta y|_{\mathfrak{p}}^{n_{\mathfrak{p}}} = |A|^{-1}.$$

Now applying the product formula (2.2) and using (2.6), we find that

$$1 = |\mathrm{Nm}(x - \theta y)| \prod_{\mathfrak{p} \in S} |x - \theta y|_{\mathfrak{p}}^{n_{\mathfrak{p}}} \prod_{\mathfrak{p} \notin S} |x - \theta y|_{\mathfrak{p}}^{n_{\mathfrak{p}}} \le$$

$$\le |x - \theta y| X^{n-1} \prod_{\theta' \ne \theta} (1 + |\theta'|) \prod_{\mathfrak{p} \in S} |x - \theta y|_{\mathfrak{p}}^{n_{\mathfrak{p}}} |A|^{-1},$$

where θ' ranges over the conjugates of θ, other than θ itself. So we have

$$|x - \theta y| \prod_{\mathfrak{p} \in S} |x - \theta y|_{\mathfrak{p}}^{n_{\mathfrak{p}}} > (2 n H_\theta)^{-n+1} X^{-n+1} |A|, \tag{2.7}$$

where H_θ is the height of θ (the height of the minimal polynomial of θ). Applying Theorem 1.1 to the equation (2.5), we obtain

$$|A| > H_\theta^{-1} c_1^{-1/c_2} X^{1/c_2},$$

and then it follows from (2.7) that

$$|x - \theta y| \prod_{\mathfrak{p} \in S} |x - \theta y|_{\mathfrak{p}}^{n_{\mathfrak{p}}} > (2n)^{-n+1} c_1^{-1/c_2} H_\theta^{-n} X^{-n+1+1/c_2}. \tag{2.8}$$

This proves inequality (2.1) for integral θ. In the general case let a be the minimal natural number for which $a\theta$ is an integer, and apply (2.8) to this number with $x = ax'$:

$$|x' - \theta y| \prod_{\mathfrak{p} \in S} |x' - \theta y|_{\mathfrak{p}}^{n_{\mathfrak{p}}} > a^{-1} \prod_{\mathfrak{p} \in S} |a|_{\mathfrak{p}}^{-n_{\mathfrak{p}}} (2n)^{-n+1} c_1^{-1/2} H_\theta^{-n} X^{-n+1-1/c_2}.$$

Since $a \le H_\theta$ and $|a|_{\mathfrak{p}} \le 1$, we again obtain (2.1). This completes the proof of the theorem, and in addition we see the way in which H_θ influences c_{24} when the field $\mathbb{K}$ remains fixed.

It is clear that inequality (2.1) contains more information than the inequality

$$|x - \theta y| > c_{25} X^{-n+1+1/c_2},$$

since such an inequality follows from (2.1) by virtue of

$$\prod_{\mathfrak{p} \in S} |x - \theta y|_{\mathfrak{p}}^{n_{\mathfrak{p}}} \leq \prod_{\mathfrak{p}} \max(1, |\theta|_{\mathfrak{p}}^{n_{\mathfrak{p}}}) = c_{26}.$$

The ineffective version of inequality (2.1) is much stronger: the exponent in the right-hand side of (2.1) is $-1 - \varepsilon$, with $\varepsilon > 0$ arbitrary [136].

3. The Greatest Prime Factor of a Binary Form

The result of the previous section is a consequence of our results concerning the solutions of the equation (1.2) in the case of fixed $p_1, \ldots, p_s$ and growing A. We now deal with the equally interesting consequences of the estimates (1.4) and (1.25), as well as (1.26), obtained for $A = 1$ and changing numbers $p_1, \ldots, p_s$ [207].

Let x, y be rational integers with $(x, y) = 1$, and let $P = P[f(x, y)]$ be the greatest prime divisor of the number $f(x, y)$. We assume that the binary form f is integral, irreducible and of degree $n \geq 3$. If $n \geq 4$, and if $n = 4$ but form f is non-exceptional, we find from Theorem 1.1 and the expressions (1.25) for the values c_1, c_2, that if $X = \max(|x|, |y|) \geq e^e$ then

$$\ln \ln X < 20(n! + 1)(s + n + 2) \ln P + c_{27} s (\ln \ln P' + \ln H_f), \qquad (3.1)$$

where c_{27} depends only on n and $P' = \max(e^e, P)$. Since s does not exceed the number of primes not exceeding P, it follows that $s \leq 2P/\ln P$, and we have

$$\ln \ln X < 40(n! + 1)P + c_{28} \frac{P}{\ln P} (\ln \ln P' + \ln H_f).$$

Hence, we obtain

$$P > \frac{\ln \ln X}{40 n! + 33} \qquad (3.2)$$

for any X exceeding a certain computable value $X_0 = X_0(n, H_f)$. Returning again to (3.1) we observe a more pithy inequality for $X > X_1(n, H_f)$, namely

$$s \ln P > \frac{\ln \ln X}{20 n! + 21}, \qquad (3.3)$$

which is especially interesting in the case of a bounded number s of prime divisors of $f(x, y)$:

$$P > (\ln X)^{1/s(20 n! + 21)}.$$

The estimate (1.26) gives slightly sharper inequalities of type (3.2), (3.3) for all forms of degree $n \geq 3$. Indeed, it shows that

$$\ln \ln X < 200 s \ln s + (n! + 1) \ln P + c_{29} s (\ln \ln P' \ln H_f),$$

where c_{29} depends only on n. As $s \leq \pi(P) \sim P/\ln P$, it follows that in the denominator of (3.2) one can put $200 + \delta$ with any $\delta > 0$, provided that $X > X_2(n, H_f, \delta)$. Accordingly, (3.3) becomes a little stronger.

It is not necessary to suppose irreducibility of the form f for the truth of an inequality similar to (3.2), but it is a necessary condition that the form f have at least three distinct roots. We shall prove that an inequality of type (3.2) holds under this assumption ([211], [190]).

Theorem 3.1 *Let f be an integral binary form with at least three distinct roots. Then for any rational integers x, y with $(x,y) = 1$, and $f(x,y) \neq 0$ with*

$$X = \max(|x|, |y|) \geq X_3(n, H_f),$$

we have the inequality

$$P[f(x,y)] > c_{30} \ln \ln X. \tag{3.4}$$

Here $X_3(n, H_f)$ is effectively expressed in terms of n and H_f, and $c_{30} > 0$ in terms of n.

Proof. By Gauss' lemma, the form f may be expressed as the product of irreducible integral forms $f = f_1 \ldots f_k$; and it is obvious that

$$P[f(x,y)] = \max_{(i)} P[f_i(x,y)] \quad (i = 1, 2, \ldots, k).$$

So if at least one of the forms f_i is of degree not less than 3, inequality (3.4) follows from (3.2). Thus we must consider the case in which all the forms f_i are linear or quadratic. Since the form f has at least three different roots, it is enough to obtain (3.4) in the following special cases:

1) f is the product of three (non-proportional) linear forms;
2) f is the product of one linear and one irreducible quadratic form;
3) f is the product of two different irreducible quadratic forms.

In the first case we have

$$(a_1 x - b_1 y)(a_2 x - b_2 y)(a_3 x - b_3 y) = p_1^{z_1} \cdots p_s^{z_s},$$

and we can assume that all $b_i \neq 0$ and that all the quotients a_i/b_i are distinct $(i = 1, 2, 3)$, whilst $p_1, \ldots, p_s$ are distinct prime numbers. Since $(x,y) = 1$, the common factors of the numbers $a_i x - b_i y$, $a_j x - b_j y$, $(i \neq j)$, divide $a_i b_j - a_j b_i \neq 0$, so that

$$a_i x - b_i y = d_i p_1^{u_{i1}} \cdots p_s^{u_{is}}, \qquad (i = 1, 2, 3), \tag{3.5}$$

where the d_i are determined by the numbers a_i, b_i, $(1 \leq i, j \leq 3)$; the u_{ij} are unknown integers, and for any j only one of the numbers $u_{1j} u_{2j} u_{3j}$ is not zero. Eliminating x and y from (3.5), we obtain

$$D_1 p_1^{u_{11}} \cdots p_s^{u_{1s}} + D_2 p_1^{u_{21}} \cdots p_s^{u_{2s}} + D_3 p_1^{u_{31}} \cdots p_s^{u_{3s}} = 0, \tag{3.6}$$

where $D_i \neq 0$ $(i = 1, 2, 3)$. For definiteness we set

$$U = \max_{(i,j)} u_{ij} = u_{11} \qquad (i = 1, 2, 3; j = 1, 2, \ldots, s).$$

Then we find from (3.6) that

$$\left| (-\frac{D_2}{D_3}) p_1^{u_{21}-u_{31}} \ldots p_s^{u_{2s}-u_{3s}} - 1 \right|_{p_1} \leq p_1^{-U}.$$

We estimate the left-hand side of this inequality from below by Lemma 8.3 of Ch.III, which gives

$$U \leq c_{31} (16(s + 2))^{12(s+2)} p_1 (\ln P)^s (\ln U)^2,$$

where P is the maximum of the primes $p_1, \ldots, p_s$, and $P' = \max(e^e, P)$. Then we find that

$$U \leq c_{32} (16(s + 2))^{12(s+2)} P(\ln P)^s (s \ln P)^2,$$

from which, recalling (3.5), we obtain

$$P \geq \frac{1}{13} \ln \ln X, \qquad X > X_4,$$

where X_4 is determined by a_i, b_j, $(i, j = 1, 2, 3)$.

The two other cases can be dealt with by a combination of such arguments with the arguments of §1. We shall meet those cases below in a more general situation (see the proof of Theorem 3.2).

We pass now from forms with integral rational coefficients to forms with integral algebraic coefficients and ask what assertion is a generalisation of Theorem 3.1 in this case? To be specific, let $\mathbb{K}$ be an algebraic number field, and let $f = f(x, y)$ be a binary form of degree n with integer coefficients from $\mathbb{K}$. If the field $\mathbb{K}$ contains units of infinite order, then for infinitely many pairs of integers x, y the totality of prime ideals in $f(x, y)$ can remain unchanged. For if, say, ξ is such a unit, then the principal ideals $(f(x, y))$ and $(f(\xi^m x, \xi^m y))$ coincide for every integer m. Therefore it is natural to ask about the growth of the norm of prime ideals dividing the number $f(x, y)$, in terms of the growth of $\max(|\operatorname{Nm}(x)|, |\operatorname{Nm}(y)|)$. Then we do come to a generalisation of Theorem 3.1 (which we shall state in the next section). In connection with this it is interesting to draw attention to the following assertion [108], strengthening a well known result of Siegel [191]:

Theorem 3.2 *Let f be a binary form with at least two different roots, and with integer coefficients in the field $\mathbb{K}$. Let ε be an arbitrary number in the interval $0 < \varepsilon < 1$, and suppose that*

$$x, y \in I_{\mathbb{K}}, \qquad 0 \neq \overline{|y|} \leq \overline{|x|}^{1-\varepsilon}, \qquad f(x, y) \neq 0.$$

Then the maximal prime number P with $(f(x, y), P) \neq 1$ satisfies

$$P > c_{33} \ln \ln \overline{|x|}, \tag{3.7}$$

provided that $\lceil x \rceil > c_{34}$, where $c_{33} > 0$ and c_{34} can be effectively determined in terms of f, $\mathbb{K}$, and ε.

Proof. Let θ_1, θ_2 be different roots of the polynomial $f(x, 1)$, and let α_0 be its leading coefficient. Set $\mathbb{L} = \mathbb{K}(\theta_1, \theta_2)$, and let $\mathfrak{p}_1, \ldots, \mathfrak{p}_s$ be the distinct prime ideals of the field $\mathbb{L}$ dividing $f(x, y)$. Then we have

$$(\alpha_0(x - \theta_1 y)) = \mathfrak{p}_1^{z_{11}} \cdots \mathfrak{p}_s^{z_{1s}}, \qquad z_{1i} \geq 0,$$
$$(\alpha_0(x - \theta_2 y)) = \mathfrak{p}_1^{z_{21}} \cdots \mathfrak{p}_s^{z_{2s}}, \qquad z_{2i} \geq 0, \qquad (i = 1, 2, \ldots s).$$

Let h be the class-number of the field $\mathbb{L}$, and set

$$\mathfrak{p}_i^h = (\rho_i), \qquad \rho_i \in I_{\mathbb{L}}, \qquad \mathrm{Nm}(\rho_i) = p_i^{f_i h}, \qquad f_i \leq d = [\mathbb{L} : \mathbb{Q}].$$

Then, denoting the fundamental units of the field $\mathbb{L}$ (or the independent units constructed by Lemma 2.1 Ch.II) by $\varepsilon_1, \ldots, \varepsilon_r$ we find

$$\alpha_0(x - \theta_1 y) = \sigma_1 \rho_1^{u_{11}} \cdots \rho_s^{u_{1s}} \varepsilon_1^{v_{11}} \cdots \varepsilon_r^{v_{1r}}, \tag{3.8}$$

$$\alpha_0(x - \theta_2 y) = \sigma_2 \rho_1^{u_{21}} \cdots \rho_s^{u_{2s}} \varepsilon_1^{v_{21}} \cdots \varepsilon_r^{v_{2r}}, \tag{3.9}$$

where, in view of Lemma 2.2 Ch.II, we may suppose that

$$\max(\lceil \sigma_1 \rceil, \lceil \sigma_2 \rceil) < c_{35}^s(p_1 \cdots p_s)^h, \qquad \lceil \rho_i \rceil < c_{36} p_i^h, \quad (i = 1, 2, \ldots, s). \tag{3.10}$$

Here σ_1 and σ_2 are in $I_{\mathbb{L}}$, and c_{35} and c_{36} are functions of the degree and the regulator of the field $\mathbb{L}$. Taking the absolute norm in the equalities (3.8), (3.9), we find

$$\max(u_{1i}, u_{2j}) < c_{37} \ln \lceil x \rceil, \qquad (i, j = 1, 2, \ldots, s), \tag{3.11}$$

and passing now to the conjugates of these equalities, we obtain

$$\max(|v_{1i}|, |v_{2j}|) < c_{38} \ln \lceil x \rceil, \qquad (i, j = 1, 2, \ldots, s), \tag{3.12}$$

where c_{37} and c_{38} are determined by α_0, θ_1 and θ_2, and the degree and regulator of $\mathbb{L}$. Consequently,

$$U = \max_{(i,j)}(u_{1i}, u_{2j}) + \max_{(i,j)}(|v_{1i}|, |v_{2j}|) < c_{39} \ln \lceil x \rceil. \tag{3.13}$$

Suppose now that τ is an isomorphism of $\mathbb{L}$ into the field of complex numbers such that $\lceil x \rceil = |x^\tau|$. Then we have

$$0 \neq \left| \frac{\alpha_0^\tau(x^\tau - \theta_1^\tau y^\tau)}{\alpha_0^\tau(x^\tau - \theta_2^\tau y^\tau)} - 1 \right| = \left| \frac{(\theta_2^\tau - \theta_1^\tau)y^\tau}{x^\tau - \theta_2^\tau y^\tau} \right| < c_{40} \lceil x \rceil^{-\varepsilon}, \tag{3.14}$$

which follows from the condition $0 \neq \lceil y \rceil \leq \lceil x \rceil^{1-\varepsilon}$. Here c_{40} is determined by f and we assume that $\lceil x \rceil$ exceeds some similar value. Further, (3.8) and (3.9) give

$$\left| \frac{\alpha_0^\tau (x^\tau - \theta_1^\tau y^\tau)}{\alpha_0^\tau (x^\tau - \theta_2^\tau y^\tau)} - 1 \right| = \left| (\frac{\sigma_1}{\sigma_2})^\tau (\rho_1^\tau)^{u_{11} - u_{21}} \cdots (\rho_s^\tau)^{u_{1s} - u_{2s}} \times \right.$$

$$\left. \times (\varepsilon_1^\tau)^{v_{11} - v_{21}} \cdots (\varepsilon_r^\tau)^{v_{1r} - v_{2r}} - 1 \right|.$$

Applying Lemma 8.2 Ch.III, we find from the estimates (3.10), (3.13) that

$$c_{40} \boxed{x}^{-\varepsilon} > \exp\{-c_{41}^s (s + d)^{200(s+d)} (\ln P)^{s+2} \ln U\}, \qquad (3.15)$$

where c_{41} is determined by f. Hence, recalling (3.13), we obtain

$$(1.005)^{-1} \ln \ln \boxed{x} < 200(s + d) \ln(s + d) + (s + 2)(\ln \ln P' + c_{42}),$$

where c_{42} depends on f and ε. Since s is not more than the number of all primes not exceeding P, we obtain (3.7) with $c_{33} = (201)^{-1}$.

The inequality (3.15) allows one to take ε decreasing together with $\boxed{x}^{-1}$, and still obtain assertions on the unbounded growth of P. For example, taking δ in the interval $0 < \delta < 1$, and setting $\varepsilon = (\ln \boxed{x})^{\delta - 1}$, we find (3.7) with $c_{33} = \delta(201)^{-1}$ for $\boxed{y} \leq \exp\{-(\ln \boxed{x})^\delta\}$. Under the even weaker condition $\boxed{y} \leq \boxed{x}(\ln \boxed{x})^{-1-\delta}$, we still find that P increases unboundedly together with $\boxed{x}$, but now as the triple logarithm of $\boxed{x}$.

4. The Generalised Thue-Mahler Equation

Let $\mathbb{K}$ be an algebraic number field and let $f = f(x, y)$ be a binary form of degree $n \geq 3$ with integer coefficients from the field $\mathbb{K}$ and irreducible over $\mathbb{K}$. Let $\alpha, \beta_1, \ldots, \beta_s$ and a be nonzero integers in $\mathbb{K}$ such that the (β_j) are powers of distinct prime ideals of $\mathbb{K}$, $(j = 1, 2, \ldots, s)$. Then the equation

$$f(x, y) = \alpha \beta_1^{z_1} \cdots \beta_s^{z_s}, \qquad (x, y)|a, \qquad (4.1)$$

in which x, y are unknown integers in $\mathbb{K}$, and $z_1 \geq 0, \ldots, z_s \geq 0$ are unknown rational integers, is a natural generalisation of (1.2). However, it has a peculiarity arising from the extension of the domain of solutions x, y from rational integers to algebraic integers in $\mathbb{K}$.

It is natural to relate the analysis of (4.1) with that of the equation

$$f(x, y) = \alpha \beta_1^{z_1} \cdots \beta_s^{z_s} \xi, \qquad (x, y)|a, \qquad (4.2)$$

where ξ is a new unknown number, which is to be some unit of $\mathbb{K}$. If the field $\mathbb{K}$ contains a unit ε of infinite order, then for every solution $x, y, z_1, \ldots, z_s, \xi$ the new set of numbers $\varepsilon x, \varepsilon y, z_1, \ldots, z_s, \xi \varepsilon^n$, also satisfies the equation, so that the number of solutions is infinite. However, it turns out that if one identifies solutions in which the pairs x, y differ only by a factor which is a unit in $\mathbb{K}$, then the number of 'solutions' will be finite, and all of them may be effectively found ([115], [116]). In particular, that means that the exponents $z_1, \ldots, z_s$ are determined by α, the coefficients of the form f and the parameters of the field $\mathbb{K}$. Since (4.2) implies (4.1), we obtain a bound for these exponents in (4.1) as well. It therefore reduces to a finite number of the generalised Thue equations (see §6 Ch.IV).

Our first objective is the following theorem.

Theorem 4.1 *For every solution of (4.2) there exists a unit $\eta \in \mathbb{K}$ such that*

$$\max(\lceil \eta x \rceil, \lceil \eta y \rceil, p_1^{z_1}, \ldots, p_s^{z_s}) < c_{43}(\mathrm{Nm}(\alpha a)H_f)^{c_{44}}, \qquad (4.3)$$

where the p_j are rational prime numbers divisible by the corresponding prime ideals of the β_j, $(j = 1, 2, \ldots, s)$, and H_f is the height of the form f, whilst c_{43} and c_{44} are effectively determined in terms of the degree, the regulator and the ideal class number of the decomposition field of the form f, and by s and $P = \max_{(j)} p_j$, $(1 \le j \le s)$.

Proof. Let

$$f(x, y) = a_0(x - \theta_1' y) \ldots (x - \theta_n' y)$$

be a decomposition of the form $f(x, y)$ in the field $\mathbb{G} = \mathbb{K}(\theta_1' \ldots \theta_n')$, and set $[\mathbb{G} : \mathbb{Q}] = g$, $[\mathbb{G} : \mathbb{K}] = d$.

By considering the decomposition into prime ideals in $I_{\mathbb{G}}$ of the left and right-hand sides of (4.2), we obtain

$$(x' - \theta_i y') = \mathfrak{a}_i \mathfrak{p}_1^{U_{1i}} \cdots \mathfrak{p}_t^{U_{ti}}, \qquad (i = 1, 2, \ldots, n),$$

where $x' = a_0 x$, $y' = y$, $\theta_i = a_0 \theta_i'$, and the $\mathfrak{p}_j$ are prime ideals contained in the numbers $\beta_1, \ldots, \beta_s$; $\mathfrak{a}_i$ is an integral ideal contained in $a_0^{n-1}\alpha$, and $t \le sd$. Setting $U_{ji} = u_{ji}h + r_{ji}$, with $(0 \le r_{ji} < h)$, where $h = h_{\mathbb{G}}$ is the class-number of $\mathbb{G}$, we find that

$$x' - \theta_i y' = \pi_i \rho_1^{u_{1i}} \cdots \rho_t^{u_{ti}} \varepsilon_i' \qquad (i = 1, 2, \ldots, n), \qquad (4.4)$$

where the $(\rho_j) = \mathfrak{p}_j^h$ and $(\pi_i) = \mathfrak{a}_i \mathfrak{p}_1^{r_{1i}} \cdots \mathfrak{p}_t^{r_{ti}}$ are principal ideals, $\rho_j, \pi_i \in I_{\mathbb{G}}$, and the ε_i' are units of $\mathbb{G}$. Lemma 2.2 of Ch.II allows one to take numbers ρ_j, π_i so that

$$\lceil \rho_j \rceil \le e^{c_{45}R}(\mathrm{N}(\mathfrak{p}_j^h))^{1/g} \le e^{c_{45}R}P^h \qquad (j = 1, 2, \ldots, t),$$

$$\lceil \pi_j \rceil \le e^{c_{45}R}(\mathrm{N}(\mathfrak{a}_i \mathfrak{p}_1^{r_{1i}} \cdots \mathfrak{p}_t^{r_{ti}}))^{1/g} < e^{c_{45}R}|\mathrm{Nm}(a_0^{n-1}\alpha)|^{1/g}P^{th} <$$
$$< e^{c_{45}R}H_f^{n-1}|\mathrm{Nm}(\alpha)|^{1/g}P^{sdh} \qquad (i = 1, 2, \ldots, n),$$

where c_{45} depends only on g, and $R = R_{\mathbb{G}}$ is the regulator of the field $\mathbb{G}$. Therefore we may assume that in (4.4) the integers ρ_j and π_i are such that

$$\lceil \rho_j \rceil \le e^{c_{45}R}P^h \qquad (j = 1, 2, \ldots, t), \qquad (4.5)$$

$$\lceil \pi_j \rceil < e^{c_{45}R}H_f^{n-1}|\mathrm{Nm}(\alpha)|^{1/g}P^{sdh} \qquad (i = 1, 2, \ldots, n). \qquad (4.6)$$

Let $U_{\mathbb{G}}$ be the group of units in $\mathbb{G}$ constructed by Lemma 2.1 Ch.II, and let $\eta_1, \ldots, \eta_r$ be its generating elements. Since $\varepsilon_i' \in \mathbb{G}$ and $\mathrm{Nm}(\varepsilon_i') = \pm 1$, then by Lemma 2.2 Ch.II there exists a unit ξ_i in $U_{\mathbb{G}}$ such that we have

$$\varepsilon_i' = \varepsilon_i'' \xi_i, \qquad \left| \overline{\varepsilon_i''} \right| < e^{c_{45} R} \qquad (i = 1, 2, \ldots, n). \tag{4.7}$$

Put

$$\varepsilon' = \varepsilon_1' \cdots \varepsilon_n' = \varepsilon_0 \xi_1 \cdots \xi_n, \qquad \varepsilon_0 = \varepsilon_1'' \cdots \varepsilon_n'',$$
$$\xi_i = \eta_1^{v_{1i}'} \cdots \eta_r^{v_{ri}'} \qquad (i = 1, 2, \ldots, n), \tag{4.8}$$
$$W_j = \sum_{i=1}^n v_{ji}' \qquad (j = 1, 2, \ldots, r).$$

Then $\varepsilon' = \varepsilon_0 \eta_1^{W_1} \cdots \eta_r^{W_r}$. Set $W_j = n w_j + v_j$, $0 \le v_j < n$ $(j = 1, 2, \ldots, r)$ and $\varepsilon = \eta_1^{-w_1} \cdots \eta_r^{-w_r}$. Then

$$\varepsilon^n \varepsilon' = \varepsilon_0 \eta_1^{v_1} \ldots \eta_r^{v_r}, \qquad \varepsilon \varepsilon_i' = \varepsilon_i'' \eta_1^{v_{1i}' - w_1} \ldots \eta_r^{v_{ri}' - w_r}.$$

Moreover, on setting

$$v_{ji} = v_{ji}' - w_j \qquad (j = 1, 2, \ldots, r; \quad i = 1, 2, \ldots, n),$$

we find from (4.8) that

$$0 \le \sum_{i=1}^n v_{ji} = v_j < n \qquad (j = 1, 2, \ldots, r). \tag{4.9}$$

On multiplying both parts of (4.4) by ε, we obtain

$$\mu_i = X - \theta_i Y = \sigma_i \rho_1^{u_{1i}} \cdots \rho_t^{u_{ti}} \eta_1^{v_{1i}} \cdots \eta_r^{v_{ri}} \qquad (i = 1, 2, \ldots, n), \tag{4.10}$$

where $X = \varepsilon x'$, $Y = \varepsilon y'$ and $\sigma_i = \varepsilon_i'' \pi_i$ whilst by (4.6), (4.7) the size of σ_i is bounded by an expression of the type (4.6) with c_{45} replaced by $2c_{45}$.

From (4.10) we obtain the system of equalities

$$(\theta_1 - \theta_2)\mu_i - (\theta_i - \theta_2)\mu_1 - (\theta_1 - \theta_i)\mu_2 = 0, \tag{4.11}$$

where now $i = 3, 4, \ldots, n$. These equalities together with (4.9) and (4.10) will yield a bound for

$$H = \max_{(k,l,i)} (u_{ki}, |v_{li}|) \qquad (k = 1, 2, \ldots, t; l = 1, 2, \ldots, r; i = 1, 2, \ldots, n)$$

in terms of the parameters of the equation (4.2).

In the sequel we shall use the connection between estimates of linear forms in the logarithms of algebraic numbers in different metrics and accordingly we now suppose that $n \ge 5$, or $n = 4$ but f is not an exceptional form over $\mathbb{K}$ (see §3 Ch.III).

We first consider how to estimate some value u_{ki}, say, u_{11}.

It is easy to see by considering the common divisor of the numbers $X - \theta_1 Y$, $X - \theta_i Y$ that if

$$h(u_{11} - 1) \geq [c_{46}(\ln H_f + \ln|\mathrm{Nm}(\alpha)| + \ln|\mathrm{Nm}(a)|)] = U_0, \qquad (4.12)$$

each of the numbers within square brackets in the equality

$$\frac{(\theta_1 - \theta_i)\mu_2}{(\theta_1 - \theta_2)\mu_i} = \left[\frac{(\theta_1 - \theta_i)\sigma_2}{(\theta_1 - \theta_2)\sigma_i} \rho_1^{u_{12}-u_{1i}}\right] \prod_{k=2}^{t} [\rho_k]^{u_{k2}-u_{ki}} \prod_{l=1}^{r} [\eta_l]^{v_{12}-v_{li}} \qquad (4.13)$$

is a $\mathfrak{p}_1$-adic unit, and we have the congruence

$$1 - \frac{(\theta_1 - \theta_i)\mu_2}{(\theta_1 - \theta_2)\mu_i} \equiv 0 \pmod{\mathfrak{p}_i^{hu_{11}-U_0}}. \qquad (4.14)$$

The multiplicative structure of the right-hand side of (4.13), and Lemma 7.1 of Ch.III, allows us to estimate from above the order of the left-hand side of (4.14) with respect to the prime ideal $\mathfrak{p}_1$. In view of (4.6) the logarithm of the height of the number $(\theta_1 - \theta_i)\sigma_2/(\theta_1 - \theta_2)\sigma_i$ is bounded by

$$c_{47}(R + \ln H_f + \ln|\mathrm{Nm}(\alpha)| + sh\ln P),$$

and the logarithms of the heights of the numbers ρ_k and η_l are respectively bounded by $c_{48}(R + h\ln P)$ and $c_{49}R$. Moreover, $T = \mathrm{Nm}(\mathfrak{p}_1^{2e})(1 - \frac{1}{\mathrm{Nm}(\mathfrak{p}_1)})$ satisfies $T < P^{2g}$.

Let q be a prime number in the range $P < q < 2P$, and denote by $|\ |_q$ the q-adic valuation of the field $\mathbb{G}$ induced by a prime ideal $\mathfrak{q}$ contained in q. To apply Lemma 7.1 of Ch.III we distinguish two cases. Either

$$\left|\frac{(\theta_1 - \theta_2)\sigma_i}{(\theta_1 - \theta_i)\sigma_2}\right|_q \neq 1, \qquad (4.15)$$

for some i, $(3 \leq i \leq n)$, or this norm is equal to 1 for every $i = 3, 4, \ldots, n$. If (4.15) holds, then under the condition

$$H \geq (c_{50}\delta^{-1}s^4(R + h\ln P)^2)^{4(sd+g+1)} P^{16g(sd+g)+18g} \times$$
$$\times (\ln|\mathrm{Nm}(\alpha a)| + \ln H_f + R + sh\ln P) \qquad (4.16)$$

we obtain

$$hu_{11} - U_0 < \delta e H, \qquad e = \mathrm{ord}_{\mathfrak{p}_1} p_1, \qquad (4.17)$$

where δ is arbitrary in the range $0 < \delta \leq (\sqrt{2}P^g)^{-1}$.

If (4.15) does not hold, then in accordance with Lemma 7.1 Ch.III we must establish the inequality

$$\left|\log\left(\frac{(\theta_1 - \theta_i)\sigma_2}{(\theta_i - \theta_2)\sigma_i}\right)^S - \sum_{k=1}^{t}(u_{k2} - u_{ki})\log(\rho_k)^S -\right.$$
$$\left.- \sum_{l=1}^{r}(v_{l2} - v_{li})\log(\eta_l^S)\right|_q > q^{-\frac{1}{5}\sigma^{-4(sd+g)-2}H}, \qquad (4.18)$$

where σ is bounded above by a value of the form

$$c_{51}\delta^{-1}s^4(R + h\ln P)^2, \qquad S = \mathrm{Nm}(\mathfrak{q}^{2e'})(1 - \frac{1}{\mathrm{Nm}(\mathfrak{q})}), \qquad e' = \mathrm{ord}_{\mathfrak{q}}q.$$

Assuming that for all $i = 3, 4, \ldots, d$ the opposite to the inequalities (4.18) holds, and arguing as in §2 Ch.IV, we obtain the relations

$$\frac{\theta_1 - \theta_i}{\theta_2 - \theta_i} \cdot \frac{\theta_2 - \theta_j}{\theta_1 - \theta_j} = \frac{1 - \zeta_i}{1 - \zeta_j}, \tag{4.19}$$

where $\zeta_i, \zeta_j \neq 1$ are distinct S-th roots of 1, $(i \neq j; \quad i, j = 3, 4, \ldots, n)$. But from the results of §3 Ch.IV we know that if $[\theta : \mathbb{K}] \geq 5$ the equalities (4.19) cannot hold, otherwise the form is exceptional. Hence (4.18) is true at any rate for some i, and again we obtain (4.17).

Therefore, under condition (4.16) we have

$$U = \max_{(k,i)} u_{ki} < \delta g H + U_0 + 1 \qquad (k = 1, 2, \ldots, t; \quad i = 1, 2, \ldots, n). \tag{4.20}$$

To bound

$$V = \max_{(l,i)} |v_{li}| \qquad (l = 1, 2, \ldots, r; \quad i = 1, 2, \ldots, n) \tag{4.21}$$

we apply Lemma 1.1 of Ch.III in the same way to obtain archimedean values of the left-hand side of the equality

$$1 - \frac{(\theta_1^\tau - \theta_2^\tau)\mu_i^\tau}{(\theta_1^\tau - \theta_i^\tau)\mu_2^\tau} = -\frac{(\theta_i^\tau - \theta_2^\tau)\mu_1^\tau}{(\theta_1^\tau - \theta_i^\tau)\mu_2^\tau},$$

which follows from (4.11); here τ denotes an isomorphism of the field $\mathbb{G}$ into the field of complex numbers. The multiplicative structure of the equality (4.13) plays the determining role again. As before, we assume that H satisfies (4.16) and distinguish two cases: For at least one i, $(3 \leq i \leq n)$ we have

$$\left| \frac{(\theta_1^\tau - \theta_i^\tau)\sigma_2^\tau}{(\theta_1^\tau - \theta_2^\tau)\sigma_i^\tau} \right|_q \neq 1,$$

or equality holds for all $i = 3, 4, \ldots, n$. In the latter case we check whether an inequality of type (4.18) holds when $\theta_1, \theta_2, \theta_i, \rho_k, \eta_l$ are replaced by their images under the isomorphism τ, and see that (4.18) holds for at least one i $(3 \leq i \leq n)$ if θ is non-exceptional. For that i we have

$$\left| \frac{(\theta_i^\tau - \theta_2^\tau)\mu_1^\tau}{(\theta_1^\tau - \theta_i^\tau)\mu_2^\tau} \right| > e^{-\delta H},$$

and then for all τ we find that

$$|\mu_1^\tau/\mu_2^\tau| > H_f^{-c_{52}} e^{-\delta H}. \tag{4.22}$$

Since $\mu_1/\mu_2 = (x' - \theta_1 y') / (x' - \theta_2 y')$, where $x', y' \in \mathbb{K}$, and θ_1 and θ_2 are conjugates over $\mathbb{K}$, the absolute norm of the number μ_1/μ_2 is equal to 1, and the product of the left-hand sides of (4.22) over all τ is 1. Hence, we find that

$$\max_{(\tau)} \left| \ln |\mu_1^\tau/\mu_2^\tau| \right| < g\delta H + c_{53} \ln H_f . \tag{4.23}$$

Turning to (4.10) and taking into account the estimates (4.5), (4.6), (4.20), we obtain for all τ :

$$\left| \sum_{j=1}^{r} (v_{j1} - v_{j2}) \ln |\eta_j^\tau| \right| < c_{54}(R + \ln H_f + \ln|\operatorname{Nm}(\alpha)| +$$

$$+ sh \ln P) + 2(\delta g H + U_0 + 1)(c_{45} R + h \ln P).$$

This shows that

$$\max_{(j)} |v_{j1} - v_{j2}| < 2\delta g H(c_{45} R + h \ln P) + c_{55}(\ln|\operatorname{Nm}(\alpha a)| +$$

$$+ \ln H_f + s)(c_{45} R + h \ln P). \tag{4.24}$$

Obviously, we can find inequalities of type (4.22) for μ_k^τ/μ_l^τ with distinct $k, l = 1, 2, \ldots, n$ in the same way. Hence,

$$V' = \max_{(j,k,l)} |v_{jk} - v_{jk}| \qquad (j = 1, 2, \ldots, r; \quad k, l = 1, 2, \ldots, n; \quad k \neq l)$$

is bounded above by a value of the form (4.24). To be specific, set $V = |v_{11}|$, where V is determined by (4.21). Then we see that the previous arguments give an upper bound for $|(n-1)v_{11} - (v_{12} + \cdots + v_{1n})|$ of the form (4.24) multiplied by n. Then (4.9) shows that V is bounded above by a value of the form (4.24). Consequently, we have an upper bound of the same type for H. In view of the supposed inequality (4.16) we must have

$$1 < (c_{45} R + h \ln P)(2\delta g + (c_{46}\delta))^{4(sd+g+1)} .$$

If we take

$$\delta = \min\left((\sqrt{2} P^g)^{-1}, (4g)^{-1}(c_{45} R + h \ln P)^{-1}, (2c_{46})^{-1}\right),$$

then the last inequality is impossible and we conclude that with such δ inequality (4.16) does not hold. Hence, it follows that

$$H \leq H_0 = (c_{57} s^4 (R + h \ln P)^3)^{4(sd+g+1)} P^{20g(sd+g+1)+22g} \times$$

$$\times (\ln|\operatorname{Nm}(\alpha a)| + \ln H_f + R + sh \ln P).$$

We now obtain from (4.10)

$$\overline{|\mu_i|} \leq \overline{|\sigma_i|} \, \overline{|\rho_1|}^{u_{1i}} \cdots \overline{|\rho_t|}^{u_{ti}} \overline{|\mu_1|}^{(g-1)|v_{1i}|} \cdots \overline{|\eta_r|}^{(g-1)|v_{ri}|} <$$

$$< H_f^n |\operatorname{Nm}(\alpha)| P^{sdh} e^{2c_{45} R} (e^{2c_{45} R} P^h)^{tH_0} e^{c_{59} R H_0} <$$

$$< \exp\{c_{59}((R + h \ln P) s H_0 + \ln|\operatorname{Nm}(\alpha)| + \ln H_f + sh \ln P)\}. \tag{4.25}$$

Hence, the absolute norm of the number $\mu_1 = X - \theta_1 Y$ is bounded by a similar value, and, since $\mu_1 = \varepsilon(X' - \theta_1 Y')$, this also deals with the norm of the number $X' - \theta_1 Y'$. Lemma 2.2 Ch.II now shows that the field $\mathbb{K}$ contains a unit η such that

$$\lceil \eta(X' - \theta_1 Y') \rceil$$

does not exceed a value of the form (4.25) multiplied by $\exp\{c_{60} R_{\mathbb{K}}\}$. By Lemma 2.3 we have $R_{\mathbb{K}} < c_{61} R$, so that we obtain

$$\max(\lceil \overline{\eta x} \rceil, \lceil \overline{\eta y} \rceil) < \exp\{c_{62} s^4 (R + h \ln P)^3\}^{4(sd+g+1)+1} \times$$
$$\times P^{20g(sd+g+1)+22g}(\ln|\operatorname{Nm}(\alpha a)| + \ln H_f + R + sh \ln P)\}. \quad (4.26)$$

This concludes the proof of the Theorem 4.1 for forms of degree $n \geq 5$ and for non-exceptional forms of degree $n = 4$. We have explicit representations for the values c_{43} and c_{44} in terms of the regulator and class-number of the field as well as g, s and P. Similar arguments, relying on direct estimates of the linear forms in logarithms of algebraic numbers give a proof for the theorem also in the remaining cases ($n = 3, 4$).

Turning to equation (4.1), we observe that from the bound (4.3) a similar bound follows for $\lceil \beta_1^{z_1} \cdots \beta_s^{z_s} \rceil$, and then by Theorem 6.1 of Ch.IV we obtain

Theorem 4.2 *All solutions of* (4.1) *satisfy*

$$\max(\lceil \overline{x} \rceil, \lceil \overline{y} \rceil, p_1^{z_1}, \ldots, p_s^{z_s}) < (\lceil \overline{a} \rceil H_f | \operatorname{Nm}(a)|)^{c_{64}}$$

where c_{63}, c_{64} *are effectively determined by the degree, regulator and ideal class number of the decomposition field of the form* f, *and by* s, P *and the maximal size of* $\beta_1^{z_1}, \ldots, \beta_s^{z_s}$.

Inequality (4.26) yields a corollary which supplements the results of the previous section.

Theorem 4.3 *Let* $x, y \in I_{\mathbb{K}}$, *with* $(x, y) = 1$, *and let* P *be the maximal prime number with* $(P, f(x, y)) \neq 1$. *Then*

$$P \geq \frac{1}{36g^2} \ln \ln N, \qquad N = \max(|\operatorname{Nm}(x)|, |\operatorname{Nm}(y)|) \geq N_0, \quad (4.27)$$

where N_0 *is effectively determined by* f *and the parameters of the decomposition field of the form* f, *and* g *is the degree of that field.*

Indeed, one applies the bound (4.26) in the case $\alpha = a = 1$ and observes that

$$|\operatorname{Nm}(x)| = |\operatorname{Nm}(\eta x)| \leq \lceil \overline{\eta x} \rceil^{[\mathbb{K}:\mathbb{Q}]};$$

and similarly for $\operatorname{Nm}(y)$. A simple computation then gives (4.27).

It is easy to see that in the analysis described above of equations (4.1) and (4.2) one can avoid supposing the irreducibility of the form f, if one applies direct estimates of the linear forms in the logarithms of algebraic numbers. It is enough to assume that the form f has three distinct roots. In particular, application of Lemmas 8.2 and 8.3 of Ch.III gives the inequality of the form (4.27) for all such forms.

5. Approximations to Algebraic Numbers by Algebraic Numbers of a Fixed Field

Theorem 4.1 allows one to obtain the deepest and strongest generalisation of the Liouville inequality on approximation to algebraic numbers by rational numbers [116]. Before we proceed with this topic directly, we simplify the assumptions of Theorem 4.1 concerning the greatest common divisor of the numbers x and y and the choice of the number a, satisfying (4.2).

Let the numbers $\beta_1, \ldots, \beta_s$ be powers of the prime ideals $\mathfrak{p}_1, \ldots, \mathfrak{p}_s$ of the ring $I_{\mathbb{K}}$: respectively, $(\beta_i) = \mathfrak{p}_i^{h_i}$, $(i = 1, 2, \ldots, s)$ with $h_i | h$, where $h = h_{\mathbb{K}}$ is the class-number of the field $\mathbb{K}$. We set $(x, y) = \mathfrak{a}\mathfrak{p}_1^{t_1'} \cdots \mathfrak{p}_s^{t_s'}$, where the ideal $\mathfrak{a}$ is relatively prime to the prime ideals $\mathfrak{p}_1, \ldots, \mathfrak{p}_s$ and $t_i' = h_i t_i + r_i$, $0 \le r_i < h_i$, $(i = 1, 2, \ldots, s)$. Then

$$(x, y) = \mathfrak{a}\mathfrak{b}\beta_1^{t_1} \ldots \beta_s^{t_s},$$
$$\mathrm{ord}_{\mathfrak{p}_i}\mathfrak{a} = 0, \qquad \mathrm{ord}_{\mathfrak{p}_i}\mathfrak{b} \le h_i - 1 \qquad (i = 1, 2, \ldots, s),$$

and we find that

$$x = x_0\beta_1^{t_1} \cdots \beta_s^{t_s}, \qquad y = y_0 \beta 1^{t_1} \cdots \beta_s^{t_s}, \qquad (x_0, y_0) = \mathfrak{a}\mathfrak{b}, \qquad x_0, y_0 \in I_{\mathbb{K}}.$$

From (4.2) we see that $\mathfrak{a} \mid \alpha$, hence, $(x_0, y_0) \mid \alpha\beta_1 \cdots \beta_s$, since $\mathfrak{b} \mid \beta_1 \cdots \beta_s$. Thus, all solutions x, y of (4.2) without the condition $(x, y) \mid a$ are obtained in the form

$$x = x_0\beta_1^{t_1} \cdots \beta_s^{t_s}, \qquad y = y_0 \beta 1^{t_1} \cdots \beta_s^{t_s}, \qquad (x_0, y_0) \mid \alpha\beta_1 \cdots \beta_s,$$

where $t_1, \ldots, t_s$ are arbitrary non-negative integers, and x_0, y_0 satisfies (4.2) with $a = \alpha\beta_1 \cdots \beta_s$. Now applying Theorem 4.1 and the properties of heights of algebraic numbers described in §1 Ch.II, we obtain:

Theorem 5.1 *Under the conditions of the previous section all solutions* $x, y \in I_{\mathbb{K}}$, $y \ne 0$, *to equation (4.2) without the condition* $(x, y) \mid a$, *satisfy*

$$h(x/y) < c_{65}(|\operatorname{Nm}(\alpha)|H_f)^{c_{66}}, \tag{5.1}$$

where $h(x/y)$ is the height of x/y, and c_{65} and c_{66} are effectively determined by the degree, regulator, and class-number of the decomposition field of the form f, and by the number $\beta_1 \cdots \beta_s$.

Now we proceed with consideration of approximation of algebraic numbers θ by algebraic numbers $\kappa \in \mathbb{K}$. We set $[\mathbb{K}(\theta) : \mathbb{K}] = d$, and $[\mathbb{K} : \mathbb{Q}] = m$.

Consider the field $\mathbb{L} = \mathbb{K}(\theta)$. Suppose it has r_1 real and r_2 complex isomorphisms into the field of complex numbers (so $r_1 + 2r_2 = md$). Let Ω denote the set of all non-equivalent valuations $| \; |_v$ on $\mathbb{L}$. For $\lambda \in \mathbb{L}$ set

$$\|\lambda\|_v = |\lambda|_v^{n_v}, \qquad n_v = [\mathbb{L}_v : \mathbb{Q}_v],$$

where $\mathbb{L}_v$, $\mathbb{Q}_v$ are the completions of $\mathbb{L}$, $\mathbb{Q}$ in the metric $| \; |_v$. With $\lambda \neq 0$ one has the 'product formula' [121]

$$\prod_{v \in \Omega} \|\lambda\|_v = 1. \tag{5.2}$$

Now let S be an arbitrary finite subset of Ω. We shall prove that

$$\prod_{v \in S} \min(1, \|\theta - \kappa\|_v) > c_{67}(h(\kappa))^{-md}, \tag{5.3}$$

in which $h(\kappa)$ is the height of $\kappa \in \mathbb{K}$, and $c_{67} > 0$ is a value determined explicitly in terms of θ, $\mathbb{K}$ and S.

Let t and k' be the leading coefficients of the minimal polynomials of θ, respectively κ. Then

$$\lceil t\theta \rceil \leq h(\theta) \deg \theta, \qquad \lceil k'\kappa \rceil \leq h(\kappa) \deg \kappa$$

(see §1 Ch.II, proof of the inequality (1.4)). Hence, setting $k = tk'$ we find that

$$\lceil k\theta - k\kappa \rceil \leq |k'| \lceil t\theta \rceil + |t| \lceil k'\kappa \rceil \leq m(d+1)h(\theta)h(\kappa). \tag{5.4}$$

Writing $\theta' = k\theta$, $\kappa' = k\kappa$ and applying (5.2) to $\lambda = \theta' - \kappa'$, we find that

$$\prod_{v \in S} \|\theta' - \kappa'\|_v \cdot \prod_{v \in \Omega \setminus S} \|\theta' - \kappa'\|_v = 1. \tag{5.5}$$

For the archimedean v we find from (5.4)

$$\|\theta' - \kappa'\|_v \leq (m(d+1)h(\theta)h(\kappa))^{n_v},$$

while for the non-archimedean values $\|\theta' - \kappa'\|_v \leq 1$, since θ' and κ' are algebraic integers. Consequently,

$$\prod_{v \in \Omega \setminus S} \|\theta' - \kappa'\|_v \leq c_{68}(h(\kappa))^{r_1' + 2r_2'}, \tag{5.6}$$

where r_1' is the number of real, and r_2' is the number of complex valuations in $\Omega \setminus S$, and

$$c_{68} = (m(d+1)h(\theta))^{r_1' + 2r_2'}. \tag{5.7}$$

Now from (5.5), (5.6) we obtain

$$\prod_{v \in S} \|k\|_v \cdot \prod_{v \in S} \|\theta - \kappa\|_v \geq c_{68}^{-1}(h(\kappa))^{-r_1' - 2r_2'}. \tag{5.8}$$

Let S_∞ be the set of archimedean valuations in S and let S_0 be the set of non-archimedean valuations in S. Then, by (5.4), for $v \in S_\infty$ we have

$$\|k\|_v \|\theta - \kappa\|_v \leq (m(d+1)h(\theta)h(\kappa))^{n_v} \min(1, \|\theta - \kappa\|_v) ,$$

$$\prod_{v \in S_\infty} \|k\|_v \cdot \prod_{v \in S_\infty} \|\theta - \kappa\|_v \leq$$

$$\leq (m(d+1)h(\theta)h(\kappa))^{r_1'' + 2r_2''} \prod_{v \in S_\infty} \min(1, \|\theta - \kappa\|_v) , \quad (5.9)$$

where r_1'' is the number of real and r_2'' is the number of complex valuations in S_∞. Finally, for $v \in S_0$ we find that if $|\theta - \kappa|_v \leq 1$ because $|k|_v \leq 1$, then

$$\|k\|_v \|\theta - \kappa\|_v \leq |\theta - \kappa|_v^{n_v} = \min(1, \|\theta - \kappa\|_v),$$

whilst if $|\theta - \kappa|_v > 1$, because $k(\theta - \kappa)$ is an integer,

$$\|k\|_v \|\theta - \kappa\|_v \leq 1 = \min(1, \|\theta - \kappa\|_v).$$

Hence, in both cases, for $v \in S_0$ we have

$$\|k\|_v \|\theta - \kappa\|_v \leq \min(1, \|\theta - \kappa\|_v),$$

and then (5.8) and (5.9) yield (5.3), since $r_1' + r_1'' + 2(r_2' + r_2'') = md$. From (5.7) and (5.9) we find explicit representation for c_{67} in (5.3).

If the numbers κ are integers, then inequality (5.3) may be improved, since then $k' = 1$, $\|k\|_v \leq (h(\theta))^{n_v}$ for $v \in S_\infty$, and it follows from (5.8) that instead of (5.3) we have

$$\prod_{v \in S} \|\theta - \kappa\|_v > c_{69}(h(\kappa))^{-md + r_1'' + 2r_2''} . \quad (5.10)$$

It is clear that inequality (5.8) may be considered as the most general form of Liouville's inequality on rational approximation to algebraic numbers. We shall now prove that Theorem 5.1 implies a strengthening of the inequalities (5.3) and (5.10).

Theorem 5.2 *Let $\mathbb{K}$ be an algebraic number field of degree $[\mathbb{K} : \mathbb{Q}] = m$, and let θ be an algebraic number of degree $[\mathbb{K}(\theta) : \mathbb{K}] = d \geq 4$ over $\mathbb{K}$. Denote by S an arbitrary finite set of valuations on $\mathbb{L} = \mathbb{K}(\theta)$. Then for all $\kappa \in \mathbb{K}$ we have*

$$\prod_{v \in S} \min(1, \|\theta - \kappa\|_v) > c_{70}(h(\kappa))^{-md + c_{71}} , \quad (5.11)$$

while for all $\kappa \in I_\mathbb{K}$ we have

$$\prod_{v \in S} \|\theta - \kappa\|_v > c_{72}(h(\kappa))^{-md + r + c_{71}} , \quad (5.12)$$

where r is the number of archimedean valuations in S (complex values are counted twice), and c_{70}, c_{71}, c_{72} are positive numbers effectively determined by the parameters of $\mathbb{L}$ and S.

Proof. Let, as before, t and k' be the leading coefficients of the minimal polynomials of θ and κ respectively, and set $k = tk'$. Write $k\kappa = l$, and

$$(\mathrm{Nm}_{\mathbb{L}/\mathbb{K}}(l - \theta k)) = \mathfrak{a}\mathfrak{p}_1^{z_1} \cdots \mathfrak{p}_s^{z_s}, \qquad (\mathfrak{a}, \mathfrak{p}_1 \cdots \mathfrak{p}_s) = 1$$

where $\mathfrak{p}_1, \dots, \mathfrak{p}_s$ are those prime ideals of the ring $I_{\mathbb{K}}$ which are divisible by prime (finite) ideals of the set S, $\mathfrak{p}_i^{h_{\mathbb{K}}} = (\beta_i)$, $(i = 1, 2, \dots, s)$. Then

$$\mathrm{Nm}_{\mathbb{L}/\mathbb{K}}(l - \theta k)) = \alpha \beta_1^{z_1'} \cdots \beta_s^{z_s'} \xi, \tag{5.13}$$

where ξ is some unknown unit in $\mathbb{K}$, and $\alpha, \beta_1, \dots, \beta_s \in I_{\mathbb{K}}$. Equation (5.13) has the form (4.2), where $f(x, y) = \mathrm{Nm}_{\mathbb{L}/\mathbb{K}}(x - \theta y)$, but without the condition $(x, y) \mid \mathfrak{a}$. Therefore we may apply Theorem 5.1, which gives

$$h(l/k) = h(\kappa) < c_{73} |\mathrm{Nm}(\alpha)|^{c_{66}}. \tag{5.14}$$

From (5.13) we find that

$$\prod_{\mathfrak{p} \dagger \mathfrak{p}_1 \cdots \mathfrak{p}_s} |\mathrm{Nm}_{\mathbb{L}/\mathbb{K}}(l - k\theta)|_{\mathfrak{p}} = \prod_{\mathfrak{p} \dagger \mathfrak{p}_1 \cdots \mathfrak{p}_s} |\alpha|_{\mathfrak{p}} = \prod_{\mathfrak{p}} |\alpha|_{\mathfrak{p}} \prod_{\mathfrak{p} \mid \mathfrak{p}_1 \cdots \mathfrak{p}_s} |\alpha|_{\mathfrak{p}}^{-1}. \tag{5.15}$$

Since $\mathrm{ord}_{\mathfrak{p}} \alpha < h_{\mathbb{K}}$ it follows that $|\alpha|_{\mathfrak{p}} > P^{-h_{\mathbb{K}}}$, where P is the maximal rational prime divisible by a prime ideal from the set $\mathfrak{p}_1, \dots, \mathfrak{p}_s$. Hence,

$$\prod_{\mathfrak{p} \mid \mathfrak{p}_1 \cdots \mathfrak{p}_s} |\alpha|_{\mathfrak{p}}^{-1} < P^{sh_{\mathbb{K}}} = c_{74} \tag{5.16}$$

Moreover $|\alpha|_{\mathfrak{p}} = \|\alpha\|_{\mathfrak{p}}^{1/n_{\mathfrak{p}}}$, where $n_{\mathfrak{p}} = [\mathbb{K}_{\mathfrak{p}} : \mathbb{Q}_{\mathfrak{p}}] \leq m$. Hence,

$$\prod_{\mathfrak{p}} |\alpha|_{\mathfrak{p}} \leq (\prod_{\mathfrak{p}} \|\alpha\|_{\mathfrak{p}})^{1/m}. \tag{5.17}$$

By the product formula we have

$$|\mathrm{Nm}(\alpha)| \prod_{\mathfrak{p}} \|\alpha\|_{\mathfrak{p}} = 1,$$

so that by (5.17) and (5.16),

$$\prod_{\mathfrak{p} \dagger \mathfrak{p}_1 \cdots \mathfrak{p}_s} |\mathrm{Nm}_{\mathbb{L}/\mathbb{K}}(l - k\theta)|_{\mathfrak{p}} < c_{76} h(\kappa)^{-1/mc_{66}} \tag{5.18}$$

If $\mathfrak{P}$ denotes a prime ideal of the ring $I_{\mathbb{L}}$, we have the well known formula

$$|\mathrm{Nm}_{\mathbb{L}/\mathbb{K}}(l - k\theta)|_{\mathfrak{p}} = \prod_{\mathfrak{P} \mid \mathfrak{p}} \|l - k\theta\|_{\mathfrak{P}}.$$

Consequently, inequality (5.18) shows that

$$\prod_{v \in \Omega_0 \setminus S_0} \|l - k\theta\|_v = \prod_{\mathfrak{p} \nmid \mathfrak{p}_1 \ldots \mathfrak{p}_s} \prod_{\mathfrak{P} | \mathfrak{p}} \|l - k\theta\|_{\mathfrak{P}} < c_{76} h(\kappa)^{-1/mc_{66}}, \qquad (5.19)$$

where Ω_0 is the set of all non-archimedean valuations in Ω.

Now, again by the product formula, and (5.19), we obtain (where Ω_∞ is the set of all archimedean valuations in Ω)

$$1 = \prod_{v \in \Omega_\infty} \|l - k\theta\|_v \prod_{v \in S_0} \|l - k\theta\|_v \prod_{v \in \Omega_0 \setminus S_0} \|l - k\theta\|_v <$$

$$< c_{76} h(\kappa)^{-1/mc_{66}} \prod_{v \in \Omega_\infty} \|l - k\theta\|_v \prod_{v \in S_0} \|l - k\theta\|_v. \qquad (5.20)$$

For $v \in \Omega_\infty$ we have

$$\|l - \theta k\|_v = \|k\|_v \|\kappa - \theta\|_v \le (m(d+1) h(\theta) h(\kappa))^{n_v} \min(1, \|\theta - \kappa\|_v),$$

while for $v \in S_0$ we have

$$\|l - \theta k\|_v = \|k\|_v \|\kappa - \theta\|_v \le \min(1, \|\theta - \kappa\|_v). \qquad (5.21)$$

So from (5.20) we obtain

$$1 < c_{77} h(\kappa)^{-1/mc_{66}+md} \prod_{v \in \Omega_\infty \cup S_0} \min(1, \|\theta - \kappa\|_v).$$

This gives (5.11).

If κ is an integer, then $k = t$ and for $v \in \Omega_\infty$

$$\|l - \theta k\|_v \le (h(\theta))^{n_v} \|\theta - \kappa\|_v.$$

Hence

$$\prod_{v \in \Omega_\infty} \|l - \theta k\|_v \le (h(\theta))^{md} \prod_{v \in \Omega_\infty} \|\theta - \kappa\|_v = c_{78} |\mathrm{Nm}_{\mathbb{L}/\mathbb{Q}}(\theta - \kappa)|.$$

But, from (5.20), (5.21) we obtain

$$|\mathrm{Nm}_{\mathbb{L}/\mathbb{Q}}(\theta - \kappa)| \prod_{v \in S_0} \min(1, \|\theta - \kappa\|_v) > c_{79} h(\kappa)^{1/mc_{66}}. \qquad (5.22)$$

Since it is obvious that

$$|\mathrm{Nm}_{\mathbb{L}/\mathbb{Q}}(\theta - \kappa)| < c_{80} (h(\kappa))^{md-r} \prod_{v \in S_\infty} \|\theta - \kappa\|_v,$$

(5.22) yields (5.12).

We conclude this chapter by 'deciphering' the meaning of the inequality (5.22) to show how far we have developed the initial ideas connected to equation (1.2). The following theorem is an easy consequence of (5.22).

Theorem 5.3 *Let $\mathbb{K}$ be an algebraic number field of degree m and let $\theta_1, \ldots, \theta_m$ be an integral basis of $\mathbb{K}$. Further, let θ be an algebraic integer of degree $d \geq 4$ over $\mathbb{K}$, let A be a rational integer, and let $p_1, \ldots, p_s$ be rational primes. Then all solutions $x_1, \ldots, x_m$ of the diophantine equation*

$$\mathrm{Nm}(\theta_1 x_1 + \ldots + \theta_m x_m + \theta) = A p_1^{z_1} \ldots p_s^{z_s}$$

satisfy the inequality

$$X = \max_{1 \leq i \leq m} |x_i| < c_{81} |A|^{c_{82}},$$

where the quantities c_{81} and c_{82} are effectively determined by the numbers $\theta_1, \ldots, \theta_m$, and θ, and the primes $p_1, \ldots, p_s$.

It is easy to see that the explicit estimates of the previous paragraph spread to the values c_{81} and c_{82}. In particular, the greatest prime factor of the number

$$\mathrm{Nm}(\theta_1 x_1 + \ldots + \theta_m x_m + \theta)$$

is bounded below by some multiple of $\ln \ln X$ as X tends to infinity (cf. also [87]).

The ineffective form of the inequalities (5.11) and (5.12) is much stronger. The exponent on the right-hand side of (5.11) is $-2 - \varepsilon$, and that on the right-hand side of (5.12) is $-1 - \varepsilon$, where $\varepsilon > 0$ is arbitrary ([136], [137]).

VI. Elliptic and Hyperelliptic Equations

The equations considered in this chapter are in essence different both from the Thue equation and from its direct generalisations. It is possible to prove the existence of an effective bound for solutions of these equations by purely arithmetic methods, by reducing them to the Thue equation or to its generalisations over relative fields. However the bounds obtained in this way are not quite satisfactory in the general case, and we again turn to exponential equations to obtain better results. We give an analysis of the hyperelliptic equation and of integer and S-integer points on elliptic curves.

1. The Simplest Elliptic Equations

In this chapter we mainly consider diophantine equations of the form

$$f(x) = y^2 \qquad (1.1)$$

where $f(x)$ is a polynomial of degree $n \geq 3$. At first we suppose that the coefficients of $f(x)$, and the unknowns x, y, are rational integers, but subsequently we shall also consider (1.1) over a finite extension of the ring of rational integers.

The famous problem on the difference between cubes and squares of natural numbers leads to an equation of the form of (1.1):

$$x^3 - y^2 = k, \qquad k \neq 0. \qquad (1.2)$$

This problem has been systematically studied since Bachet (1621) and has its own deep history [144]. Nevertheless, many natural questions concerning (1.2) remain unclear. In particular, it is not known for which numbers k the equation is solvable. Mordell [141] was the first to prove that (1.2) has only finitely many solutions; he did so by directly connecting solutions of the equation with the solutions of a finite number of Thue equations. Mordell observes that all the solutions of the diophantine equation

$$X^2 + kY^2 = Z^3, \qquad (X, Z) = 1, \qquad (1.3)$$

are given by the formulae

$$X = \frac{1}{2}G(u,v), \qquad Y = f(u,v), \qquad Z = H(u,v), \qquad (1.4)$$

where u, v are integer parameters, and $f(u,v)$ is a certain integral binary cubic form of discriminant $4k$; $G(u,v)$ and $H(u,v)$ are the cubic and quadratic covariants of this form. Consequently, if we consider solutions x, y of (1.2) with $(x,y) = 1$, then (1.3) and (1.4) imply $f(u,v) = 1$ for some binary cubic form of discriminant $4k$. However, it is known that there exist only a finite number of classes of integral binary cubic forms with given discriminant, where a class consists of forms obtained one from another by means of unimodular integral transformations of the variables. Since equivalent forms (forms of the same class) represent the same numbers, we may consider the form (1.4) to be chosen from a fixed finite set of representatives of the classes of forms with discriminant $4k$. Thus we obtain a finite number of Thue equations $f(u,v) = 1$. From these we first determine all possible values for u, v and then, by the formulae $x = H(u,v)$ and $y = G(u,v)/2$, we find all x, y satisfying (1.2) and the condition $(x,y) = 1$.

Similar arguments allow one to remove the condition $(x,y) = 1$. Let $f(u,v)$ be a binary cubic form

$$f(u,v) = u^3 - 3xuv^2 - 2yv^3. \qquad (1.5)$$

It follows from (1.2) that the discriminant of this form is $108k$. By the reduction theory of binary cubic forms it is not difficult to show that $f(u,v)$ is equivalent to a form

$$g(u',v') = f(\alpha u' + \beta v', \gamma u' + \delta v'), \qquad \alpha\delta - \beta\gamma = \pm 1,$$

of height at most $(108|k|)^{1/2}$ (cf. [14]). On noting that

$$g(\delta u - \beta v, \alpha v - \gamma u) = \pm f(u,v) \qquad (1.6)$$

and comparing the coefficients of u^3 on the two sides of this equality we obtain the Thue equation to determine δ and γ:

$$g(\delta, -\gamma) = \pm 1. \qquad (1.7)$$

Recalling that the form $f(u,v)$ does not contain a term in $u^2 v$, we now obtain α and β from (1.6) and the equality $\alpha\delta - \beta\gamma = \pm 1$. Finally, we see from (1.5) and (1.6) that x and y are represented by polynomials in α, β, γ and δ and the coefficients of the form $g(u',v')$. Therefore, for fixed $k \neq 0$, evidently $\max(|x|, |y|)$ is bounded.

It is obvious from the foregoing arguments that to obtain the explicit bound for solutions of the equation (1.2) it suffices to have a bound for the solutions of the Thue equation (1.7). Thus Baker [14] obtained the inequality

$$\max(|x|, |y|) < \exp\{(10^{10}|k|)^{10^4}\},$$

improved by Stark [227] to the inequality

$$\max(|x|, |y|) < \exp\{(c_1|k|)^{1+\varepsilon}\},\qquad(1.8)$$

where $\varepsilon > 0$ is arbitrary, and $c_1 = c_1(\varepsilon)$ is effectively determined by ε. It is clear that bounding the solutions of the equation (1.7) in the light of inequality (4.4) of Chapter IV yields

$$\max(|\delta|, |\gamma|) < \exp\{(c_2|k|)(\ln(|k| + 1))^6\},\qquad(1.9)$$

where c_2 is a computable absolute constant. To derive this estimate, we recall that the field $\mathbb{K}$, is generated by a root of the form $f(u, v)$, and therefore the discriminant of $\mathbb{K}$ divides the discriminant $108k$ of a root of the polynomial. Hence the $|D|$ appearing in (4.4) of Chapter IV, is bounded above by a quantity of order $|k|$, implying (1.9). If the form $g(u', v')$ is reducible, the bound for the solutions of (1.7) is essentially stronger than (1.9). In that way we obtain the following:

Theorem 1.1 *All solutions of the diophantine equation (1.2) satisfy*

$$\max(|x|, |y|) < \exp\{(c_3|k|)(\ln(|k| + 1))^6\},$$

where c_3 is a computable absolute constant.

A similar idea underlies Mordell's proof [142] of the finiteness of the number of solutions of (1.1) in the case of an integral polynomial $f(x)$ without multiple roots.

Let $f(x) = ax^3 + bx^2 + cx + d$, where all the coefficients are rational integers, and $a \neq 0$. On multiplying (1.1) by $(27a)^2$ we obtain

$$s^3 - \frac{1}{4}g_2 s - \frac{1}{4}g_3 = t^2,\qquad(1.10)$$

where

$$s = 9ax + 3b, \quad t = 27ay, \quad g_2 = 108(b^2 - 3ac), \quad g_3 = 108(9abc - 2b^3 - 27a^2 d),$$

and both the discriminant of the polynomial on the left of (1.10), and the discriminant of $f(x)$ are non-zero: $\Delta = g_2^3 - 27g_3^2 \neq 0$. Mordell proved that all the solutions of the diophantine equation

$$X^3 - g_2 XY^2 - g_3 Y^3 = Z^2, \qquad (X, Y) = 1,\qquad(1.11)$$

are given by the formulae

$$X = H(u, v), \qquad Y = f(u, v),\qquad(1.12)$$

where $f(u, v)$ is a certain integral binary biquadratic form with invariants g_2, g_3, and $H(u, v)$ is the Hessian of this form. Since there exist only a finite number of classes of integral biquadratic forms with given invariants, it is sufficient to examine the finite number of forms $f(u, v)$ in the formulae (1.12), after taking representatives of the classes. Equation (1.10) is obtained from

(1.11) on setting $Y = 1$, which leads by (1.12) to a finite number of Thue equations $f(u,v) = 1$. On determining u, v from these equations we find s and t, and then x and y. Thus we obtain the finiteness of the number of solutions of (1.1) with $f(x)$ a cubic polynomial without multiple roots.

There exists another direct method to analyse (1.10), relying on the theory of binary biquadratic forms [15]. One associates a solution s, t of the equation with a binary form

$$f(u,v) = u^4 - 6su^2v^2 + 8tuv^3 + (g_2 - 3s^2)v^4,$$

with invariants g_2, g_3. The form is equivalent to a form $g(u', v')$, one of the finite number of representatives of all the classes of forms with invariants g_2, g_3. Consequently, similarly to (1.6), (1.7), one obtains a Thue equation

$$g(\delta, -\gamma) = 1, \tag{1.13}$$

from which to determine δ and γ. Since the coefficients of $f(u,v)$ are polynomials in $\alpha, \beta, \gamma, \delta$ and the coefficients of the form $g(u', v')$, one can first find α, β, and then s and t.

Baker [15] was the first to obtain a bound for solutions of (1.1) with a cubic polynomial $f(x)$ without multiple roots. He proved that (1.1) implies

$$\max(|x|, |y|) < \exp\{(10^6 H)^{10^6}\},$$

where H is the height of the polynomial $f(x)$ (the maximum of the absolute values of its coefficients). To derive this estimate, Baker used Mordell's arguments above, supplementing them with his own estimate for the value of solutions of the equation (1.13) and with a lemma that any form $f(u,v)$ with invariants g_2, g_3 is equivalent to a form $g(u', v')$ of height at most $10^4 \max(|g_2|^2, |g_3|^{2/3})$.

An application of inequality (4.4) of Chapter IV allows one to obtain the following result.

Theorem 1.2 *Suppose that in (1.10) g_2, g_3 are arbitrary integers with $\Delta = g_2^3 - 27g_2^2 \neq 0$. Then all solutions of this equation satisfy*

$$\max(|s|, |t|) < \exp\{c_4 [|\Delta|^{1/2} (|\Delta|^{1/2} + \ln G)]^{1+\varepsilon}\}, \tag{1.14}$$

where $G = \max(|g_2|, |g_3|)$; $\varepsilon > 0$ is arbitrary, and c_4 is effectively determined by ε. Therefore, all solutions of (1.1) with a cubic polynomial $f(x)$ without multiple roots, satisfy the inequality

$$\max(|x|, |y|) < \exp\{c_5 [|a||D|^{1/2} (|a||D|^{1/2} + \ln H)]^{1+\varepsilon}\}. \tag{1.15}$$

Here D is the discriminant of the polynomial $f(x)$ and H is its height; $\varepsilon > 0$ is arbitrary, and c_5 effectively determined by ε.

We leave the detailed proof of the inequalities (1.14) and (1.15) to the reader. We remark only that to prove (1.14) on the basis of the estimate (4.4) of Chapter IV, one must recall that in the most important case, that of irreducibility of the form $g(u', v')$, a root of this form generates a field coinciding with that generated by a root of the form $f(x, y)$. Hence the discriminant of this field is at most $|\Delta|$. To prove (1.15) one utilises (1.14) and the identity

$$27Da^2 = 4(b^2 - 3ac)^3 - (9abc - 2b^3 - 27a^2 d)^2 .$$

Thus all the solutions of the diophantine equation (1.1) with a cubic polynomial without multiple roots, satisfy an inequality of the form

$$\max(|x|, |y|) < \exp\{c_6 H^{6+\varepsilon}\}, \tag{1.16}$$

with an arbitrary $\varepsilon > 0$, and c_6 effectively determined by ε.

The ideas described above allow one to analyse S-integer solutions of the corresponding equations as well. Indeed, consider the basic equation (1.10). Obviously, if (s, t) is a rational point on the curve (1.10), then if $s = p/q$, with $(p, q) = 1$, we find that q is a square of a rational integer: $q = r^2$. On using the formulae (1.11), (1.12) we come to equations of the form

$$f(u, v) = r^2 , \tag{1.17}$$

where a binary form may be regarded as a representative of one of the finite number of form classes with invariants g_2 and g_3 — in particular, we may consider its height to be bounded by Baker's lemma. If we consider S-integer points on the curve (1.10), then (1.17) is a Mahler equation of type (1.1) of Chapter V. For in this case the collection of prime divisors of the number r must be fixed, and relative primality of u and v follows from (1.12) and the homogeneity of the polynomials $H(u, v)$ and $f(u, v)$. On employing Theorem 1.2, Chapter V to estimate solutions of (1.17), we obtain the following assertion:

Theorem 1.3 *Let (s, t) be a rational point on the curve (1.10), with $s = p/q$ and $(p, q) = 1$, and with $P \geq 2$ and $l \geq 1$ denoting the greatest prime divisor and the number of different prime divisors of q respectively. Finally, set $h(s) = \max(|p|, |q|)$. Then*

$$h(s) < \exp\{(c_7 P)^{201l+832} G^{6l+13}\}, \quad G = \max(|g_3|^2, |g_2|^{3/2}), \tag{1.18}$$

where c_7 is a computable absolute constant.

In particular, we see from inequality (1.18) that as the height $h(s)$ grows the magnitude of P increases as the iterated logarithm of $h(s)$. Similarly we can use the formulae (1.3) and (1.4), and Theorem 1.2 of Ch.V to bound the solutions of the equation

$$x^3 - y^2 = k p_1^{z_1} \cdots k p_l^{z_l}, \quad (x, y) = 1, \tag{1.19}$$

where $p_1, \ldots, p_l$ are fixed prime numbers, and $x, y, z_1 \geq 0, \ldots z_l \geq 0$ are unknown integers (cf. also [48]). As a consequence we find that the greatest prime divisor of the difference $x^3 - y^2, (x, y) = 1$, increases with growing $X = \max(|x|, |y|)$ as the square root of the iterated logarithm of X. It is interesting to note, that these results are completely consistent with those on the greatest prime divisor of binary forms (cf. §3 Ch.V). The effective analysis of equation (1.19) turned out to be useful for the explicit determination of elliptic curves with given conductor [48].

We have only sketched these results because the arguments just now described are very specific and admit no generalisations to polynomials $f(x)$ of degree greater than 3 nor to polynomials $f(x)$ of third degree with algebraic coefficients. So in the general case we have to analyse equation (1.1) in an essentially different way. Below we shall describe in detail a method of analysis leading to inequalities of type (1.15), (1.16), (1.18) in the general case.

2. The General Hyperelliptic Equation

The first step in analysis of the general equation (1.1) was made by Siegel [192] who proved the finiteness of the number of it's solutions under the assumption that the polynomial $f(x)$ has at least three simple roots.

We notice that in the general case, for (1.1) to have just a finite number of solutions it is necessary to suppose that $f(x)$ has at least three different roots of odd degree. After that one has no trouble in turning to the case of three simple roots.

Let α_1, α_2, α_3 be three simple roots of the polynomial $f(x)$ and let $\mathbb{K}$ denote the algebraic number field $\mathbb{Q}(\alpha_1, \alpha_2, \alpha_3)$. On factoring $x - \alpha_j$ as a product of prime ideals in $\mathbb{K}$, (1.1) yields

$$x - \alpha_i = \lambda_i \xi_i^2 \qquad (i = 1, 2, 3), \tag{2.1}$$

where the *lambda*$_i$ and ξ_i lie in the field $\mathbb{K}$, and the λ_i belong to a fixed set of numbers depending on the parameters of $\mathbb{K}$. The ξ_i are unknown integers (cf. below, §4). Eliminating x from the equalities (2.1) we obtain

$$\alpha_j - \alpha_i = \lambda_i \xi_i^2 - \lambda_j \xi_j^2 \qquad (1 \le i, j \le 3). \tag{2.2}$$

Consequently, the numbers $\xi_i \sqrt{\lambda_i} - \xi_j \sqrt{\lambda_j}$, have absolute norms independent of unknowns ξ_i, and then on multiplying by a suitable unit of the field $\mathbb{L} = \mathbb{K}(\sqrt{\lambda_1}, \sqrt{\lambda_2}, \sqrt{\lambda_3})$ they may be reduced to the finite set of fixed numbers (Lemma 2.2 Chapter II). We obtain

$$\xi_i \sqrt{\lambda_i} - \xi_j \sqrt{\lambda_j} = \mu_{ij} \eta_{ij} \qquad (i \ne j; 1 \le i, j \le 3), \tag{2.3}$$

where the μ_{ij} lie in $\mathbb{L}$ and have heights bounded by parameters depending on the field $\mathbb{L}$ and the η_{ij} are units of $\mathbb{L}$. On expressing η_{ij} in terms of a basis of the group of units of the field $\mathbb{L}$, we obtain the representation $\eta_{ij} = \nu_{ij} \zeta_{ij}^3$, where the ν_{ij} lie in the fixed set and the ζ_{ij} are unknown units. Therefore (2.3) takes the form

$$\xi_i \sqrt{\lambda_i} - \xi_j \sqrt{\lambda_j} = \nu'_{ij}\zeta_{ij} \qquad (i \neq j; 1 \leq i, j \leq 3), \tag{2.4}$$

whence we find

$$\nu'_{12}\zeta_{12}^3 + \nu'_{23}\zeta_{23}^3 + \nu'_{31}\zeta_{31}^3 = 0.$$

Setting $\sigma = \zeta_{12}\zeta_{31}^{-1}, \tau = \zeta_{23}\zeta_{31}^{-1}$, we observe that the latter equation is of the form

$$\alpha\sigma^3 + \beta\tau^3 = \gamma, \tag{2.5}$$

where α, β, γ are non-zero numbers of the field $\mathbb{L}$, and σ, τ are unknown integers of this field. Thus we have a generalised Thue equation.

Developing Thue's method, Siegel [191] obtained the generalisation of the Thue theorem on rational approximations to algebraic numbers in that case, when algebraic numbers are approximated by algebraic numbers of a fixed field. This allowed him to prove the finiteness of the number of solutions of the generalised Thue equation, in particular, the equation (2.5).

Having proved that there are only finitely many choices for σ and τ, it follows from (2.4) that there are only finitely many choices for the quotient $\xi_1\xi_3^{-1}$, and then by (2.2) we observe that there are only finitely many possibilities for ξ_3. Finally, from (2.1) with $i = 3$ we see that x admits only a finite number of possibilities. This completes the proof of the finiteness of the number of solutions of equation (1.1).

In this discussion we do not use the rationality of the polynomial $f(x)$, so our argument proves that (1.1) has only finitely many solutions for polynomials with algebraic coefficients having at least three simple roots and with the unknowns x and y lying in the ring of integers of an arbitrary fixed algebraic number field of finite degree over $\mathbb{Q}$.

Similarly we can establish the finiteness of the number of solutions in S-integers on the basis of $\mathfrak{p}$-adic analogues of the Siegel approximation theorems.

Baker's analysis [15] of Thue's equation in the form of (2.5) yields an explicit bound for the solutions of that equation. Inserting that into the above argument allowed him to prove that all solutions of (1.1), with $f(x)$ a polynomial with integer coefficients and with three simple roots, satisfy the inequality

$$\max(|x|, |y|) < \exp\exp\exp(n^{10n}H)^{n^2}, \tag{2.6}$$

where n is the degree of the polynomial $f(x)$ and H is its height.

By applying Theorem 6.1 Ch.IV (we need only the estimate (6.12)) and the theory of differents of algebraic number fields, it is not difficult to obtain a sharpening of inequality (2.6) of the type $\exp H^{c_8}$, where c_8 depends only on n. In addition, (1.1) may be considered over a finite extension of $\mathbb{Q}$ (in this case H denotes the maximal size of the coefficients of the polynomial $f(x)$, and the bound is for $\max(\overline{|x|}, \overline{|y|})$. Similarly, one can analyse S-integer solutions, based on Theorem 4.1 Ch.V, utilising the explicit expressions for the values c_{43}, c_{44} (the inequality (4.26)). However, the bounds found in this way for solutions of (1.1) are considerably less sharp than those obtained by the method of the following paragraphs. That method does not reduce (1.1) to

Thue equations but appeals directly to exponential equations and the theory of differents of algebraic number fields ([210], [218]).

3. Linear Dependence of Three Algebraic Units

Let $\mathbb{L}$ be an algebraic number field, set $l = [\mathbb{L} : \mathbb{Q}]$, denote by r the rank of the group of units of the field $\mathbb{L}$ and let $R = R_{\mathbb{L}}$ be the regulator of $\mathbb{L}$. By Lemma 2.1 Ch.II there exists a maximal system of independent units $\eta_1, \ldots, \eta_r$ in $\mathbb{L}$ such that

$$\prod_{i=1}^{r} \ln \overline{|\eta_i|} < c_9 R, \tag{3.1}$$

where c_9 is a function of l. We denote by $U_{\mathbb{L}}$ the group of units generated by the units $\eta_1, \ldots, \eta_r$.

Lemma 3.1 *Let α, β, γ be non-zero numbers in the field $\mathbb{L}$, with*

$$\max(h(\alpha), h(\beta), h(\gamma)) \leq T, \tag{3.2}$$

where $T > e$.
Then all the solutions x, y in units of the group $U_{\mathbb{L}}$ of the equation

$$\alpha x + \beta y = \gamma \qquad (x, y \in U_{\mathbb{L}}) \tag{3.3}$$

satisfy

$$\max(\overline{|x|}, \overline{|y|}) < \exp\{c_{10}(R \ln T)^{1+\varepsilon}\}, \tag{3.4}$$

where $\varepsilon > 0$ is arbitrary and c_{10} is effectively determined in terms of l and ε.

Proof. (cf. also [210], [218]). Let

$$x = \eta_1^{a_1} \cdots \eta_r^{a_r}, \qquad y = \eta_1^{b_1} \cdots \eta_r^{b_r}, \tag{3.5}$$

where $a_1, \ldots a_r$ and $b_1, \ldots, b_r$ are unknown integers with

$$A = \max_i |a_i|, \qquad B = \max_i |b_i|.$$

Substituting the expressions (3.5) in the equations (3.3) and employing the isomorphism $\sigma : \mathbb{L} \to \mathbb{C}$ to the obtained equations, we have

$$|(\eta_1^{\sigma})^{a_1} \cdots \eta_r^{\sigma})^{a_r}| = \left| \left(\frac{\gamma}{\alpha}\right)^{\sigma} \left(1 - \left(\frac{\beta}{\gamma}\right)^{\sigma} (\eta_1^{\sigma})^{b_1} \cdots (\eta_r^{\sigma})^{b_r}\right) \right|.$$

Applying Lemma 8.2 Ch.III to the right hand side of the latter equality and recalling (3.1), (3.2), we obtain

$$|(\eta_1^{\sigma})^{a_1} \cdots (\eta_r^{\sigma})^{a_r}| > e^{-D},$$

where

$$D = c_{11} \ln T \ln(B+1) R \ln(R+1). \qquad (3.6)$$

Here c_{11} depends on l. Thus, $\min_\sigma |x^\sigma| > e^{-D}$, and because $x \in U_\mathbb{L}$ we have

$$\overline{|x|} = \max_\sigma |x^\sigma| \le (\min_\sigma |x^\sigma|)^{-l+1} < e^{lD}. \qquad (3.7)$$

Therefore

$$\max_\sigma \left| \sum_{j=1}^r a_j \ln |\eta_j^\sigma| \right| < lD,$$

and we obtain from the inequality (3.1) that

$$A < c_{12} \ln T \ln(B+1) R \ln(R+1). \qquad (3.8)$$

By symmetry we have, similarly to the foregoing, that

$$B < c_{13} \ln T \ln(A+1) R \ln(R+1). \qquad (3.9)$$

Using the inequality (3.9) in (3.8), and the inequality (3.8) in (3.9), we obtain

$$\max(A, B) < c_{14} (R \ln T)^{1+\varepsilon'}, \qquad (3.10)$$

where $\varepsilon' > 0$ is arbitrary and c_{14} depends on l and ε'. Turning now to (3.6) we have

$$D < c_{15} (R \ln T)^{1+\varepsilon}, \qquad (3.11)$$

which by (3.7) yields inequality (3.4) for $\overline{|x|}$. Because of the symmetry of x and y in (3.3), the same holds for $\overline{|y|}$.

In fact, the inequalities (3.8), (3.9) show that $(R \ln T)^\varepsilon$ in (3.10), (3.11) may be replaced by iterated logarithms. We do not do that to shorten this discussion, because such a replacement does not seem to be of importance. The exponent $1 + \varepsilon$ in (3.4) and, consequently, in (3.10), (3.11) is best possible (to within ε). $\square$

We shall apply Lemma 3.1 in the proof of the central theorem of this chapter, Theorem 4.1. Here we make some remarks on variants and generalisations of the lemma.

Inequality (3.1) shows that the group $U_\mathbb{L}$ is a subgroup of finite index, bounded by a value depending on l, in the group $E_\mathbb{L}$ of all units of the field L. Therefore, in considering solutions $x, y \in E_\mathbb{L}$ in (3.3), we may utilise the representation (3.5), in which a_i, b_j are rational numbers with a denominator bounded by a function of l. Our proof of Lemma 3.1 is still applicable with slight changes. We again obtain an inequality like (3.4) with only c_{10} changed.

Lemma 3.2 *Under the hypotheses of the previous lemma, all solutions x, y in units of the group $E_\mathbb{L}$ of (3.3) satisfy an inequality of the form (3.4).*

In order to estimate the number of solutions of diophantine equations it is sometimes useful to have an estimate for the number of solutions of (3.3) in units of the groups $U_\mathbb{L}$ or $E_\mathbb{L}$. Obviously (3.10) bounds the number of possibilities for the exponents a_i, b_j in (3.5), and we obtain, for instance:

Lemma 3.3 *The number of pairs $x, y \in E_{\mathbb{L}}$, satisfying (3.3), does not exceed $c_{16}(R \ln T)^{2r+\varepsilon}$, where $\varepsilon > 0$ is arbitrary and $c_{16} = c_{16}(l, \varepsilon)$.*

In certain cases it is advisable to consider (3.3), with x belonging to the group of units of some field, while y is a unit of some other field. It is easy to see that in this case the bounds are similar to the previous case, with a product of regulators of corresponding fields replacing R. The coefficients α, β, γ may lie in some third field; the values of the constants now depend on the degrees of all three fields.

4. Main Theorem

We now come to the proof of a theorem on the solutions of the diophantine equation

$$f(x) = Ay^2, \tag{4.1}$$

where $f(x)$ is an integral polynomial of degree n with at least three simple roots and $A \neq 0$ is an integer. By multiplying by A and substituting $y' = Ay$, this equation is reduced to (1.1), but this transition smoothes out the influence of A on the bound and that is undesirable for our purposes. The estimate for solutions of the equation (4.1) in terms of A for a fixed polynomial is of special interest in view of the fact that in this way we have an opportunity to estimate from below a magnitude of the square-free kernel of the values of a polynomial $f(x)$ at integer points. In the sequel that result will be essential in the description of the parametric constructions of algebraic number fields with large class number (cf. §2 of Ch.VIII).

We shall suppose that the leading coefficient a_0 of the polynomial $f(x)$ is 1. If not, we multiply by a_0^{n-1} and substitute $x' = a_0 x, A' = a_0^{n-1} A$.

Theorem 4.1 *Let $f(x)$ be a monic polynomial with integer coefficients and of degree n and having three simple roots $\alpha = \alpha_0, \alpha_1, \alpha_2$. Let f_j be the minimal polynomials and D_j the discriminants respectively of the numbers α_j. Set $\mathbb{F} = \mathbb{Q}(\alpha_0, \alpha_1, \alpha_2)$, $d_j = [\mathbb{F} : \mathbb{Q}(\alpha_j)]$, $s = [\mathbb{F} : \mathbb{Q}]$ $(j = 0, 1, 2)$.*

Then all integer solutions x, y of (4.1) satisfy

$$\max(|x|, |y|) < \exp\{c_{17}|A|^{4(d_0+d_1+d_2)+\varepsilon} B^{4+\varepsilon} (\ln H)^{1+\varepsilon}\},$$

where $H > e$ is an upper bound for the height of the polynomial $f(x)$ and

$$B = \prod_{0 \leq j \leq 2} |R(f', f_j)| \prod_{0 \leq j \leq 2} |D_j|^{2s+3d_j}.$$

The $R(f', f_j)$ are resultants of the stated polynomials, $\varepsilon > 0$ is arbitrary and c_{17} is effectively determined in terms of n and ε.

We suppose here and in §7 that the discriminant of a rational number is 1.

Before turning to the proof of the theorem we make a general remark:

Lemma 4.1 *Let $\mathbb{L}$ be an algebraic number field and suppose that $\mathfrak{a}_1, \ldots, \mathfrak{a}_m$ are integer ideals of $I_{\mathbb{L}}$, with $[\mathfrak{a}_1, \ldots, \mathfrak{a}_m]$ their least common multiple; $(\mathfrak{a}_i, \mathfrak{a}_j)$ is the greatest common divisor of ideals $\mathfrak{a}_i, \mathfrak{a}_j$ $(i, j = 1, 2, \ldots, m)$. Then*

$$\prod_{i=1}^{m} \mathfrak{a}_i \mid [\mathfrak{a}_1, \ldots, \mathfrak{a}_m] \prod_{1 \le i < j \le m} (\mathfrak{a}_i, \mathfrak{a}_j).$$

Proof. Let $\mathfrak{p}$ be a prime ideal in $I_{\mathbb{L}}$ and $\mathfrak{p}^{\delta_i}$ the exact power of $\mathfrak{p}$ in $\mathfrak{a}_i$. Then the exact powers of $\mathfrak{p}$ in $(\mathfrak{a}_i, \mathfrak{a}_j)$ and in $[\mathfrak{a}_1, \ldots, \mathfrak{a}_m]$ are respectively $\min(\delta_i, \delta_j)$ and $\max_{(i)} \delta_i$. To prove the lemma it is sufficient to obtain the inequality

$$\sum_{1 \le i \le m} \delta_i \le \max_{(i)} \delta_i + \sum_{1 \le i < j \le m} \min(\delta_i, \delta_j). \tag{4.2}$$

But this inequality is independent of the indexing of the numbers δ_i, so we may suppose that $\delta_1 \le \delta_2 \le \cdots \le \delta_m$. Then (4.2) follows from the observation that

$$\sum_{1 \le i < j \le m} \min(\delta_i, \delta_j) = \sum_{1 \le i < j \le m} \delta_i = \sum_{1 \le i < m} (m - i)\delta_i \ge \sum_{1 \le i < m} \delta_i.$$

$\square$

Proof of Theorem 4.1. Let $\mathbb{K} = \mathbb{Q}(\alpha)$. We see from (4.1) that we have the following ideal equation in $\mathbb{K}$: $(x - \alpha) = \mathfrak{a}\mathfrak{p}^2$, where $\mathfrak{a} \mid Af'(\alpha)$. Similarly for the conjugate ideals: $\mathfrak{a}^{(i)} \mid Af'(\alpha^{(i)})$. Then $(\mathfrak{a}^{(i)}, \mathfrak{a}^{(j)})$ divides $(x - \alpha^{(i)}, x - \alpha^{(j)})$, which divides $\alpha^{(i)} - \alpha^{(j)}$. Hence

$$\prod_{1 \le i < j \le m} (\mathfrak{a}^{(i)}, \mathfrak{a}^{(j)}) \mid D_\alpha, \tag{4.3}$$

where $m = [\mathbb{K} : \mathbb{Q}]$ and D_α is the discriminant of α. Further, $\mathfrak{a}^{(i)} \mid AR(f', f_\alpha)$, where f_α is the minimal polynomial for α. Therefore

$$[\mathfrak{a}^{(1)}, \ldots, \mathfrak{a}^{(m)}] \mid AR(f', f_\alpha). \tag{4.4}$$

By Lemma 4.1 we find from (4.3), (4.4) that

$$\mathfrak{a}^{(1)} \cdots \mathfrak{a}^{(m)} \mid AD_\alpha R(f', f_\alpha),$$

and as $\mathfrak{a}^{(1)} \cdots \mathfrak{a}^{(m)} = (N(\mathfrak{a}))$, we obtain

$$N(\mathfrak{a}) \le |AD_\alpha R(f', f_\alpha)|. \tag{4.5}$$

If $\mathfrak{p}'$ is a prime ideal in the field $\mathbb{K}$, belonging to the class inverse to that of $\mathfrak{p}$ and with norm at most $|D_{\mathbb{K}}|^{1/2}$, then $\mathfrak{p}\mathfrak{p}'$ is a principal ideal (ξ'), $x - \alpha = \gamma' {\xi'}^2$, where $(\gamma') = \mathfrak{a}(\mathfrak{p}')^{-2}$, $g\gamma'$ is an integer and $g = \mathrm{Nm}(\mathfrak{p}')^2 \leq |D_{\mathbb{K}}|$. By (4.5) we have

$$|\mathrm{Nm}(\gamma')| \leq |AD_\alpha R(f', f_\alpha)| \, . \tag{4.6}$$

In what follows the values $c_{18}, \ldots, c_{21}$ depend on n only. By Lemma 2.2 Ch.II applied to the field $\mathbb{K}$ and the number $\gamma' \neq 0$, there exists a unit θ in $\mathbb{K}$ with

$$\overline{|\gamma'\theta|} < |\mathrm{Nm}(\gamma')|^{1/m} e^{c_{18} R_{\mathbb{K}}} \, . \tag{4.7}$$

Then θ may be represented in the form $\theta = \theta_1 \theta_2^2$, where $\theta_1, \theta_2 \in U_{\mathbb{K}}$ and θ_1 is a product of distinct fundamental units of the group $U_{\mathbb{K}}$. Setting $\gamma = \gamma'\theta\theta_1^{-1}$, $\xi = \xi'\theta_2^{-1}$, we find $x - \alpha = \gamma\xi^2$, that $g\gamma$ is an integer, and by (4.7)

$$\overline{|\gamma|} \leq \overline{|\gamma'\theta|} \, \overline{|\theta_1^{-1}|} < |\mathrm{Nm}(\gamma')|^{1/m} e^{c_{19} R_{\mathbb{K}}} \, .$$

Since $|R(f', f_\alpha)| < c_{20} H^{m+n-1}$ and $|D_\alpha| < c_{21} H^{2m-2}$, we have from (4.6)

$$\overline{|\gamma|} < c_{22} \left| A H^{3m+n-3} \right|^{1/m} e^{c_{19} R_{\mathbb{K}}} \, . \tag{4.8}$$

Similarly, on setting $\mathbb{K}_1 = \mathbb{Q}(\alpha_1)$, $\mathbb{K}_2 = \mathbb{Q}(\alpha_2)$, we obtain $x - \alpha_i = \gamma_i \xi_i^2$, where γ_i, ξ_i lie in the fields $\mathbb{K}_i$ respectively, and both ξ_i and $g_i \gamma_i$ are integers, $g_i \leq |D_i|$, and inequalities of the type (4.8) hold for γ_i, $(i = 1, 2)$.

The relations so obtained imply

$$\alpha_j - \alpha_i = \gamma_i \xi_i^2 - \gamma_j \xi_j^2 \qquad (0 \leq i, j \leq 2), \tag{4.9}$$

where $\alpha_0 = \alpha$, $\gamma_0 = \gamma$, and $\xi_0 = \xi$. Let $\mathbb{L} = \mathbb{Q}(\alpha, \alpha_1, \alpha_2, \sqrt{\gamma}, \sqrt{\gamma_1}, \sqrt{\gamma_2})$. We see from (4.9) that the integers $\xi_i \sqrt{\gamma_i} - \xi_j \sqrt{\gamma_j}$ lying in $\mathbb{L}$ have absolute norms with moduli at most $c_{20} H^{8n^3}$. Hence, by Lemma 2.2 Ch.II there exist units $\varepsilon_1, \varepsilon_2, \varepsilon_3$ in the group of units $U_{\mathbb{L}}$, such that

$$\xi\sqrt{\gamma} - \xi_1\sqrt{\gamma_1} = \varepsilon_1 \lambda_1, \quad \xi\sqrt{\gamma} - \xi_2\sqrt{\gamma_2} = \varepsilon_2 \lambda_2, \quad \xi_2\sqrt{\gamma_2} - \xi_0\sqrt{\gamma_1} = \varepsilon_3 \lambda_3, \tag{4.10}$$

where the λ_i are integers in $\mathbb{L}$ with

$$\max_{(i)} \overline{|\lambda_i|} < (c_{20} H^{8n^3})^{1/l} e^{c_{21} R} \, , \tag{4.11}$$

and $l = [\mathbb{L} : \mathbb{Q}]$, $R = R_{\mathbb{L}}$. From (4.10) we obtain

$$\lambda_1 \delta_1 - \lambda_2 \delta_2 = lambda_3, \quad \delta_1 = \varepsilon_1 \varepsilon_3^{-1}, \quad \delta_2 = \varepsilon_2 \varepsilon_3^{-1} \, .$$

This equation has the form (3.3) and, applying Lemma 3.1 with regard to the estimate (4.11), we have

$$\max(\overline{|\delta_1|}, \overline{|\delta_2|}) < \exp\{c_{22} R^{1+\varepsilon} (R + \ln H)^{1+\varepsilon}\} \, , \tag{4.12}$$

where $\varepsilon > 0$ is arbitrary and $c_{22} = c_{22}(n, \varepsilon)$.

Setting $\mu_1 = \delta_1 \lambda_1, \mu_2 = \delta_2 \lambda_2$, we find from (4.10) that

$$\xi\sqrt{\gamma} - \xi_1\sqrt{\gamma_1} = \mu_1 \varepsilon_3, \quad \xi\sqrt{\gamma} - \xi_2\sqrt{\gamma_2} = \mu_2 \varepsilon_3 \tag{4.13}$$

and (4.11), (4.12) imply

$$\max(\lceil\overline{\mu_1}\rceil, \lceil\overline{\mu_2}\rceil) < \exp\{c_{23}R^{1+\varepsilon}(R+\ln H)^{1+\varepsilon}\}, \tag{4.14}$$

where $c_{23} = c_{23}(n,\varepsilon)$. Eliminating ε_3 from the relations (4.13), we obtain

$$\xi_2 = \sigma\xi + \tau\xi_1, \quad \sigma = \frac{(\mu_1 - \mu_2)\sqrt{\gamma}}{\mu_1\sqrt{\gamma_2}}, \quad \tau = \frac{\mu_2\sqrt{\gamma_1}}{\mu_1\sqrt{\gamma_2}}. \tag{4.15}$$

By the estimate (4.8), similar bounds for $\lceil\overline{\gamma_1}\rceil$ and $\lceil\overline{\gamma_2}\rceil$ and the bound (4.14), we find that

$$\max(h(\sigma), h(\tau)) < \exp\{c_{24}R^{1+\varepsilon}(R+\ln H)^{1+\varepsilon} + c_{25}(\ln A + R^\star)\}, \tag{4.16}$$

where $R^\star = \max(R_{\mathbb{K}}, R_{\mathbb{K}_1}, R_{\mathbb{K}_2})$, $c_{24} = c_{24}(n,\varepsilon)$, and $c_{25} = c_{25}(n)$.

Now taking two of the equations (4.9) we have

$$\alpha_1 - \alpha = \gamma\xi^2 - \gamma_1\xi_1^2, \quad \alpha_2 - \alpha = \gamma\xi^2 - \gamma_1(\bar\sigma\xi + \bar\tau\xi_1)^2,$$

where $\bar\sigma = \sigma\sqrt{\gamma_2/\gamma_1}$ and $\bar\tau = \tau\sqrt{\gamma_2/\gamma_1}$. Eliminating ξ_1 from this system, we see that ξ satisfies

$$\zeta_1\xi^4 + \zeta_2\xi^2 + \zeta_3^2 = 0 \tag{4.17}$$

where

$$\zeta_1 = 4\bar\sigma^2\bar\tau^2\gamma_1\gamma - (\gamma - \bar\tau^2\gamma - \bar\sigma^2\gamma_1)^2,$$
$$\zeta_2 = 4\bar\sigma^2\bar\tau^2\gamma_1(\alpha - \alpha_1) + 2(\gamma - \bar\tau^2\gamma - \bar\sigma^2\gamma_1)(\bar\tau^2(\alpha - \alpha_1) - \alpha + \alpha_2),$$
$$\zeta_3 = \bar\tau^2(\alpha - \alpha_1) - \alpha + \alpha_2.$$

At least one of the numbers ζ_3, ζ_2 is non-zero, since otherwise $\sigma = 0$, which would imply $\alpha_1 = \alpha_2$. Consequently, (4.17) is non-trivial. Using (4.8) to find similar bounds for $\lceil\overline{\gamma_1}\rceil$, $\lceil\overline{\gamma_2}\rceil$, and (4.16), we obtain from (4.17) a bound for $|\xi|^2$ in a form similar to the right hand side of (4.16), though with other constants. In view of $x - \alpha = \gamma\xi^2$ and the bound (4.8) we have for $|x|$ that

$$|x| < \exp\{c_{26}R^{1+\varepsilon}(R+\ln H)^{1+\varepsilon} + c_{27}(\ln|A| + R^\star)\}, \tag{4.18}$$

where $c_{26} = c_{26}(n,\varepsilon)$ and $c_{27} = c_{27}(n)$.

To complete the proof of the theorem it remains to bound R and $R^\star$ from above.

Since the fields $\mathbb{K}$, $\mathbb{K}_1$ and $\mathbb{K}_2$ are contained in $\mathbb{L}$, by Lemma 2.3 Ch.II we have $R^\star < c_{28}R$, where $c_{28} = c_{28}(n)$. To bound R we employ Lemma 2.4 Ch.II:

$$R < c_{29}|D_{\mathbb{L}}|^{1/2}(\ln|D_{\mathbb{L}}|)^{l-1}. \tag{4.19}$$

Now it remains to bound the discriminant $D_{\mathbb{L}}$. We view $\mathbb{L}$ as a field obtained from $\mathbb{Q}$ by means of successive adjunction of the integers α, α_1, α_2, $g\sqrt{\gamma}$, $g_1\sqrt{\gamma_1}$, and $g_2\sqrt{\gamma_2}$. Because of the multiplicative property of relative fields differents, we see that the different of the field $\mathbb{L}$ is a divisor of the product of the differents of the successively adjoined numbers. Therefore the discriminant $D_{\mathbb{L}}$, being the norm of the different, divides the absolute norm of the number

$$\Delta = f'_\alpha(\alpha) f'_{\alpha_1}(\alpha_1) f'_{\alpha_2}(\alpha_2) 2g\sqrt{\gamma} 2g_1\sqrt{\gamma_1} 2g_2\sqrt{\gamma_2} .$$

But we have that $\mathrm{Nm}(\Delta)$ is given by

$$D_\alpha^{l/m} D_{\alpha_1}^{l/m_1} D_{\alpha_2}^{l/m_2} (8gg_1g_2)^l (\mathrm{Nm}(\gamma))^{l/me} (\mathrm{Nm}(\gamma_1))^{l/m_1e_1} (\mathrm{Nm}(\gamma_2))^{l/m_2e_2} ,$$

where $m = [\mathbb{K} : \mathbb{Q}]$, $e = [\mathbb{K}(\sqrt{\gamma}) : \mathbb{K}]$ and m_1, m_2, e_1, e_2 are defined analogously. Because of $g \leq |D_{\mathbb{K}}|$, noting that $|\mathrm{Nm}(\gamma)|$ coincides with $|\mathrm{Nm}(\gamma')|$ and is bounded by the inequality (4.6), and because of similar inequalities for g_1, g_2, $|\mathrm{Nm}(\gamma_1)|$, and $|\mathrm{Nm}(\gamma_2)|$, we find that

$$|\mathrm{Nm}(\Delta)| \leq 8^l |A|^{4(d+d_1+d_2)} |D_\alpha|^{l+12d} S^{4d} |D_{\alpha_1}|^{l+12d_1} S_1^{4d_1} |D_{\alpha_2}|^{l+12d_2} S_2^{4d_2} ,$$

where $S = |R(f', f'_\alpha)|$, S_1 and S_2 are similarly defined and we have noted that $|D_{\mathbb{K}}| \leq |D_\alpha|$; similarly for $|D_{\mathbb{K}_1}|$ and $|D_{\mathbb{K}_2}|$. Consequently, from the inequality $|D_{\mathbb{L}}| \leq |\mathrm{Nm}(\Delta)|$ and the estimate (4.19), we find that

$$R < c_{30} |A|^{2(d+d_1+d_2)} B_1 (\ln |A| + \ln B_1)^{l-1} ,$$

where $B_1 = S^{2d} S_1^{2d_1} S_2^{2d_2} |D_\alpha|^{l/2+6d} |D_{\alpha_1}|^{l/2+6d_1} |D_{\alpha_2}|^{l/2+6d_2}$. Turning back to (4.18) we see that

$$|x| < \exp\{c_{31} |A|^{4(d+d_1+d_2)+\varepsilon} B_1^{2+\varepsilon} (\ln H)^{1+\varepsilon}\} ,$$

and this completes the proof of the theorem.

We notice that if the polynomial $f(x)$ has no multiple roots, B in the statement of Theorem 4.1 may be replaced by

$$B' = |D_f|^{6s+4(d+d_1+d_2)} ,$$

where D_f is the discriminant of the polynomial $f(x)$.

Of course $R(f', f_\alpha)$, $D_\alpha, \ldots$ are divisors of the discriminant D_f. If $f(x) = f_\alpha(x) g(x)$, then

$$D_f = D_\alpha D_g R^2(f_\alpha, g), \qquad D_f = R(f', f_\alpha) R(f', g) .$$

It is not difficult to exhibit the bound B in terms of H. In any case

$$B < c_{32} H^{12(n^2-1)(n-1)(n-2)} , \tag{4.20}$$

if one takes into account that $d_j \leq (n-1)(n-2)$, $s \leq n(n-1)(n-2)$,

$$|R(f', f_j)| < c_{33} H^{2(n-1)}, \qquad |D_j| < c_{34} H^{2(n-1)}$$

(all the values c_{32}, c_{33}, c_{34} are explicitly determined by n). Certainly, if special properties of the polynomial $f(x)$ are known, all the estimates may be improved in an essential way. For example, if $f(x)$ is a normal polynomial (that is, the field generated by a root of the polynomial is normal) then $d_j = 1$, $s = n$ and the bound for solutions of (4.1) assumes the form

$$\max(|x|, |y|) < \exp\{c_{35}|A|^{12+\varepsilon}|D_f|^{24(n+2)+\varepsilon}(\ln H)^{1+\varepsilon}\}.$$

It is interesting here that the exponent of the power of $|A|$ is an absolute constant. Thus for integer $x \neq 0$ the square-free kernel of the number $f(x)$ is estimated from below by the value $c_{36}(\ln|x|)^{1/12-\varepsilon}$, while in general case, for integer $x \neq 0$ and $f(x) \neq 0$ this kernel is bounded from below by

$$c_{37}(\ln|x|)^{(1-\varepsilon)/3(n-1)(n-2)}.$$

Here c_{36} and c_{37} are positive, effectively determined quantities depending on the coefficients and degree of $f(x)$, and on an arbitrary $\varepsilon > 0$.

5. Estimate for the Number of Solutions

In Chapter VIII we will need a bound for the number of solutions of equation (4.1) when the polynomial $f(x)$ is fixed and the number A is variable. The arguments and the estimates of the previous section, together with Lemma 3.3, allow us to obtain such a bound easily. In fact, we obtain a bound for the number of solutions which also takes into account varying coefficients of $f(x)$. It will be see from the nature of arguments below that for the number of solutions of all the equations under review in this monograph (except for the equations of Chapter IX) similar estimates for their number of solutions may be obtained in the same way. However, the accuracy of these results is barely satisfactory and special methods or significant changes of argument are necessary to approach the true situation (in connection with this, see the papers by Evertse [61–64]).

Theorem 5.1 *Under the hypotheses of Theorem 4.1 the number of solutions of* (4.1) *is at most*

$$c_{38}(|A|^{d+d_1+d_2}B)^{64s-6+\varepsilon}H^\varepsilon \prod_{0 \le j \le 2} |D_j|^{1/2},$$

where c_{38} is effectively determined in terms of n and ε.

Proof. We follow the arguments given in the proof of Theorem 4.1 and estimate the number of possibilities for the equations determining a solution x of the original equation (4.1).

Since the ideal $\mathfrak{a}$ divides $AR(f', f_\alpha)$, it admits at most $c_{39}(|A|H)^\varepsilon$ possibilities. Similarly the number of possibilities for the ideal $\mathfrak{p}$ is at most the number of integer ideals of the field $\mathbb{K}$ with norm at most $|D_\mathbb{K}|^{1/2}$; thus it is bounded by $c_{40}|D_\mathbb{K}|^{1/2+\varepsilon}$. Hence the number of non-associated γ' does not exceed $c_{39}c_{40}(|A|H)^\varepsilon|D_\mathbb{K}|^{1/2+\varepsilon}$, and the number of different γ (analogously, γ_1, γ_2) is bounded by a quantity of the same type.

We fix the field $\mathbb{L} = \mathbb{Q}(\alpha, \alpha_1, \alpha_2, \sqrt{\gamma}, \sqrt{\gamma_1}, \sqrt{\gamma_2})$ and consider the number of possibilities for λ_1 (similarly for λ_2, λ_3). By (4.9) the absolute norm of λ_1

is fixed, so the number of principal ideals (λ_1) is bounded by $c_{41}H^\varepsilon$. All λ_1 generating the same ideal (λ_1), differ by just a unit of the field $\mathbb{L}$, and, because $U_\mathbb{L}$ is a subgroup of $E_\mathbb{L}$ of the index at most c_{42}, all these λ_1 fall into at most $c_{41}c_{42}H^\varepsilon$ classes, where we take a class to consist of those λ_1 which differ just by factors from the subgroup $U_\mathbb{L}$. In view of (4.11) it suffices to count the number of those units $\eta \in U_\mathbb{L}$, for which we have for fixed λ_1^*

$$\overline{|\lambda_1^* \eta|} < (c_{20}H^{8n^3})^{1/l}e^{c_{21}R}. \tag{5.1}$$

Assuming $\eta = \eta_1^{t_1} \cdots \eta_r^{t_r}$, we find from (5.1) that

$$\max_{(\sigma)} \left| \sum_{j=1}^r t_j \ln \left| \eta_j^\sigma \right| \right| < c_{43}(R + \ln H),$$

whence by (3.1) it follows that

$$\max_{(j)} |t_j| < c_{44}(R + \ln H).$$

Therefore the number of possible choices of λ_1 is bounded by $c_{45}R^r H^\varepsilon$.

By Lemma 3.3 the number of δ_1, δ_2, which satisfy $\lambda_1\delta_1 - \lambda_2\delta_2 = \lambda_3$, does not exceed

$$c_{46}[R(R + \ln H)]^{2r+\varepsilon}.$$

It is obvious that the values $\gamma, \gamma_1, \gamma_2, \lambda_1, \lambda_2, \lambda_3, \delta_1, \delta_2$ uniquely determine (4.17). It gives at most four values for ξ (in fact, at most two, because $\zeta_1 = 0$). For R we use the estimate obtained at the close of the previous paragraph and then we have the following upper bound for the number of solutions of (4.1):

$$c_{45}(|A|^{d+d_1+d_2}B)^{64s-6+\varepsilon}H^\varepsilon|D_\mathbb{K}D_{\mathbb{K}_1}D_{\mathbb{K}_2}|^{1/2}.$$

Since $D_\mathbb{K}$ divides D_α and a similar statement is true for $D_{\mathbb{K}_1}, D_{\mathbb{K}_2}$, this implies the assertion of the theorem.

In particular, for a fixed polynomial $f(x)$ the number of solutions of (4.1) does not exceed

$$c_{46}|A|^{64n^5}, \tag{5.2}$$

where c_{46} is given in terms of the coefficients and degree of the polynomial.

6. Linear Dependence of Three S-units

As in §3 let $\mathbb{L}$ be an algebraic number field and suppose that $l = [\mathbb{L} : \mathbb{Q}]$, and that Ω denotes the set of all non-equivalent absolute values of the field $\mathbb{L}$. Let Ω_∞ be the set of all the archimedean (infinite) absolute values in Ω, and let S be a finite subset of Ω containing Ω_∞. The numbers $\lambda \in \mathbb{L}$ with

$$|\lambda|_v = 1 \text{ for all } v \notin S \tag{6.1}$$

form the group $E_{\mathbb{L}}(S)$ with respect to multiplication and are called S-units. Since (6.1) holds for the units of the field $\mathbb{L}$, plainly $E_{\mathbb{L}}(S) \supset E_{\mathbb{L}}$ and we are dealing with a generalisation of ordinary algebraic units. This generalisation has proved its utility for the solution of many problems in algebraic number theory.

Let $\mathfrak{p}_1, \ldots, \mathfrak{p}_s$ be prime ideals of the ring $I_{\mathbb{L}}$ corresponding to non-archimedean values in S, and let $P_1, \ldots, P_s$ be the absolute norms of these ideals. Denoting by $h = h_{\mathbb{L}}$ the class number of the field $\mathbb{L}$, we have

$$\mathfrak{p}_i^h = (\pi_i) \qquad (i = 1, 2, \ldots, s), \tag{6.2}$$

where the π_i are integers of $\mathbb{L}$, which by Lemma 2.2 Ch.II may be chosen so that

$$\overline{|\pi_i|} < P_i^{h/l} e^{c_{47} R} \qquad (i = 1, 2, \ldots, s). \tag{6.3}$$

Denote by $U_{\mathbb{L}}(S)$ the free group generated by the units $\eta_1, \ldots, \eta_r$, constructed in the light of Lemma 2.1 Ch. II, and by the numbers $\pi_1, \ldots, \pi_s$. It is easy to see that this group has finite index in $E_{\mathbb{L}}(S)$. On taking $\lambda \in E_{\mathbb{L}}(S)$ and considering the factorisation of the principal ideal (λ) as a product of powers of prime ideals in $I_{\mathbb{L}}$, we observe that by (6.1) only the $\mathfrak{p}_1, \ldots, \mathfrak{p}_s$ appear:

$$(\lambda) = \mathfrak{p}_1^{z_1} \cdots \mathfrak{p}_s^{z_s},$$

where the z_i are integers. By reducing each z_i modulo h, we can write,

$$(\lambda) = \mathfrak{a}\pi_1^{z_1'} \cdots \pi_s^{z_s'}, \text{ with } \mathfrak{a} = \mathfrak{p}_1^{r_1} \cdots \mathfrak{p}_s^{r_s} \qquad (0 \le r_i < h).$$

The ideal $\mathfrak{a}$, being a quotient of principal ideals, is a principal ideal (σ) say, a generating element of which, by Lemma 2.2 Ch. II, may be chosen so that

$$\overline{|\sigma|} < |\mathrm{Nm}(\sigma)|^{1/l} e^{c_{48} R}. \tag{6.5}$$

From (6.4), (6.5) we evidently obtain

$$\overline{|\sigma|} < (P_1 \cdots P_s)^{(h-1)/l} e^{c_{48} R}, \tag{6.6}$$

σ being an integer. Thus each number $\lambda \in E_{\mathbb{L}}(S)$ may be represented in the form

$$\lambda = \xi \sigma \pi_1^{z_1'} \cdots \pi_s^{z_s'}, \qquad \xi \in E_{\mathbb{L}},$$

where σ satisfies the inequality (6.6). Finally we notice that from (3.1) it follows each unit $\xi \in E_{\mathbb{L}}$ may be represented in the form $\xi = \xi_0 \eta$, where $\eta \in U_{\mathbb{L}}$, and $\overline{|\xi_0|} < \exp\{c_{49} R\}$. This is clear from geometrical considerations and also follows from Lemma 2.2 Ch.II. We may replace the number ξ_0 by σ in order to keep the notation and obtain:

Lemma 6.1 *Each number* $\lambda \in E_{\mathbb{L}}(S)$ *may be represented in the form*

$$\lambda = \sigma\pi, \qquad \pi \in U_{\mathbb{L}}(S),$$

where $\sigma \in I_{\mathbb{L}}$ satisfies an inequality of the type (6.6) *with c_{48} depending only on l.*

Now let α, β, γ be non-zero elements of the field $\mathbb{L}$ satisfying the condition (3.2). We consider the equation

$$\alpha x + \beta y = \gamma \tag{6.7}$$

in unknowns $x, y \in E_{\mathbb{L}}(S)$ and prove a generalisation of Lemma 3.2 (cf. also [117]).

By Lemma 6.1 we have

$$x = \sigma \eta_1^{a_1} \cdots \eta_r^{a_r} \pi_1^{x_1} \cdots \pi_s^{x_s}, \qquad y = \tau \eta_1^{b_1} \cdots \eta_r^{b_r} \pi_1^{y_1} \cdots \pi_s^{y_s}, \tag{6.8}$$

where a_i, b_j, x_k, y_l are unknown integers; where $\sigma, \tau \in I_{\mathbb{L}}$, and where if P is the greatest prime number dividing one of the numbers $P_1, \ldots, P_s$, we have

$$\max(\overline{|\sigma|}, \overline{|\tau|}) < P^{hs} e^{c_{49} R}. \tag{6.9}$$

On writing

$$\max_{(i)} |a_i| = A, \quad \max_{(j)} |b_j| = B, \quad \max_{(k)} |x_k| = X, \quad \max_{(l)} |y_l| = Y$$

and employing (6.3), (6.7)–(6.9), (3.1) for archimedean values v, by Lemma 8.2 Ch.III, we find

$$|x|_v = \left| \frac{\gamma}{\alpha} \left(1 - \frac{\beta}{\gamma} y \right) \right|_v > e^{-D}, \tag{6.10}$$

$$D = c(\varepsilon)(c_{50}(s+l))^{200(s+l)} [(R+h \ln P)^s (R+sh \ln P) R \ln T]^{1+\varepsilon} \ln \max(B, Y), \tag{6.11}$$

where $\varepsilon > 0$ is arbitrary and $c_{50} = c_{50}(l)$. Similarly, for non-archimedean $v \in S$ by Lemma 8.3 Ch.III, we obtain the inequality of the form (6.10) with D' instead of D, where

$$D' = (c_{501}(s+l))^{12(s+l)} P^l (R+h \ln P)^s (R+sh \ln P) R \ln T (\ln \max(B, Y))^2, \tag{6.12}$$

with $c_{51} = c_{51}(l)$. Consequently, from (6.11), (6.12), for all $v \in S$ we have an inequality of the form (6.10) with D'' instead of D, where

$$D'' = c(\varepsilon)(c_{52}(s+l))^{200(s+l)} P^l [(R+h \ln P)^s (R+sh \ln P) R \ln T]^{1+\varepsilon} (\ln Z)^2, \tag{6.13}$$

here $Z = \max(A, B, X, Y)$.

Because $x^{-1}, yx^{-1} \in E_{\mathbb{L}}(S)$ and $\gamma x^{-1} - \beta y x^{-1} = -\alpha$, we can apply the previous argument to x^{-1} instead of x, which by virtue of the multiplicative representation (6.8), yields a system of inequalities of the form

$$\left| \ln |\sigma|_v + \sum_{i=1}^{r} a_i \ln |\eta_i|_v + \sum_{j=1}^{s} x_j \ln |\pi_j|_v \right| < D'', \tag{6.14}$$

with v running through all the values in S. For non-archimedean $v \in S$ we have $\ln |\eta_i|_v = 0$ $(i = 1, 2, \ldots, r)$ and since every such valuation is $\mathfrak{p}$-adic corresponding to prime ideals $\mathfrak{p}_1, \ldots, \mathfrak{p}_s$, we obtain the subsystem

$$\left| \ln |\sigma|_{\mathfrak{p}_j} + x_j \ln |\pi_j|_{\mathfrak{p}_j} \right| < D'' \qquad (j = 1, 2, \ldots, s). \tag{6.15}$$

From (6.2) it follows that $|\pi_j|_{\mathfrak{p}_j} = p_j^{-h}$, where p_j is the rational prime divided by $\mathfrak{p}_j$, and, as $\sigma \subset I_{\mathbb{L}}$, the product formula implies

$$\overline{|\sigma|}^{-l} \leq |\sigma|_{\mathfrak{p}_j} \leq 1 \qquad (j = 1, 2, \ldots, s).$$

Therefore we obtain from (6.15) and (6.9)

$$X = \max_{(j)} |x_j| < \frac{1}{h \ln 2} (D'' + R + hs \ln P) \qquad (j = 1, 2, \ldots, s). \tag{6.16}$$

Returning to the system (6.14) and now considering the archimedean valuations v, because of (6.3), (6.9), (6.16), we obtain

$$\max_{v \in \Omega_\infty} \left| \sum_{i=1}^r a_i \ln |\eta_i|_v \right| < D''', \tag{6.17}$$

where D''' is a value of the type (6.13). By (3.1) it follows that the elements of the inverse matrix of the inequalities comprising the system (6.17) are bounded above by a quantity depending only on l. Hence, $A = \max_{(i)} |a_i|$ is bounded in terms of a quantity similar to D''.

Given all that, we notice that similar arguments may be advanced with respect to y instead of x. In this way we find that Z is bounded by a quantity of the type D''. This shows that

$$Z < c_{53}(c_{52}(s + l))^{200(s+l)} P^l [(R + h \ln P)^s (R + sh \ln P) R \ln T]^{1+\varepsilon}, \tag{6.18}$$

where $c_{53} = c_{53}(\varepsilon)$, with $\varepsilon > 0$ arbitrary. In view of (6.14) a similar estimate is true for $\max |\ln |x|_v|$ with $v \in S$, and because of the symmetry of x and y in (6.7), also for $\max |\ln |y|_v|$ with $v \in S$. Thus we obtain the following main assertion on the equation (6.7).

Lemma 6.2 *Let* α, β, γ *be non-zero elements of the field* $\mathbb{L}$ *with heights at most* $T > e$. *Then all solutions of equation (6.7) in* S-units x, y *of the group* $E_{\mathbb{L}}(S)$ *satisfy*

$$\max_{v \in S} |\ln |x|_v| + \max_{v \in S} |\ln |y|_v| <$$

$$< c_{54}(c_{52}(s + l))^{200(s+l)} P^l [(R + h \ln P)^s (R + sh \ln P) R \ln T]^{1+\varepsilon}, \tag{6.19}$$

where s *is the number of non-archimedean valuations in* S, P *is the greatest prime with* $|P|_v < 1$ *for* $v \in S$, *and* l, R *and* h *are the degree, regulator and class number of the field* $\mathbb{L}$ *respectively. The constant* c_{52} *is effectively determined by* l, *and the constant* c_{54} *by* l *and* ε, *where* $\varepsilon > 0$ *is arbitrary.*

For $s = 0$ this lemma yields a formally worse estimate than Lemma 3.2 (it has R^2 instead of R) but, actually in this case one should replace the factors P^l and $R + sh \ln P$ by 1, since those terms appear in the argument only if $s \neq 0$.

From (6.19) it is easy to turn to estimates for the heights of the numbers x, y in any variant of definition of height. Since we shall proceed to work with the ordinary notion of height, we employ inequality (1.5) of Ch.II to estimate the heights of the numbers x, y. For that, it is necessary to have upper bounds for the sizes $\overline{|x|}$, $\overline{|y|}$ and for the minimal natural numbers a_0, b_0 (the 'denominators') such that $a_0 x$, $b_0 y$ are integers. Because S includes all the archimedean valuations, we obtain the bounds for $\overline{|x|}$ and $\overline{|y|}$ from (6.19). Further we observe that for non-archimedean $v \in S$ and corresponding prime ideals $\mathfrak{p}$

$$|x|_v = |x|_{\mathfrak{p}} = p^{-\frac{1}{e} \operatorname{ord}_{\mathfrak{p}} x}, \qquad e = e_{\mathfrak{p}} = \operatorname{ord}_{\mathfrak{p}} p,$$

where p is the prime divided by $\mathfrak{p}$. Since $1 \le e_{\mathfrak{p}} \le l$, then the product over all $\mathfrak{p}$ for which $\operatorname{ord}_{\mathfrak{p}} x < 0$

$$a_0 = \prod_{\mathfrak{p}} |x|_{\mathfrak{p}}^l \tag{6.20}$$

is a natural number and $a_0 x$ is an integer. Analogously, the natural number $b_0 = \prod_{\mathfrak{p}} |y|_{\mathfrak{p}}^l$ is such that $b_0 y$ is an integer. Therefore, applying the inequality (6.19) we obtain:

Lemma 6.3 *Under the hypotheses of Lemma 6.2 the heights of the numbers x, y satisfy the inequalities*

$$\ln \max(h(x), h(y)) <$$
$$< c_{55}(s + l)(c_{52}(s + l))^{200(s+l)} P^l [(R + h \ln P)^s (R + sh \ln P) R \ln T]^{1+\varepsilon},$$

where c_{55} is effectively determined in terms of l and ε, with $\varepsilon > 0$ arbitrary.

We make use of this lemma to obtain further results on the bounds for solutions of diophantine equations in S-integers. We also notice that it allows one to obtain an effective analysis of the general ternary exponential equation [222], that is of the equation

$$\alpha \alpha_1^{x_1} \cdots \alpha_l^{x_l} + \beta \beta_1^{y_1} \cdots \beta_m^{y_m} + \gamma \gamma_1^{z_1} \cdots \gamma_n^{z_n} = 0, \tag{6.21}$$

where α, β, γ, and the α_i, β_j, γ_k are non-zero algebraic numbers, and the x_i, y_j, z_k are non-negative integer unknowns. If this equation has a finite number of solutions, an effective upper bound for them is determined in terms of the parameters of the equation. If there are infinitely many solutions, then the bound for the generating elements of the of the solution family is determined. It is the problem of exceptional interest and importance to achieve a similar result for equations of the type (6.21) with a greater number of terms on its left-hand side.

7. Solutions in S-integers

We continue with the notation of the previous paragraph. Let $I_{\mathbb{L}}(S)$ be the set of all numbers $\lambda \in \mathbb{L}$ with

$$|\lambda|_v \leq 1 \quad \text{for all } v \notin S. \tag{7.1}$$

It is clear that $I_{\mathbb{L}}(S)$ is a ring containing $I_{\mathbb{L}}$. Its elements are called the S-integers of the field $\mathbb{L}$.

We shall deal with solutions of equation (4.1) under the assumption that the coefficients of the polynomial $f(x)$ are integers of $\mathbb{L}$, that A is also an integer of this field, and that the unknowns x, y lie in $I_{\mathbb{L}}(S)$. Again, as in §4, we suppose that the polynomial $f(x)$ is monic of degree n has three simple roots $\alpha, \alpha_1, \alpha_2$.

Suppose that x and y are in $I_{\mathbb{L}}(S)$ and satisfy (4.1). From the condition (7.1) it follows that the factorisations of the principal ideals $(x), (y)$ as products of prime ideals of $\mathbb{L}$, have only powers of those prime ideals $\mathfrak{p}_1, \ldots, \mathfrak{p}_s$ determining non-archimedean valuations from S in their denominators. Let $\mathfrak{d}$ be an integral ideal of $\mathbb{L}$ from the ideal class inverse to that of the denominator of (x) and with norm at most $|D_{\mathbb{L}}|^{1/2}$. Then we have a representation

$$x = X/z, \qquad X \in I_{\mathbb{L}}, \qquad z \in I_{\mathbb{L}}, \qquad (X, z) = \mathfrak{d}, \tag{7.2}$$

where the ideal (z) contains only the prime ideals $\mathfrak{p}_1, \ldots \mathfrak{p}_s$ and divisors of $\mathfrak{d}$. On defining a binary form $f(X, z)$ by $f(X, z) = z^n f(X/z)$, we find from that (4.1) becomes

$$f(X, z) = Ay^2 z^n. \tag{7.3}$$

Let $\mathbb{K} = \mathbb{L}(\alpha)$, where α is a root of $f(x)$. From (7.3) we see that every prime ideal of the field $\mathbb{K}$ which is not contained in $\mathfrak{d}Af'(\alpha)$ divides $X - \alpha z$ with even exponent. Hence, the factorisation of $X - \alpha z$ as a product of prime ideals of $\mathbb{K}$ is of the shape

$$(X - \alpha z) = \mathfrak{a}\mathfrak{p}^2, \qquad \mathfrak{a} \mid \mathfrak{d}Af'(\alpha).$$

On arguing further as in §4 we find a representation $X - \alpha z = \gamma \xi^2$, where $\gamma, \xi \in \mathbb{K}$, and g, γ, ξ are integers. We obtain bounds for g and $\overline{|\gamma|}$ of the same kind as in §4. Setting $\mathbb{K}_1 = \mathbb{L}(\alpha_1)$, $\mathbb{K}_2 = \mathbb{L}(\alpha_2)$, we obtain similar equalities $X - \alpha_i z = \gamma_i \xi_i^2$ $(i = 1, 2)$ in these fields, whence

$$(\alpha_j - \alpha_i)z = \gamma_i \xi_i^2 - \gamma_j \xi_j^2 \qquad (0 \leq i, j \leq 2), \tag{7.4}$$

where $\alpha_0 = \alpha$, $\gamma_0 = \gamma$, and $\xi_0 = \xi$. It is now appropriate to write $\mathbb{M} = \mathbb{L}(\alpha, \alpha_1, \alpha_2, \sqrt{\gamma}, \sqrt{\gamma_1}, \sqrt{\gamma_2})$, and to deal with (7.4) by working in $\mathbb{M}$, observing in particular that the integers $\xi_i \sqrt{\gamma_i} \pm \xi_j \sqrt{\gamma_j}$ belong to $\mathbb{M}$. This means working with all the archimedean valuations of this field and all the continuations of the non-archimedean valuations of the set S from $\mathbb{L}$ to $\mathbb{M}$. By the same arguments, again as in §4, but using Lemma 6.3 instead of (3.4), we obtain an explicit bound for solutions of the equation (4.1) in $x, y \in I_{\mathbb{L}}(S)$.

In this way L. A. Trelina [239] has obtained the following result.

Theorem 7.1 *Let $f(x)$ be a polynomial of degree n with coefficients from $I_{\mathbb{L}}$, with leading coefficient a, and with three simple roots $\alpha_j (j = 0, 1, 2)$. Let f_j be the minimal polynomials and D_j the discriminants of the numbers $a\alpha_j$ respectively, and set $l_j = [\mathbb{L}(\alpha_j) : \mathbb{L}](j = 0, 1, 2)$. Then all solutions of (4.1), provided $A \in I_{\mathbb{L}}$ and $x, y \in I_{\mathbb{L}}(S)$, satisfy the inequality*

$$\max(h(x), h(y)) < H_0^{c_{56}} \exp\{(c_{57}(l + s))^{65l'(l+s)} |\mathrm{Nm}(A)|^{32(l')^2(l+s)} BP^{3l'l}\},$$

where H_0 is the height of the number α_0, $l' = 4l_0 l_1 l_2$, and

$$B = \left| D_{\mathbb{L}}^3 (D_0 D_1 D_2)^{l+3} \mathrm{Nm}(aR_0 R_1 R_2) \right|^{32(l')^2(l+s)}.$$

The R_j are the resultants of the polynomials $a^{n-1}f(x/a)$ and the $f_j(x)(j = 0, 1, 2)$, whilst P is the greatest prime number divided by a prime ideal from S. The constants c_{56} and c_{57} are effectively determined in terms of l and n.

The interesting feature of this theorem is the nature of the influence of the height of the polynomial $f(x)$ on the bound for the solutions. This is achieved by means of the following improvement of the arguments described in §4. We set

$$\xi_i \sqrt{\gamma_i} - \xi_j \sqrt{\gamma_j} = \sigma_{ij}, \qquad \xi_i \sqrt{\gamma_i} + \xi_j \sqrt{\gamma_j} = \tau_{ij}. \tag{7.5}$$

Then from (7.4) we observe the equality

$$\sigma_{ij}\tau_{ij} = (\alpha_j - \alpha_i)z$$

and, by summing (7.5), we find the relations

$$2\xi\sqrt{\gamma} = \sigma_{01} + \tau_{01}, \tag{7.6}$$

$$\sigma_{01} + \sigma_{12} + \sigma_{20} = 0, \qquad \tau_{01} - \tau_{12} + \sigma_{20} = 0. \tag{7.7}$$

Both latter equalities are relations of linear dependence between three S^*-units of the field $\mathbb{M}$, in which the coefficients are ± 1, that is they are independent of the height of $f(x)$, which is the advantage in comparison with the previous argument. Application of Lemma 6.3 to the equalities (7.7) shows that $\sigma_{01} = \sigma_{20}\sigma = \tau_{01}\tau$, where the heights of numbers σ, τ are bounded in terms of n, the parameters of the field $\mathbb{M}$ and s and P. Now we observe from (7.6) that

$$2\xi\sqrt{\gamma} = (1 + \tau^{-1})\sigma_{01} = (1 + \tau)\tau_{01},$$

$$4\xi^2\gamma = (1 + \tau)(1 + \tau^{-1})(\alpha_1 - \alpha)z.$$

Because of (7.2) this implies the equality

$$x = \alpha + \frac{1}{4}(1 + \tau)(1 + \tau^{-1})(\alpha_1 - \alpha),$$

whence we obtain the bound for the height of x.

The main corollary from Theorem 7.1 is a bound of the type

$$P > c_{58} \ln \ln \max(h(x), h(y)) \tag{7.8}$$

for the greatest absolute norm of prime ideals which divide the denominators of the numbers x, y, satisfying the equation (4.1) and lying in the field $\mathbb{L}$. Here $c_{58} > 0$ and is effectively determined in terms of $f(x)$, A and $\mathbb{L}$. Another consequence is a bound from below for the absolute norm of the square-free kernel of the values taken by the polynomial $f(x)$ for $x \in I_\mathbb{L}$, by a quantity like $c_{59}(\ln h(x))^{1/32(l+s)l'^2}$, with $c_{59} = c_{59}(f, \mathbb{L}) > 0$. It seems probable that the above mentioned feature of Theorem 7.1 concerning the influence of the height of the polynomial $f(x)$ on the bounds for the solutions also has interesting effects, but those may be deeply concealed (cf. [240]).

8. *S*-integer Points on Elliptic Curves

According to the well-known theorem of Siegel [193], there are only finitely many integer points on any given algebraic curve of genus greater than zero. Siegel's arguments were founded on two fundamental results obtained in different but adjoining branches of number theory and based on distinct but similar ideas. These were firstly obtained by Siegel himself [191] applying improvements and generalisations of the results and method of Thue on rational approximations to algebraic numbers by algebraic numbers of a fixed field (or even by algebraic numbers of bounded degree) and, secondly, applying what is now the well-known Weil's theorem (cf. [120]) on the existence of a finite basis of the group of rational points on an algebraic curve, a generalisation of the initial result of Mordell [143] on curves of genus 1 (elliptic curves). Both results were obtained by ineffective methods and did not yield any bounds for all the integer points of the given curve. Later, Mahler [132] obtained a finiteness theorem for *S*-integer points generalising the arguments of Siegel and inheriting the ineffectiveness of their conclusions.

For a long time even the case of elliptic curves remained insuperable for effective arguments, till Baker and Coates at last found a new proof of this case of Siegel's theorem and obtained an explicit bound for all integer points on such curves. Their arguments were based on the construction in explicit form of a birational transformation of the curve to a canonical form under which the coefficients of the equation and the integer points of the curve are sent to points with bounded 'denominator' in some algebraic number field, effectively determined in terms of the coefficients and degree of the equation determining a curve. Since the canonical equation is of the form of (1.1) with a cubic polynomial $f(x)$ without multiple roots this gives an opportunity to construct a bound for solutions of the initial equation (transition from the initial equation of the curve to the canonical equation is realised without serious loss and the most essential contribution to the values of the bounds for the solutions is made by the bounds for the canonical equation).

Let $F(x,y)$ be an absolutely irreducible polynomial with integer coefficients, determining an elliptic curve

$$F(x,y) = 0; \tag{8.1}$$

H and n are its height and degree respectively. Baker and Coates [26] proved that all the integer points x, y on the curve (8.1) satisfy the inequality

$$\max(|x|, |y|) < \exp\exp\exp\{(2H)^{10^{n^{10}}}\}, \tag{8.2}$$

in accordance with the bound (2.6) previously obtained by Baker [15] for solutions of (1.1).

Here we sketch the arguments devised by Baker and Coates on the above mentioned reduction of the curve to canonical form. By the method described in this chapter, this yields the same improvement on the inequality (2.6). In addition, there are no obstacles to consideration of S-integer points on the curve.

By using these ideas S. V. Kotov and L. A. Trelina [117] obtained the following result.

Theorem 8.1 *Let S be a finite set of rational primes and let P be the greatest prime in S. Then all S-integer rational points x, y on the curve (8.1) satisfy the inequality*

$$\max(h(x), h(y)) < \exp\{(c_{60}(s+1)^{8^{4+n^6}} H^{8^6 + 4n^6 + n^8})^{(s+2)} P^{8^{2+n^6}}\};$$

where c_{60} is effectively determined in terms of n.

Denote by $\overline{\mathbb{Q}}$ the field of all algebraic numbers and let $\overline{\mathbb{Q}}(x)$ denote the field of rational functions of x over $\overline{\mathbb{Q}}$. Let $\mathfrak{R}$ be the finite algebraic extension of $\overline{\mathbb{Q}}(x)$ generated by a root of (8.1). Let $Q_1, \ldots, Q_r$ be all the absolute values of $\mathfrak{R}$ extending the valuation of $\overline{\mathbb{Q}}(x)$ determined by x^{-1}, and let $e_1, \ldots, e_r$ be the corresponding ramification indices.

In the vector space $\mathfrak{M}$, formed by all the elements g of the field $\mathfrak{R}$ with $\mathrm{ord}_{Q_1}(g) \geq -3$ and $\mathrm{ord}_U(g) \geq 0$ for all the other valuations U of $\mathfrak{R}$, one may construct a basis of the form

$$x'^h \omega_j \quad (1 \leq j \leq l, 0 \leq h \leq \lambda_j) \tag{8.3}$$

with certain integers $l, \lambda_1, \ldots, \lambda_l$, where $x' = (x - b)^{-1}$, b is an integer in the interval $0 \leq b \leq n^3$, and where $w_1, \ldots, w_l$ are elements of $\mathfrak{R}$ with the following properties: w_j has a Puiseux expansion at Q_i of the form

$$w_j = (x')^{-\nu_i/e_i} \sum_{k=0}^{\infty} w_{ijk} x'^{k/e_i} \quad (1 \leq i \leq r, 1 \leq j \leq l), \tag{8.4}$$

where $\nu_i = 3$ for $i = 1$, $\nu_i = 0$ for $i \neq 1$. All the coefficients w_{ijk} belong to the field $\mathbb{K} = \mathbb{Q}(\theta)$ with an algebraic integer θ of degree at most 8^{n^6} and of size

$|\theta| \le (2H)^\mu$ where $\mu = 8^{n^8}$. There exists a natural number Δ, such that all the numbers $\Delta^{k+1} w_{ijk}$ are algebraic integers of $\mathbb{K}$ and

$$\max(\Delta^{k+1}, \Delta^{k+1}\overline{|w_{ijk}|}) \le (2H)^{\mu(k+1)} \qquad (k = 0, 1, \ldots). \qquad (8.5)$$

In addition, each basis element (8.3) may be represented as a rational function of x, y over $\mathbb{K}$. By the Riemann-Roch Theorem the dimension of the space $\mathfrak{M}$ is equal to the degree of the divisor determined by $\mathfrak{M}$; that is, it is 3. This shows that $l = 3$ and that one may take $\lambda_j = 0$, $(0 \le j \le 3)$.

Among the linear combinations of the three functions w_1, w_2, w_3 one can choose two, X_1 and X_2, say, such that $\mathrm{ord}_{Q_1}(X_1) = -2$, $\mathrm{ord}_{Q_1}(X_2) = -3$, and $\mathrm{ord}_U(X_1) \ge 0$, $\mathrm{ord}_U(X_2) \ge 0$ for all the other valuations U of the field $\mathfrak{R}$. These functions will have an expansions of the type (8.4) and, roughly speaking, will satisfy the inequalities

$$\max(\Delta^{k+1}, \Delta^{k+1}\overline{|w_{ijk}|}) \le (2H)^{\mu(k+1)} \qquad (k = 0, 1, \ldots). \qquad (8.5)$$

$$X_j = (x')^{-\nu_i/e_i} \sum_{k=0}^{\infty} u_{ijk} x'^{k/e_i} \qquad (j = 1, 2), \qquad (8.6)$$

where $u_{ijk} \in \mathbb{K}$, $\Delta^{k+1} u_{ijk} \in I_\mathbb{K}$, and

$$\max(\Delta^{k+1}, \Delta^{k+1}\overline{|u_{ijk}|}) \le (3H)^{2\mu(k+1)} \qquad (k = 0, 1, \ldots). \qquad (8.7)$$

On these grounds, one establishes by direct computation that X_1 satisfies an equation

$$X_1^m + P_1(x)X_1^{m-1} + \cdots + P_m(x) = 0, \qquad (8.8)$$

where the $P_j(x)$ are quadratic polynomials

$$P_j(x) = q_{j0}x^2 + q_{j1}x + q_{j2}, \qquad \Delta^{n^5} q_{jk} \in I_\mathbb{K}. \qquad (8.9)$$

It follows from (8.7) that

$$\max(\Delta^{n^5}, \Delta^{n^5}\overline{|q_{jk}|}) \le (2H)^{\mu^5} \qquad (1 \le j \le m; k = 1, 2).$$

We now consider the seven functions

$$1, X_1, X_2, X_1^2, X_2^2, X_1^3, X_1 X_2.$$

The inequalities (8.6), (8.7) imply that at Q_1 these functions have expansions of the type

$$(x')^{-6/l_i} \sum_{k=0}^{\infty} v_{jk} x'^{k/l_i} \qquad (j = 1, 2, \ldots, 7)$$

respectively, where $\Delta^{6(k+1)} v_{jk} \in I_\mathbb{K}$, and

$$\max(\Delta^{6(k+1)}, \Delta^{6(k+1)}\overline{|v_{jk}|}) \le (4H)^{12\mu(k+1)}.$$

As they have no other poles, except for the one at Q_1 of order at most 6, they are linearly dependent by the Riemann-Roch Theorem. The coefficients

$x_j (j = 1, 2, \ldots, 7)$ of their linear dependence may be chosen in $I_{\mathbb{K}}$ to satisfy the inequalities

$$\overline{|x_j|} \leq \{7(4H)^{84\mu}\}^6 \leq (2H)^{\mu^2}. \tag{8.10}$$

Thus

$$x_1 + x_2 X_1 + x_3 X_2 + x_4 X_1^2 + x_5 X_2^2 + x_6 X_1^3 + x_7 X_1 X_2 = 0, \tag{8.11}$$

where it is not difficult to observe that $x_5 \neq 0$. Now on assuming that

$$X' = X_1, \qquad Y' = 2x_5 X_2 + x_7 X_1 + x_3,$$
$$a' = -4x_5 x_6, \qquad b' = x_7^2 - 4x_4 x_5,$$
$$c' = 2x_3 x_7 - 4x_2 x_5, \qquad d' = x_3^2 - 4x_1 x_5,$$

we obtain the equation

$$Y'^2 = a' X'^3 + b' X'^2 + c' X' + d'. \tag{8.12}$$

We can convince ourselves that the polynomial on the right of (8.12) has no multiple roots. Indeed, if the equation is reduced to

$$(Y'/(X' - \alpha))^2 = a'(X' - \beta)$$

with certain α, β, then, since $X' - \beta$ has only a pole at Q_1, $Y'/(X' - \alpha)$ also has a pole only at Q_1. But $x_5 \neq 0$ and $\mathrm{ord}_{Q_1}(X_1) = -2$, $\mathrm{ord}_{Q_1}(X_2) = -3$, so that order of the pole $Y'/(X' - \alpha)$ at Q_1 is exactly 1. This is incompatible with the assumption of multiple roots because the genus of the field $\mathfrak{R}$ is 1.

After all that, we suppose that

$$X = \Delta^{2n^5} X', \qquad Y = \Delta^{3n^5} Y',$$
$$a = a', \qquad b = \Delta^{2n^5} b', \qquad c = \Delta^{4n^5} c', \qquad d = \Delta^{6n^5} d',$$

which by (8.12) implies

$$Y^2 = aX^3 + bX^2 + cX + d. \tag{8.13}$$

This is the desired canonical form of the equation (8.1). We notice that the coefficients of (8.13) are algebraic integers of $\mathbb{K}$ and that

$$\max(\overline{|a|}, \overline{|b|}, \overline{|c|}, \overline{|d|}) \leq (2H)^{\mu^6}. \tag{8.14}$$

If x, y is an S-integer solution of (8.1), it follows from (8.8), (8.9) that the corresponding value of X_1 is an algebraic number containing in its denominator, besides divisors of Δ^{n^5}, only prime ideals formed by the prime numbers from S. Equation (8.11) shows that the same is true with respect to $x_5 X_2$. Since X_1, X_2 are rational functions of x, y over $\mathbb{K}$ we observe that X, Y are S^*-integers in $\mathbb{K}$ for the set of prime ideals of the field $\mathbb{K}$ dividing the prime numbers from S. Equation (8.13), which has on its right a polynomial without multiple roots, together with condition (8.14), gives an opportunity to obtain an explicit bound for the height of the solutions of this equation in S^*-integers of $\mathbb{K}$ (for instance, by applying Theorem 7.1). Now from the equations (8.8), (8.9), we obtain the bound for the height of the initial number x, and from (8.1) we have a bound for the height of y. This completes the arguments.

It follows from Theorem 8.1 that the greatest prime number contained in the denominator of a rational point on an elliptic curve is bounded from below as the iterated logarithm of the height of this point; that is, an inequality of the type (7.8) is valid. By a different method using the Mordell Theorem on the finiteness of the rank of the group of rational points on an elliptic curve, the results of Lang [123] and Masser [140] on the arithmetic nature of the values of Weierstrass elliptic functions and the analogues of these results in p-adic metrics, Bertrand [29–30] has proved that in the case of the curves (8.1) with complex multiplication one has the inequality

$$ P > c_{61} \ln \max(h(x), h(y))^{0,01r^{-2}} . $$

Here r is the rank of the group of rational points on the curve and c_{61} depends only on the coefficients and degree of the equation determining the curve (the value of c_{61} is ineffective). This deep inequality is the first and as yet unique case in which an essentially stronger estimate than an iterated logarithm of the height of the argument has been obtained in problems on the arithmetic nature of numerical values of polynomials and rational points on algebraic curves.

VII. Equations of Hyperelliptic Type

We apply the methods and considerations described in the previous chapter to more general situations so as to obtain new facts generalising our former results. Then we proceed to a new type of equations in which at least one of the unknowns is a power of an unknown integer. Our aim is to bound the unknown exponent so as to reduce these new equations to those of the kind considered before. In this way we determine, for example, that any integral polynomial having at least two simple roots represents only a finite number of powers of integers with exponents greater than 2. We also give an analysis of S-integer solutions of the Catalan equation.

1. Equations with Fixed Exponent

The subject of this chapter is the equation of the form

$$f(x) = Ay^m, \qquad A \neq 0, \quad m \geq 3, \tag{1.1}$$

in which $f(x)$ is a polynomial with at least two simple roots. We shall suppose that the coefficients of the polynomial and the unknowns x, y are rational integers, but this supposition is not essential, and just as in the discussion in the previous chapter we shall be able to consider (1.1) and its solutions in finite extensions of the field of rational numbers. In this paragraph we assume that the exponent m is fixed. This makes the equation close to (4.1) of Ch. VI; to analyse it we apply the arguments similar to those given in §4 Ch. VI. Those methods for analysing (1.1) were based on reducing it to a generalised Thue equation (over a finite extension of the field of rational numbers). By this route Baker [16] found the first effective bound for its solutions. He proved that all solutions satisfy the inequality

$$\max(|x|, |y|) < \exp\exp\{(5m)^{10}(n^{10n}H)^{n^2}\},$$

where n and H are respectively the degree and the height of the polynomial $f(x)$. The next theorem gives an essential improvement to this bound and is similar to the Theorem 4.1 Ch.VI (cf. [219]).

Theorem 1.1 *Suppose $m \geq 3$ is a natural number and that $f(x)$ is an integral polynomial of degree $n \geq 3$, with height at most $H > e$, with leading coefficient*

1, and having two simple roots α, α_1. Let f_α, f_{α_1} be the minimal polynomials and D_α, D_{α_1} the discriminants of α and α_1 respectively. Denote by R_α, R_{α_1} the resultants of the polynomials f' and f_α and of f' and f_{α_1} respectively. Let $\mathbb{K}$, $\mathbb{K}_1$, and $\mathbb{F}$ denote the fields of algebraic numbers $\mathbb{Q}(\alpha)$, $\mathbb{Q}(\alpha_1)$, and $\mathbb{Q}(\alpha, \alpha_1)$ respectively, and set

$$d = [\mathbb{F} : \mathbb{K}], \quad d_1 = [\mathbb{F} : \mathbb{K}_1], \quad s = [\mathbb{K} : \mathbb{Q}], \quad s_1 = [\mathbb{K}_1 : \mathbb{Q}].$$

Then all integer solutions of the equation (1.1) satisfy

$$\max(|x|, |y|) < \exp\{c_1|A|^{m(m-1)\varphi(m)(d+d_1)+\varepsilon}(B \ln H)^{1+\varepsilon}\}, \qquad (1.2)$$

where

$$B = |D_\alpha D_{\alpha_1}|^{ss_1 m^3(m-1)\varphi(m)/2} |D_\alpha^d D_{\alpha_1}^{d_1}|^{m(m-1)\varphi(m)} |R_\alpha^d R_{\alpha_1}^{d_1}|^{m(m-1)^2\varphi(m)}.$$

Here c_1 is effectively expressible in terms of n, m and ε; $\varphi(m)$ is the Euler function, and $\varepsilon > 0$ is arbitrary.

Proof. We assume that A does not contain prime divisors with exponents greater than $m - 1$, as we may without loss of generality by separating any mth power from A and joining it to y^m, noting that this does not weaken inequality (1.2).

Suppose the rational integers x, y satisfy (1.1). In the field $\mathbb{K}$ the ideal $(x - \alpha)$ may be represented in the form $(x - \alpha) = \mathfrak{a}\mathfrak{b}^m$ where $\mathfrak{a}$ and $\mathfrak{b}$ are integral ideals, and $\mathfrak{a}$ is free of mth powers. Since

$$f(x) = (x - \alpha)(f'(\alpha) + \tfrac{1}{2}f''(\alpha)(x - \alpha) + \ldots),$$

then if $\mathfrak{p}$ is a prime ideal of the field $\mathbb{K}$ and $(f'(\alpha), \mathfrak{p}) = 1$, we have

$$\operatorname{ord}_\mathfrak{p} f(x) = \operatorname{ord}_\mathfrak{p}(x - \alpha).$$

Therefore

$$\operatorname{ord}_\mathfrak{p} A \equiv \operatorname{ord}_\mathfrak{p} \mathfrak{a} \pmod{m}$$

and, consequently, $\operatorname{ord}_\mathfrak{p} \mathfrak{a} \le \operatorname{ord}_\mathfrak{p} A$. If $\mathfrak{p} \mid f'(\alpha)$, then at any rate $\operatorname{ord}_\mathfrak{p} \mathfrak{a} \le m-1$. Hence

$$\mathfrak{a} \mid A(f'(\alpha))^{m-1} \qquad (1.3)$$

and since $f'(\alpha)$ divides R_α, then $\mathfrak{a} \mid AR_\alpha^{m-1}$. For the conjugate ideals $\mathfrak{a}^{(i)}$ we shall also have $\mathfrak{a}^{(i)} \mid AR_\alpha^{m-1}$. Consequently the least common multiple $[\mathfrak{a}(1), \ldots, \mathfrak{a}^{(s)}]$ divides AR_α^{m-1}.

Further, $(\mathfrak{a}^{(i)}, \mathfrak{a}^{(j)})$ divides $(x - \alpha^{(i)}, x - \alpha^{(j)})$, which in its turn divides $\alpha^{(i)} - \alpha^{(j)}$, and then

$$\prod_{1 \le i, j \le s} (\mathfrak{a}^{(i)}, \mathfrak{a}^{(j)}) \mid D_\alpha.$$

By Lemma 4.1 Ch.VI we find

$$\mathfrak{a}^{(1)} \ldots \mathfrak{a}^{(s)} \mid AR_\alpha^{m-1} D_\alpha.$$

Hence,

$$N(\mathfrak{a}) \leq |AR_\alpha^{m-1}D_\alpha|. \tag{1.4}$$

Let $\mathfrak{b}'$ be an integral ideals of the field $\mathbb{K}$, lying in the class opposite to the class of the ideal $\mathfrak{b}$ and having norm not exceeding $|D_\mathbb{K}|^{1/2}$. Then $\mathfrak{b}\mathfrak{b}'$ is an integral principal ideal (ξ'), and $x - \alpha = \gamma'(\xi')^m$, where $(\gamma') = \mathfrak{a}(\mathfrak{b}')^{-m}$, with $g\gamma'$ an integer, $g = \mathrm{Nm}(\mathfrak{b}')^m \leq |D_\mathbb{K}|^{m/2}$. In view of (1.4) we find

$$|\mathrm{Nm}(\gamma')| \leq N(\mathfrak{a}) \leq |AR_\alpha^{m-1}D_\alpha|. \tag{1.5}$$

By Lemma 2.2 Ch.II applied to the field $\mathbb{K}$ and the number $\gamma \neq 0$, there is a unit θ in $\mathbb{K}$ satisfying

$$\overline{|\gamma'\theta|} < |\mathrm{Nm}(\gamma')|^{1/s}e^{c_2 R_\mathbb{K}}, \tag{1.6}$$

where c_2 depends only on s (hence, c_2 may be estimated in terms of n; in what follows, $c_3, c_4, \ldots$ are similar quantities). It is possible to represent θ in the form $\theta = \theta_1\theta_2^m$, where $\theta_1, \theta_2 \in U_\mathbb{K}$, and θ_1 is made up of basic units of the group $U_\mathbb{K}$ in integral non-negative powers not exceeding $m - 1$. Setting $\gamma = \gamma'\theta\theta_1^{-1}$, and $\xi = \xi'\theta_2^{-1}$ we find

$$x - \alpha = \gamma\xi^m, \tag{1.7}$$

where ξ and $g\gamma$ are integers, and then it follows from (1.6) that we have

$$\overline{|\gamma|} \leq \overline{|\gamma'\theta|}\,\overline{|\theta_1^{-1}|} < |\mathrm{Nm}(\gamma')^{1/s}|e^{c_3 R_\mathbb{K}}$$

(to estimate $\overline{|\theta_1^{-1}|}$ we use the inequality (2.1) Ch.II and Lemma 1.1 Ch.II). Consequently, it follows from (1.5) that

$$\overline{|\gamma|} < |AR_\alpha^{m-1}D_\alpha|^{1/s}e^{c_3 R_\mathbb{K}} \tag{1.8}$$

Since

$$|R_\alpha| < c_4 H^{s+n-1}, \quad |D_\alpha| < c_5 H^{2s-2}$$

it follows from (1.8) that

$$\overline{|\gamma|} < c_6|AH^{(m-1)(s+h-1)+2s-2}|^{1/s}e^{c_3 R_\mathbb{K}}. \tag{1.9}$$

Similarly for a different root α_1 of the polynomial,

$$x - \alpha_1 = \gamma_1\xi_1^m, \tag{1.10}$$

where γ_1, ξ_1 lie in the field $\mathbb{K}_1 = \mathbb{Q}(\alpha_1)$, and $\xi_1, g_1\gamma_1$ are integers; g_1 is a natural number not exceeding $|D_{\mathbb{K}_1}|^{m/2}$ and for γ_1 we have an inequality of the type (1.9), in which s is replaced by s_1 and $R_\mathbb{K}$ by $R_{\mathbb{K}_1}$.

From (1.7), (1.10) we obtain

$$\alpha_1 - \alpha = \gamma\xi^m - \gamma_1\xi_1^m \tag{1.11}$$

which may be considered as a generalised Thue equation in unknown integers ξ, ξ_1 lying in the fixed fields $\mathbb{K}$ and $\mathbb{K}_1$. To estimate $\max(\overline{|\xi|}, \xi_1)$ one can apply

Theorem 6.1 Ch.IV, but we shall go another route and use the ideas described in §4 of Ch.VI.

Set

$$\mathbb{L} = \mathbb{Q}(\alpha, \alpha_1, \sqrt[m]{\gamma}, \sqrt[m]{\gamma_1}, \zeta),$$

where ζ is a primitive mth root of 1. Equation (1.11) shows that the numbers

$$\beta_i = \xi \sqrt[m]{\gamma} - \zeta_i \xi_1 \sqrt[m]{\gamma_1} \quad (i = 1, 2, \ldots, m; \quad \zeta_i = \zeta^i) \tag{1.12}$$

lie in the field $\mathbb{L}$ and have absolute norms of absolute value not exceeding $c_7 H^{n^2 m \varphi(m)}$. The numbers β_i are integers, since $\gamma \xi^m$, $\gamma_1 \xi_1^m$ and, hence, $\xi \sqrt[m]{\gamma}$, $\xi_1 \sqrt[m]{\gamma_1}$ are integers. By Lemma 2.2 Ch.II there are such units $\varepsilon_1, \varepsilon_2, \varepsilon_3$ in the group of units $U_\mathbb{L}$ so that the numbers λ_i defined by $\beta_i = \varepsilon_i \lambda_i$ $(i = 1, 2, 3)$ are subject to the restriction

$$\max(\lceil \lambda_1 \rceil, \lceil \lambda_2 \rceil, \lceil \lambda_3 \rceil) < (c_7 H^{n^2 m \varphi(m)})^{1/l} e^{c_8 R} \tag{1.13}$$

where $R = R_\mathbb{L}$, $l = [\mathbb{L} : \mathbb{Q}]$, and c_8 is given in terms of l.

Eliminating in (1.12) the values $\xi \sqrt[m]{\gamma}$, $\xi_1 \sqrt[m]{\gamma_1}$ we obtain, in particular,

$$(\zeta_3 - \zeta_2)\lambda_1 \delta_1 + (\zeta_1 - \zeta_3)\lambda_2 \delta_2 = (\zeta_1 - \zeta_2)\lambda_3, \tag{1.14}$$

where $\delta_1 = \varepsilon_1 \varepsilon_3^{-1}$, $\delta_2 = \varepsilon_2 \varepsilon_3^{-1}$ are units of the group $U_\mathbb{L}$. The relation (1.14) may be considered as an equation of the form (3.3) Ch.VI with respect to unknown units $\delta_1, \delta_2 \in U_\mathbb{L}$. Applying Lemma 3.1 Ch.VI, we find from (1.13), (1.14) that

$$\max(\lceil \delta_1 \rceil, \lceil \delta_2 \rceil) < \exp\{c_9 [R(R + \ln H)]^{1+\varepsilon}\}, \tag{1.15}$$

if one takes into account that the heights of the numbers

$$(\zeta_3 - \zeta_2)\lambda_1, \quad (\zeta_1 - \zeta_3)\lambda_2, \quad (\zeta_1 - \zeta_2)\lambda_3$$

are bounded by

$$T = \exp\{c_{10}(R + \ln H)\}.$$

In the inequality (1.15), $\varepsilon > 0$ is arbitrary, and c_9 is effectively computable in terms of l and ε.

Set $\mu_1 = \delta_1 \lambda_1$, $\mu_2 = \delta_2 \lambda_2$. Then from (1.13), (1.15) an estimate of the form (1.15) follows for $\max(\lceil \mu_1 \rceil, \lceil \mu_2 \rceil)$ with another value for c_9. Returning to (1.12) we find

$$\xi_1 = \frac{(\mu_2 - \mu_1)\varepsilon_3}{(\zeta_1 - \zeta_2)\sqrt[m]{\gamma_1}}, \quad \xi = \frac{(\mu_1 \zeta_1^{-1} - \mu_2 \zeta_2^{-1})\varepsilon_3}{(\zeta_1^{-1} - \zeta_2^{-1})\sqrt[m]{\gamma}}. \tag{1.16}$$

Now (1.11) gives

$$\alpha_1 - \alpha = \left(\left(\frac{\mu_1 \zeta_1^{-1} - \mu_2 \zeta_2^{-1}}{\zeta_1^{-1} - \zeta_2^{-1}} \right)^m - \left(\frac{\mu_2 - \mu_1}{\zeta_1 - \zeta_2} \right)^m \right) \varepsilon_3^m$$

Since $\alpha \neq \alpha_1$ we obtain an equation for ε_3 which gives an estimate for $\overline{|\varepsilon|}$ of the type (1.15) with a different value c_9. From (1.16) and the estimate (1.9) we obtain that

$$|\xi| < |A|^{c_{11}} \exp\{c_{12}[R(R + \ln H)]^{1+\varepsilon}\}, \tag{1.17}$$

since by Lemma 2.3 Ch.II the regulator $R_{\mathbb{K}}$ is estimated by $c_{13}R$; here $c_{11} = c_{11}(n,m)$, $c_{12} = c_{12}(n,m,\varepsilon)$, $c_{13} = c_{13}(l)$, and $\varepsilon > 0$ is arbitrary. It is obvious that (1.7) and inequality (1.9) give an estimate of the type (1.17) for $|x|$.

To complete the proof of the theorem it remains to bound the regulator R. By Lemma 2.4 Ch.II it is enough to estimate the magnitude of the discriminant $D = D_{\mathbb{L}}$.

We consider the field $\mathbb{L}$ as a field obtained from $\mathbb{Q}$ by the successive adjunction of the integers α, α_1, $g \sqrt[m]{\gamma}$, $g_1 \sqrt[m]{\gamma_1}$ and ζ.

We estimate the discriminant as the norm of the different of the field $\mathbb{L}$ which, in virtue of the multiplicative property of differents in relative extensions, will occur as a divisor of the product Δ of the differents of the successively adjoined numbers:

$$\Delta = f'_\alpha(\alpha) f'_{\alpha_1}(\alpha_1) m(g \sqrt[m]{\gamma})^{m-1} m(g_1 \sqrt[m]{\gamma_1})^{m-1} P'_m(\zeta)$$

where P_m is the mth cyclotomic polynomial. We find that

$$\mathrm{Nm}_{\mathbb{L}/\mathbb{Q}}(\Delta) = D_\alpha^{l/s} D_{\alpha_1}^{l/s_1} m^{2l} (gg_1)^{l(m-1)} \times$$
$$\times (\mathrm{Nm}_{\mathbb{K}/\mathbb{Q}}(\gamma))^{(m-1)l/st} (\mathrm{Nm}_{\mathbb{K}_1/\mathbb{Q}}(\gamma_1))^{(m-1)l/s_1 t_1} D_\zeta^{l/\varphi(m)},$$

where $t = [\mathbb{K}(\sqrt[m]{\gamma}) : \mathbb{K}]$, $t_1 = [\mathbb{K}_1(\sqrt[m]{\gamma_1}) : \mathbb{K}_1]$. Since

$$g \leq |D_{\mathbb{K}}|^{m/2}, \qquad g_1 \leq |D_{\mathbb{K}_1}|^{m/2},$$

and $D_{\mathbb{K}}, D_{\mathbb{K}_1}$ are the divisors of D_α, D_{α_1} respectively, then

$$g \leq |D_\alpha|^{m/2}, \qquad g_1 \leq |D_{\alpha_1}|^{m/2}.$$

Inequality (1.5) shows that

$$|\mathrm{Nm}_{\mathbb{K}/\mathbb{Q}}(\gamma)| \leq |AR_\alpha^{m-1} D_\alpha|$$

and a similar inequality holds for the norm of γ_1. Therefore we have

$$\mathrm{Nm}_{\mathbb{L}/\mathbb{Q}}(\Delta) <$$
$$c_{14}|A|^{(l/st + l/s_1 t_1)(m-1)} |D_\alpha|^{l/s + lm(m-1)/2 + (m-1)l/st} \times$$
$$\times |R_\alpha|^{l(m-1)^2/st} |D_{\alpha_1}|^{l/s_1 + lm(m-1)/2 + (m-1)l/s_1 t_1} |R_{\alpha_1}|^{l(m-1)^2/s_1 t_1}.$$

Since $l/s \leq dm^2\varphi(m)$, $l/st \leq dm\varphi(m)$, and similar inequalities hold if one replaces s, t by s_1, t_1 then the inequality

$$|D_{\mathbb{L}}| \leq |\mathrm{Nm}_{\mathbb{L}/\mathbb{Q}}(\Delta)|$$

and Lemma 2.4 Ch.II give

$$R < c_{15}|A|^{m(m-1)\varphi(m)(d+d_1)/2}B_1(\ln|A| + \ln B_1)^{t-1},$$

where we write

$$B_1 = |D_\alpha D_{\alpha_1}|^{ss_1 m^3(m-1)\varphi(m)/4}|D_\alpha^d D_{\alpha_1}^{d_1}|^{m(m-1)\varphi(m)/2}|R_\alpha^d R_{\alpha_1}^{d_1}|^{m(m-1)^2\varphi(m)/2}.$$

Since an estimate of the form (1.17) holds for $|x|$ we obtain that

$$|x| < \exp\{c_{16}|A|^{m(m-1)\varphi(m)(d+d_1)+\varepsilon}B_1^{2+\varepsilon}(\ln H)^{1+\varepsilon}\},$$

where $c_{16} = c_{16}(n, m, \varepsilon)$, and $\varepsilon > 0$ is arbitrary. This coincides with the inequality (1.2) for $|x|$. The inequality for $|y|$ follows now directly from (1.1), and the proof of the theorem is completed.

Given the conditions of Theorem 1.1 and that the polynomial $f(x)$ has no multiple root, the magnitude of B may be replaced by a simpler expression

$$B' = |D_f|^{m^2(m-1)\varphi(m)(ss_1 m+d+d_1)},$$

where D_f is the discriminant of the polynomial $f(x)$. This follows from the relations $D_\alpha \mid D_f, R_\alpha \mid D_f$ and similar relations for $D_{\alpha_1}, R_{\alpha_1}$ (see §4 Ch.VI). It is also possible to give an estimate for B in terms of H. In any case

$$B < c_{17}H^{2(n-1)m^2(m-1)\varphi(m)(ss_1 m+d+d_1)}, \qquad c_{17} = c_{17}(n, m).$$

It is obvious that the transition from (1.2) to such estimates is a considerable roughing and in concrete cases it is expedient to use the suppositions of the theorem fully.

We note that a slight change in the described arguments allows one to obtain extra information on the solutions of (1.1): any x satisfying (1.1) may be represented in the form $x = \alpha + \mu$ with

$$|\mu| < \exp\{c_{18}|A|^{m(m-1)\varphi(m)(d+d_1)+\varepsilon}B^{1+\varepsilon}\}, \tag{1.18}$$

where $c_{18} = c_{18}(m, n, \varepsilon)$ and $\varepsilon > 0$ is arbitrary. This is obtained because of the fact that the absolute norms of the numbers β_i defined by the equalities (1.12), are the divisors of the norm $\text{Nm}_{\mathbb{L}/\mathbb{Q}}(\alpha - \alpha_1)$ which in its turn is a divisor of the norm $\text{Nm}_{\mathbb{L}/\mathbb{Q}}(f'(\alpha))$. Consequently, it divides $R_\alpha^{l/s}$. The inequality (1.13) now may be replaced by the inequality

$$\max(house\lambda_1, house\lambda_2, house\lambda_3) < |R_\alpha|^{1/s}e^{c_{19}R}$$

which leads to the bound (1.15) with $\ln B$ instead of $\ln H$. An analogous change may be applied to (1.17), and then (1.7) shows that $x = \alpha + \mu$ with μ satisfying (1.18).

This remark shows, in particular, that with fixed discriminant $D_f \neq 0$ the magnitude of the solutions depends only trivially on the height of the polynomial. It is apparent that this phenomenon concerns also (4.1) Ch. VI (all this due to Trelina).

It is easy to understand that the described methods may be applied to the case of (1.1) over a finite extension of the field of rational numbers, as well as to the case of solutions of this equation in S-integers. The ideas described here and in §7 Ch.VI, and an analogue of Lemma 6.3 allowed Trelina [239] to obtain the following result.

Theorem 1.2 *Let $f(x)$ be a polynomial of degree n with coefficients in I_L and with leading coefficient a. Suppose that f has two simple roots α_j $(j = 1, 2)$, with respective minimal polynomial f_j and discriminant D_j. Further, set $l_i = [L(\alpha_j) : L]$ $(j = 1, 2)$. Then all solutions x, y of (1.1), in which $A \in I_L$ and $x, y \in I_L(S)$ have $\max(h(x), h(y))$ bounded above by*

$$H_0^{c_{20}} \exp\{(c_{21}(l + s))^{50(l+s)M} |\mathrm{Nm}(A)|^{6M^2(\frac{1}{l_1}+\frac{1}{l_2})(l+s)} BP^{2Ml}\},$$

where H_0 is the height of the number α_1; $M = l_1 l_2 m(m - 1)\varphi(m)$, and

$$B = \left| D_{\mathbb{K}}^{(m+3)/2}(D_1 D_2)^{2+ml/2} \mathrm{Nm}(A^{n_1}(R_1 R_2)^{m-1})^{\frac{1}{l_1}+\frac{1}{l_2}} \right|^{6M^2(l+s)}.$$

Here $n_1 = \min(n, m - 1)$, the R_j are the resultants of the polynomials $a^{n-1}f(x/a)$ and f_j $(j = 1, 2)$, and P is the maximal prime number which is divisible by a prime ideal from S. The constant c_{20} is effectively determined by l and n, and c_{21} by l, n and m.

The theorem yields a lower bound for P as an iterated logarithm of the maximum of the heights of the numbers x, y satisfying (1.1). There is a similar bound for the maximum of the norms of prime ideals occurring in the denominators of x, y when the point (x, y) lies on the curve (1.1) and $x, v \in L$. Apart from that, it follows from the theorem that the maximal norm of prime ideals of the ring I_L, occurring in the number $ax^n + by^m$ with $a \neq 0$, $b \neq 0$, when x and y are numbers in I_L and n and m are natural numbers with $n \geq 2$, $m \geq 3$, is bounded below by a quantity of the form

$$(\ln \ln X \cdot \ln \ln \ln X)^{1/2}, \qquad X = \max(h(x), h(y)) > X_0.$$

This was first proved by Kotov [112], by reducing the problem to the analysis of a generalised Thue-Mahler equation.

2. Equations with indefinite exponent

We now turn to equation (1.1) in which the exponent m is not fixed and, hence, is to be considered as a new unknown number. This problem was first considered by Schinzel and Tijdeman [172]. They proved that (1.1) with $A = 1$ and with unknown numbers x, y, where $|y| > 1$ and $m \geq 3$, has only finitely many solutions and all of them may in principle be determined in terms of the coefficients and the degree of $f(x)$. Assuming that the polynomial $f(x)$ is fixed, we obtain a bound for the exponent m as a function of A. It is obvious that Theorem 1.1 then implies a bound for x and y as well. Our considerations differ from those of Schinzel and Tijdeman not just in technical details, even though the principal argument remains the same.

Theorem 2.1 *Let $f(x)$ be an integral polynomial with at least two simple roots. Then the equation (1.1) can have an integer solution x, y with $|y| > 1$, only if m satisfies the inequality*

$$m < c_{22}(\ln(|A| + 1))^{1+\varepsilon}, \tag{2.1}$$

where c_{22} is effectively determined in terms of ε, and the coefficients and degree of the polynomial $f(x)$; $\varepsilon > 0$ is arbitrary.

Proof. It is apparent that we can assume that the leading coefficient of $f(x)$ is 1. It is possible to also suppose that the exponents of the powers of prime divisors of A are less than m, if $A \neq \pm 1$. Indeed, such exponents do not exceed $\ln|A| / \ln 2$, and if m is less than one of them, we obtain immediately a stronger inequality than (2.1). Therefore, we suppose that A is free of mth powers.

We follow the arguments commencing the proof of Theorem 1.1, and assume the notation introduced there. We find that $x - \alpha = \gamma(\xi')^m$, where $(\gamma) = \mathfrak{a}(\mathfrak{p}')^{-m}$, and ξ' is an integer, whilst in view of (1.3) the ideal $\mathfrak{a}$ divides $A(f'(\alpha))^{m-1}$.

Suppose $\mathfrak{p}_1, \ldots, \mathfrak{p}_s$ are the prime ideals of the field occurring in $f'(\alpha)$, say

$$f'(\alpha) = \mathfrak{p}_1^{a_1} \ldots \mathfrak{p}_s^{a_s}.$$

Then the ideal $\mathfrak{a}$ can be represented in the form

$$\mathfrak{a} = \mathfrak{a}_0 \mathfrak{p}_1^{z_1} \ldots \mathfrak{p}_s^{z_s}, \quad (\mathfrak{a}, \mathfrak{p}_1 \ldots \mathfrak{p}_s) = 1, \quad \mathfrak{a}_0 \mid A,$$

where all the exponents z_i lie in the intervals

$$0 \leq z_i \leq a_i(m - 1) + b_i, \quad b_i = \mathrm{ord}_{\mathfrak{p}_i} A \quad (1 \leq i \leq s). \tag{2.2}$$

Set $z_i = u_i h + z_i'$, where $0 \leq z_i' < h$ $(1 \leq i \leq s)$, and $h = g_{\mathbb{K}}$ is the class-number of $\mathbb{K}$. Then $\mathfrak{p}_i^h = (\pi_i)$ are principal ideals,

$$\mathfrak{a} = \mathfrak{a}_1 \pi_1^{u_1} \ldots \pi_s^{u_s}, \quad \mathfrak{a}_1 \mid A(f'(\alpha))^{h-1}, \tag{2.3}$$

where the exponents u_i satisfy

$$0 \leq u_i \leq a(m - 1) + \ln(|A| + 1)/\ln 2, \quad a = \max_{(i)} a_i. \tag{2.4}$$

Further, $(\mathfrak{p}')^h = (\beta)$ is a principal ideal, and setting $m = m_1 h + m_0$ for $0 \leq m_0 < h$, we find from (2.3) that

$$(\gamma) = \mathfrak{a}_1(\mathfrak{p}')^{-m_0} \beta^{-m_1} \pi_1^{u_1} \ldots \pi_s^{u_s}.$$

Consequently $\mathfrak{a}_1(\mathfrak{p}')^{-m_0} = (\beta_0)$ is a principal ideal,

$$\gamma = \beta_0 \eta \beta^{-m_1} \pi_1^{u_1} \ldots \pi_s^{u_s}, \tag{2.5}$$

where η is a unit of $\mathbb{K}$, $g_0\beta_0$ is an integer — $(g_0 = N(\mathfrak{p}')^{m_0})$ — and by Lemma 2.2 Ch.II, β_0 may be taken in such a way that

$$\overline{|\beta_0|} < c_{23}|A|, \tag{2.6}$$

if one takes into account that $\mathfrak{a}_1$ divides $A(f'(\alpha))^{h-1}$ (c_{23} depends on the field $\mathbb{K}$ and on $f(x)$ only). Thus, we find that

$$x - \alpha = \beta_0 \eta \beta^{-m_1} \pi_1^{u_1} \dots \pi_s^{u_s} (\xi')^m. \tag{2.7}$$

On representing the unit η in terms of the basis $\eta_1, \dots, \eta_k$ of the group of units in $\mathbb{K}$, we obtain

$$\eta = \zeta \eta_1^{v_1} \dots \eta_k^{v_k} (\eta')^m, \qquad 0 \le v_i < m \quad (1 \le i \le k), \tag{2.8}$$

where ζ is a root of unity and η' is a unit in $\mathbb{K}$. On replacing β_0 by $\zeta\beta_0$, which does not change the estimate (2.6), we find from (2.7), setting $\xi'\eta' = \xi$,

$$x \div \alpha = \beta_0 \eta_1^{v_1} \dots \eta_k^{v_k} \pi_1^{u_1} \dots \pi_s^{u_s} \beta^{-m_1} \xi^m.$$

Similarly, in the field $\mathbb{K}_1$, we obtain

$$x - \alpha_1 = \beta_{10} \eta_{11}^{v_{11}} \dots \eta_{1k_1}^{v_{1k_1}} \pi_{11}^{u_{11}} \dots \pi_{1s}^{u_{1s}} \beta_1^{-m_{11}} \xi_1^m.$$

We find from these equalities that

$$\alpha_1 - \alpha = \beta_0 \eta_1^{v_1} \dots \eta_k^{v_k} \pi_1^{u_1} \dots \pi_s^{u_s} \beta^{-m_1} \xi^m - $$
$$- \beta_{10} \eta_{11}^{v_{11}} \dots \eta_{1k_1}^{v_{1k_1}} \pi_{11}^{u_{11}} \dots \pi_{1s}^{u_{1s}} \beta_1^{-m_{11}} \xi_1^m. \tag{2.9}$$

In view of the obvious symmetry in that last equality we may assume that $\overline{|\xi|} \ge \overline{|\xi_1|}$. Without changing notation, we also assume that $|\xi| = \overline{|\xi|}$, since we may pass in (2.9) to the conjugate equalities. Now from (2.9) we obtain:

$$|\alpha_1 - \alpha| |\beta_0^{-1} \eta_1^{-v_1} \dots \eta_k^{-v_k} \pi_1^{-u_1} \dots \pi_s^{-u_s} \beta^{m_1}| \, \overline{|\xi|}^{-m} = $$
$$= \left| 1 - \frac{\beta_{10}}{\beta_0} \eta_1^{-v_1} \dots \eta_k^{-v_k} \eta_{11}^{v_{11}} \dots \eta_{1k_1}^{v_{1k_1}} \pi_1^{-u_1} \dots \right.$$
$$\left. \dots \pi_s^{-u_s} \pi_{11}^{u_{11}} \dots \pi_{1s}^{u_{1s}} \beta^{m_1} \beta_1^{-m_{11}} \left(\frac{\xi_1}{\xi} \right)^m \right|. \tag{2.10}$$

We apply Lemma 8.2 Ch.III to estimate the right-hand side of this equality from below. Recalling the inequalities (2.4), (2.6), (2.8) and similar inequalities for the u_{1i}, β_{10}, and the v_{1i} we find that the right-hand side of (2.10) is not less than

$$\exp\left\{ -c_{24} \ln \overline{|\xi|} \ln m [\ln(|A| + 1)]^{1+\varepsilon} \right\}, \tag{2.11}$$

where c_{24} is determined in terms of ε, the coefficients and the degree of the polynomial $f(x)$; here $\varepsilon > 0$ is arbitrary. Since the left-hand side of (2.10) is bounded from above by the value

$$|A|^{c_{25}} c_{26}^m \overline{|\xi|}^{-m}, \tag{2.12}$$

where c_{25} and c_{26} are determined in terms of the coefficients and the degree of $f(x)$, we obtain from the comparison of (2.11) and (2.12) that

$$|\xi|^m < |A|^{c_{25}} c_{26}^m |\xi|^{c_{24} \ln m [\ln(|A|+1)]^{1+\varepsilon}}. \tag{2.13}$$

We may assume that $m \geq \ln|A| + 1$, since otherwise we have more than (2.1). Hence, we obtain from (2.13) that

$$|\xi| < e^{c_{25}} c_{26} |\xi|^{c_{24} \frac{\ln m}{m} [\ln(|A|+1)]^{1+\varepsilon}}. \tag{2.14}$$

Suppose first that

$$|\xi|^{1/2} > e^{c_{25}} c_{26}. \tag{2.15}$$

Then we find from (2.14) that

$$\frac{m}{\ln m} < 2c_{24}[\ln(|A|+1)]^{1+\varepsilon}.$$

If (2.15) does not hold, then we consider (2.9) as a relation of linear dependence of three S-units in the field $\mathbb{L}$, containing the fields $\mathbb{K}$ and $\mathbb{K}_1$, and for the set of those prime ideals of the ring $I_{\mathbb{L}}$ occurring in $\pi_1, \ldots, \pi_s, \beta$ and the ξ_j. We take the numbers β_0, β_{10} and $\alpha_1 - \alpha$ as the coefficients of this linear relation. Lemma 6.3 Ch.VI shows that the logarithm of the height of the number

$$\rho = \eta_1^{v_1} \ldots \eta_k^{v_k} \pi_1^{u_1} \ldots \pi_s^{u_s} \beta^{-m_1} \xi^m$$

is bounded from above by the value $c_{27}[\ln(|A|+1)]^{1+\varepsilon}$, where c_{27} is determined by $f(x)$ and ε, with $\varepsilon > 0$ arbitrary. If we have $\ln h(\rho) > m/\ln m$, then we obtain (2.1). If the opposite inequality holds, then we find from the equality $x - \alpha = \beta_0 \rho$ that

$$|x| < c_{28}(|A|e^{m/\ln m})^{c_{29}},$$

and the condition $|y| > 1$ shows now that $m \leq c_{30}(\ln|A| + 1)$. By this the proof of Theorem 2.1 is completed.

On studying the proof of Theorem 1.1 it is easy to ascertain that the magnitude of c_1 as a function of m does not grow more rapidly than m^{4m^3}. Then (1.2) shows that for large $|x|$ we have

$$\ln\ln|x| < \frac{ss_1}{2} m^5 \ln|D_\alpha D_{\alpha_1}| + c_{31} m^4 (1 + \ln|A|),$$

where c_{31} depends only on $f(x)$. By Theorem 2.1 we can estimate m in terms of A, which gives

$$\ln\ln|x| < c_{32}[\ln(|A|+1)]^{5+\varepsilon}. \tag{2.16}$$

This inequality leads to an interesting arithmetic statement.

Let m, N be natural numbers. Denote by $A_m[N]$ the m-free part of N, that is, represent N in the form $N = N_1 N_2^m$, where all the prime divisors of N_1 occur with exponents less than m, and set $A_m[N] = N_1$. Then it follows from Theorem 1.1 that for $m \geq 3$ and an integer x with $x \neq 0$, $f(x) \neq 0$ we have

$$A_m[f(x)] > c_{33}(\ln|x|)^{1/m(m-1)\varphi(m)(d+d_1)+\varepsilon},$$

where c_{33} is effectively determined in terms of $f(x)$ and m and $\varepsilon > 0$. We see that with any fixed m the magnitude of $A_m[f(x)]$ grows unboundedly with the growth of $|x|$, but the estimate of this growth is the worse the bigger m is. Meanwhile, the inequality (2.16) shows that the magnitude

$$A_{\min}[f(x)] = \min_{m \geq 3} A_m[f(x)] \quad (f(x) \neq 0)$$

grows unboundedly with $|x|$:

$$A_{\min}[f(x)] > \exp\{c_{34}(\ln\ln|x|)^{1/5-\varepsilon}\},$$

where c_{34} is determined by $f(x)$ and ε. Thus, independently of m we have a guaranteed estimate for the speed of growth of $A_m[f(x)]$ (cf. also [236]).

3. The Catalan Equation

In 1844 Catalan [44] conjectured that the only solution of the equation

$$x^u - y^v = 1; \tag{3.1}$$

where all the unknowns x, y, u, v are integers not less than 1, is given by $3^2 - 2^3 = 1$. In 1953 Cassels [38], independently, put forward a weaker conjecture that the equation (3.1) has only a finite number of solutions.

Over many years the equation (3.1) attracted the attention of mathematicians who considered it for small values of u or v and either with fixed x, y and unknown u, v or with fixed u, v and unknown x, y (see [145] for a review). Recently Tijdeman [232] obtained the principal solution to this problem by proving the existence of a computable bound c_{35} to the solutions of the equation (3.1):

$$\max(x, y, u, v) < c_{35}.$$

In particular, the truth of Cassels' conjecture follows.

Following Van der Poorten [157] we discuss here the solution of (3.1) in S-integers, which is the same as to analyse the following problem:

Let S be a fixed set of prime numbers $p_1, \ldots, p_s$. We consider the equation

$$x^u - y^v = (p_1^{w_1} \ldots p_s^{w_s})^{[u,v]}, \quad (x, y) = 1, \tag{3.2}$$

in which $x > 1$, $y > 1$, $u > 1$, $w_1 \geq 0, \ldots, w_s \geq 0$ are unknown integers (excluding $u = v = 2$), and $[u, v]$ is the least common multiple of the numbers u and v. It is obvious that (3.2) implies the Catalan equation (3.1).

Theorem 3.1 *All the solutions to the equation (3.2) satisfy*

$$\max(x, y, u, v, w_1, \ldots, w_s) < c_{36} \tag{3.3}$$

where c_{36} is effectively determined by S.

Proof. We shall use Lemmas 8.2 and 8.3 Ch.III with the intention of first bounding the exponents u, v in (3.2).

We start with the case of $u = v$ where the desired result follows quickly. It is obvious, that if (3.2) has a solution, then not all of the numbers w_j are equal to zero, and we can assume that all of them are not zero without change in notation. Let p be a prime from the set S. By Lemma 8.3 Ch.III we find that

$$|x^u - y^u|_p = |x^u y^{-u} - 1|_p > e^{-c_{37}(p) \ln x (\ln u)^2}.$$

Turning to (3.2) we find

$$x^u - y^u = \prod_{p \in S} |x^u - y^u|_{p^{-1}} < x^{c_{38}(\ln u)^2},$$

where c_{38} is the sum of all the values $c_{37}(p)$ over the $p \in S$. It is apparent, that

$$x^u - y^u \geq x^u - (x - 1)^u > (x - 1)^{u-1},$$

so we find $u < c_{39}(\ln u)^2$, from which it follows that $u < c_{40}$ where the magnitude of c_{40} may be determined in explicit form in terms of the prime numbers of S. Now we come to the equation of Mahler (see §3 Ch.V):

$$x^u - y^u = p_1^{z_1} \ldots p_s^{z_s}, \qquad (x, y) = 1$$

from which the bound for x and y is determined. This completes the discussion of the case $u = v$ in (3.2).

Yet one more simple case corresponds to $v = 2$. Here it is enough to consider the equation

$$x^u = y^2 + z^2, \quad z = p_1^{z_1} \ldots p_s^{z_s}, \quad (x, y) = 1.$$

In the Gaussian field $\mathbb{Q}(i)$ we find that the numbers $y + iz$ and $y - iz$ have maximal common divisor dividing 2. Hence,

$$z = \alpha \sigma^u + \beta \tau^u, \qquad (\sigma, \tau) = 1 \tag{3.4}$$

for fixed nonzero numbers α, β and unknowns σ, τ, $|\sigma| > 1$, $|\tau| > 1$ (all the numbers lie in the Gaussian field). We analyse this equation relying on the ideas which were applied above. In view of the symmetry of (3.4) we may assume that $|\sigma| \geq |\tau|$. Applying Lemma 8.2 Ch.III, we obtain

$$\frac{z}{|\alpha||\sigma|^u} = \left|1 - \left(-\frac{\beta}{\alpha}\right)\left(\frac{\tau}{\sigma}\right)^u\right| > e^{-c_{41} \ln |\sigma| \ln u}, \tag{3.5}$$

if one takes into account that the height of an integer in the Gaussian field coincides with its norm. In the case $z = 1$, from (3.5) it follows that $u < c_{42}$, which is just what we need. If $z > 1$ then for a prime divisor $\mathfrak{p} \mid z$ which does not occur in τ we obtain by Lemma 8.3 Ch.III that

$$|\alpha \sigma^u + \beta \tau^u|_p = |\beta|_p \left|\left(-\frac{\alpha}{\beta}\right)\left(\frac{\sigma}{\tau}\right)^u - 1\right|_p > |\beta|_p e^{-c_{42}(p) \ln |\sigma|(\ln u)^2}.$$

If $\mathfrak{p} \mid z$ and $\mathfrak{p} \mid \tau$ then $\mathfrak{p} \mid \alpha$. Consequently, applying the 'product formula' we obtain

$$z^2 = \mathrm{Nm}(z) \leq \prod_{\mathfrak{p}\mid z} |\alpha\sigma^u + \beta\tau^u|_{\mathfrak{p}}^{-2} < e^{c_{42} \ln |\sigma|(\ln u)^2} \prod_{\mathfrak{p}} |\alpha\beta|_{\mathfrak{p}}^{-2}.$$

Since the inequality (3.5) yields $z > |\alpha||\sigma|^{u-c_{41} \ln u}$, we again obtain $u < c_{44}$. Thus, we come to (3.4) with a bounded exponent u i.e. to the generalised equation of Mahler which was discussed in §4 Ch.V. We may suppose from now on that u is odd.

Now we proceed with the main case, that is $u \neq v$, $v \geq 3$. We argue as before, but now a direct application of Lemmas 8.2 and 8.3 Ch.III is not sufficient and first of all we derive a few auxiliary facts on the arithmetic structure of the solutions of the equation (3.2).

Note that one may suppose that the numbers u, v are prime. Then we have $[u, v] = uv$. We set

$$z = p_1^{w_1} \dots p_s^{w_s}.$$

First of all we verify that

$$\left(y + z^u, \frac{y^v + z^{uv}}{y + z^u}\right) = 1 \text{ or } v. \tag{3.6}$$

If p is a prime number and p^α divides both $y + z^u$ and $(y^v + z^{uv})/(y + z^u)$, then

$$y \equiv -z^u \pmod{p^\alpha}$$
$$0 \equiv y^{v-1} - y^{v-2}z^u + \dots \equiv (-1)^v v z^{u(v-1)} \pmod{p^\alpha}$$

Since $(x, y) = 1$ then $(z, p) = 1$, and then p^α divides v, hence $\alpha = 1$, $p = v$, which proves (3.6).

Now from the equality

$$x^u = (y + z^u)\frac{y^v + z^{uv}}{y + z^u}$$

we conclude that $y + z^u$ is, up to a power of v, a uth power of a natural number. To determine the exponent to which v may occur in $y + z^u$, we consider the congruence

$$y^v + z^{uv} = ((y + z^u) - z^u)^v + z^{uv} \equiv$$
$$\equiv v(y + z^u)z^{u(v-1)} - \tfrac{1}{2}v(v - 1)(y + z^u)^2 z^{u(v-2)} \pmod{(y + z^u)^3}.$$

It follows that if $v \mid (y + z^u)$, then

$$\frac{y^v + z^{uv}}{y + z^u} \equiv v z^{u(v-1)} \pmod{v^2},$$

i.e. v may occur in $(y^v + z^{uv})/(y + z^u)$ with the exponent equal to 1. Therefore v^{u-1} divides $y + z^u$ and we find that y can be represented in the form

$$y = v_0 Y^u - z^u \tag{3.7}$$

where $v_0 = 1$ or v^{-1} and Y is a natural number.

Similarly one verifies that

$$x = u_0 X^v + z^v, \tag{3.8}$$

where u_0 or u^{-1} and X is a natural number.

Now the equality (3.2) takes the form

$$(u_0 X^v + z^v)^u - (v_0 Y^u - z^u)^v = z^{uv}, \tag{3.9}$$

and and it is to it that we apply the analytic ideas indicated above in special cases. From now on we assume that $v < u$ though there is no full symmetry with respect to u and v in the equality (3.9); nevertheless we shall be able to meet the case $u < v$ with similar arguments.

Let p be a prime occurring in z and set $w = \mathrm{ord}_p z$. Note that it follows from (3.7),(3.8) that if p coincides with u or v, then the corresponding number u_0 or v_0 is 1. From the equalities (3.2), (3.7), (3.8) we obtain correspondingly

$$|x^u y^{-v} - 1|_p = p^{-wuv}, \tag{3.10}$$

$$|yv_0^{-1}Y^{-u} - 1|_p = p^{-wu}, \tag{3.11}$$

$$|xu_0^{-1}X^{-v} - 1|_p = p^{-wv}. \tag{3.12}$$

Substituting (3.11) and (3.12) into (3.10) we obtain

$$|u_0^u v_0^{-v}(X/Y)^{uv} - 1| \leq \max(p^{-wv}, p^{-wu}, p^{-wuv}) = p^{-wv} \tag{3.13}$$

We set $\Lambda = u_0^u v_0^v (X/Y)^{uv} - 1$ and observe that $\Lambda \neq 0$. Indeed, otherwise we have $(x - z^v)^u = (y + z^u)^v$ so that $(x - z^v)^u > y^v + z^{uv}$, which is inconsistent with the main equality $x^u = y^v + z^{uv}$.

Suppose first that $X < Y$. Applying Lemma 8.3 Ch.III, we obtain

$$|\Lambda|_p > e^{-c_{45}(p)(\ln u)^4 \ln Y}, \tag{3.14}$$

where the value $c_{45}(p)$ depends only on $p \in S$. Comparing (3.13) and (3.14), we obtain

$$p^{wv} < e^{c_{45}(p)(\ln u)^4 \ln Y}$$

and taking the product of all such inequalities over all p dividing z, we find

$$z^v < Y^{c_{46}(\ln u)^4}. \tag{3.15}$$

From this inequality we shall obtain that

$$v < c_{47}(\ln u)^4, \tag{3.16}$$

but first we suppose that $v > 3c_{46}(\ln u)^4$. Then it follows from (3.15) that $z < Y^{1/3}$, which implied for $u > c_{48}$ that

$$y = v_0 Y^u - z^u > Y^{3u/4} - Y^{u/3} > Y^{2u/3}. \tag{3.17}$$

Now from the equation (3.9) we see that

$$(u_0 X^v + z^v)^u > Y^{2uv/3},$$

and then we have

$$u_0 X^v > Y^{2v/3} - z^v > Y^{v/2}. \tag{3.18}$$

In view of (3.15) and (3.18) we find from (3.8), (3.7):

$$|x/(u_0 X^v) - 1| = |z^v/(u_0 X^v)| < \tfrac{1}{2} u^{-1} Y^{c_{49}(\ln u)^4 - \frac{1}{2}v},$$

$$|y/(v_0 Y^u) - 1| = |z^u/(v_0 Y^u)| < \tfrac{1}{2} v^{-1} Y^{(c_{50}(\ln u)^4 - v)u/v}.$$

Similarly from (3.17) and the initial equation (3.2) we obtain

$$|x^u/y^v - 1| = |z^{uv}/y^v| < \frac{1}{2} Y^{(c_{51}(\ln u)^4 - \frac{1}{2}v)u}.$$

Consequently, passing to the logarithms, we obtain

$$|u \ln x - u \ln(u_0 X^v)| < Y^{c_{49}(\ln u)^4 - \frac{1}{2}v},$$

$$|v \ln y - v \ln(v_0 Y^u)| < Y^{(c_{50}(\ln u)^4 - v)u/v},$$

$$|u \ln x - v \ln y| < Y^{(c_{51}(\ln u)^4 - \frac{1}{2}v)u},$$

from which we find

$$|u \ln u_0 - v \ln v_0 + uv \ln(X/Y)| < Y^{-\frac{1}{2}(v - c_{52}(\ln u)^4)}.$$

Applying now Lemma 8.2 Ch.III, we obtain for the left-hand side of the last inequality a lower bound of the form

$$e^{-c_{53}(\ln u)^4 \ln Y} = Y^{-c_{53}(\ln u)^4}.$$

A comparison with the former estimate gives (3.16).

Our initial supposition that $X < Y$ has the following justification. It follows from (3.9) that

$$(u_0 X^v/z^v)^u \le (u_0 X^v/z^v + 1)^u - 1 = (v_0 Y^u/z^u - 1)^v < (v_0 Y^u/z^u)^v,$$

so at any rate we have

$$X \le u_0^{-1/v} v_0^{1/u} Y < 2Y,$$

if one assumes, as we did, that $v > 3c_{46}(\ln u)^4$ and $u > c_{48}$. Therefore, also in case of $X \ge Y$ we may apply the arguments described above with only inessential changes.

From the inequalities (3.10), (3.11) we obtain for a prime p dividing z:

$$|v_0^{-v}(x/Y^v)^u - 1|_p \le p^{-wu}. \tag{3.19}$$

Observe that $x < Y^v$. Because, from (3.2), (3.7) we obtain

$$(x/z^v)^u = (v_0 Y^u/z^u - 1)^v + 1 < v_0^v (Y^v/z^v)^u, \tag{3.20}$$

and since $v_0 \leq 1$, then $x < Y^v$. More than that, the left-hand side of (3.19) is not equal to zero, as the equality $x^u - v_0^v Y^{uv} = 0$ contradicts (3.20). We estimate the left-hand side of (3.19) from below using Lemma 8.3 Ch.III, which gives the bound

$$e^{-c_{54}(p)(\ln u)^3 v \ln Y} = Y^{-c_{54}(p)v(\ln u)^3}. \tag{3.21}$$

Comparing (3.19), (3.20) we obtain

$$p^{wu} < Y^{c_{54}(p)v(\ln u)^3},$$

and taking the product of all such inequalities over all p occurring in z, we obtain

$$z^u < Y^{c_{55}v(\ln u)^3},$$

where c_{55} is determined by S. Now taking into account (3.16), we find from (3.7) that $y = v_0 Y^u - z^u > Y^{u/2}$, so that we have

$$|x^u/y^v - 1| = |z^{uv}/y^v| < \frac{1}{2} Y^{(c_{56}v(\ln u)^3 - \frac{1}{2}u)v},$$

$$|y/(v_0 Y^u) - 1| = |z^u/(v_0 Y^u)| < \frac{1}{2} v^{-1} Y^{c_{57}v(\ln u)^3 - u}.$$

From these inequalities we find

$$|u \ln x - v \ln y| < Y^{(c_{56}v(\ln u)^3 - \frac{1}{2}u)v},$$

$$|v \ln y - v \ln(v_0 Y^u)| < Y^{c_{57}v(\ln u)^3 - u},$$

and it follows then that

$$|u \ln(x/Y^v) - v \ln v_0| < Y^{c_{58}v(\ln u)^3 - u}. \tag{3.22}$$

We estimate the left-hand side of this inequality by Lemma 8.2 Ch.III, which gives for it a lower bound of the form

$$e^{-c_{59}v(\ln u)^2 \ln Y} = Y^{-c_{59}v(\ln u)^2}.$$

Comparing this with the upper bound (3.22) we obtain

$$u < c_{60}v(\ln u)^3 < c_{61}(\ln u)^7,$$

if we use (3.16).

The last inequality implies that $u < c_{62}$ and, since $v < u$ we see that both numbers u, v are bounded by the effectively computable value c_{62} which depends only on S. It is easy to understand that our initial supposition $v < u$ is not a serious restriction, and the case of $v > u$ is dealt with quite similarly.

Thus, having proved the boundedness of the values u, v in the equality (3.2), we come to a finite number of equations with fixed values for $u, v - u \neq v, v \geq 3$. Since, recalling the comments after Theorem 1.2, the greatest prime divisor of $x^u - y^v$ increases unboundedly with the growth of $\max(x, y)$, we have to conclude that (3.2) entails the boundedness of x, y. All the constants involved in our estimates are effectively computable, so we obtain the assertion of the theorem.

Recently Inkeri and Van der Poorten [102] proved by similar arguments that all the solutions of the equation

$$x^p + (x + k)^p = z^p, \qquad (x, k) = 1$$

in which natural numbers x, z and the odd prime number p are unknowns, have a bound effectively determined in terms of k. This result is close to the still unproven conjecture by Markoff [139] asserting that the equation

$$x^p + y^p = (y + 1)^p$$

has no solutions in natural numbers x, y and odd primes p. The solution of this problem would allow one to fill a gap in Abel's arguments on the impossibility of the existence, of a rational point with numerator a power of an (unknown) prime number, on the Fermat curve [1]. In §4 Ch.IX we give an effective description of all such points on algebraic curves (there are infinitely many of them on some curves).

VIII. The Class Number Value Problem

It was ascertained in previous chapters that upper bounds for the solutions of diophantine equations under our consideration depend essentially on the regulators of certain algebraic number fields related to the equation. Now we concentrate our attention on this phenomenon and relate it to the general problem of the magnitude of ideal class numbers. We show that algebraic number fields with 'small' regulator (hence 'large' class number) occur very frequently and in some sense constitute the majority of fields. Bounds for the solutions of the corresponding diophantine equations, e.g., Thue equations, are much better than the general bounds.

1. Influence of the Value of the Class Number on the Size of the Solutions

The problem of determining the size of the ideal class number of algebraic number fields, takes its origin in the works of Gauss, Dirichlet and Kummer, and remains one of the central and most difficult problems of contemporary number theory. Though explicit formulæ expressing the class number in terms of the main parameters of the field exist for fields of specific types, it is quite difficult to obtain from them satisfactory information either on the value or on the arithmetic structure of the class numbers. Because of that one meets with considerable difficulties in solving problems on the magnitude or the structure of class numbers, and the need to develop special methods becomes evident.

The methods used in previous chapters to investigate diophantine equations rely in essence on the theory of algebraic units and lead to the construction of bounds for the solutions of the equations which essentially depend on the magnitude of the regulators of algebraic number fields, and thus in virtue of the Siegel-Brauer formula (see Lemma 2.4 Ch.II and the discussion after it), on the ideal class numbers of these fields. In this chapter we discuss the main aspects of the influence of the ideal class number on the magnitude of the solutions of diophantine equations and see that there is a two-sided influence: the larger the ideal class numbers of the algebraic number fields related to the equation, the better the bounds for the solutions of the equation. That is, the smaller the maximal height of a solution. Vice versa, good bounds for the solutions of specific types of diophantine equations lead to the construction of families of algebraic number fields with quite large class numbers ([218], [219]).

The phenomenon is clearly seen in the simplest non-linear diophantine equations, and we discuss several examples.

Suppose B is a square-free natural number and

$$x^2 - By^2 = 1 \qquad (1.1)$$

is Pell's equation. All the solutions of the equation are given by powers of the fundamental unit ε_0 of the field $\mathbb{K} = \mathbb{Q}(\sqrt{B})$. Denoting $h = h_\mathbb{K}$ and $R = \ln \varepsilon_0$ the class number and the regulator of $\mathbb{K}$ respectively, we have by Siegel's formula:

$$\ln(hR) \sim \frac{1}{2} \ln D \quad (D \to \infty) \qquad (1.2)$$

where $D = B$ or $4B$ is the discriminant of $\mathbb{K}$. Hence, for any $\delta > 0$ and large B we have

$$e^{B^{1/2-\delta} h^{-1}} < \varepsilon_0 < e^{B^{1/2+\delta} h^{-1}},$$

which implies similar inequalities for the components x_0, y_0 of the fundamental solution $\varepsilon_0 = x_0 + y_0 \sqrt{B}$ of (1.1). We see that the magnitude of h has essential influence on the magnitude of the fundamental solution and, in particular, we have $\varepsilon_0 < e^{B^{\delta'}}$ with small $\delta' > 0$ only when $h > B^{1/2-\delta''}$ with small enough $\delta'' > 0$.

In (1.1) the value of h influences the magnitude of the fundamental solution. If we turn to equations of higher degree, we see the same phenomenon in that or other similar form with respect to all solutions.

Consider the Delaunay equation

$$x^3 + By^3 = 1, \qquad (1.3)$$

where B is a cube-free natural number. If the equation has a solution x_0, $y_0 \neq 0$, then it is unique and defines the fundamental unit $\eta_0 = x_0 + y_0 \sqrt[3]{B}$ of the ring $\mathbb{Z}[\sqrt[3]{B}]$ with $0 < \eta_0 < 1$. This unit is not necessarily the fundamental unit ε_0 of the field $\mathbb{K} = \mathbb{Q}(\sqrt[3]{B})$, but in any event, it is a power of the latter. Because of that $|\ln \eta_0| \geq R_\mathbb{K}$ and hence

$$\max(|x_0|, |y_0|) \geq (1 + \sqrt[3]{B})^{-1} e^{R_\mathbb{K}}. \qquad (1.4)$$

We shall return to this inequality below, but notice now that if B is square-free and $B \not\equiv 1 \pmod 9$ then the ring $\mathbb{Z}[\sqrt[3]{B}]$ coincides with the ring of all integers of $\mathbb{K}$. Then $\eta_0 = \varepsilon_0$ and the Siegel-Brauer formula takes the form (1.2), where h, R, and D are now the class number, regulator and discriminant of the field $\mathbb{K}$ respectively. This produces the inequalities

$$e^{B^{1-\delta} h^{-1}} < \eta_0 < e^{B^{1+\delta} h^{-1}},$$

which imply

$$e^{B^{1-\delta'} h^{-1}} < \max(|x_0|, |y_0|) < e^{B^{1+\delta'} h^{-1}},$$

with any $\delta' > 0$ for all sufficiently large B (for which the equation (1.3) is solvable).

A similar result holds also for the Nagell equation

$$x^4 - By^4 = \pm 1,$$

the only solution of which is given by the fundamental unit of the ring $\mathbb{Z}[\sqrt[4]{-4B}]$, and also on other equations of the form

$$x^{2n} - By^{2n} = 1 \qquad (n \geq 3),$$

all solutions of which are determined by the first or second power of the fundamental unit of the ring $\mathbb{Z}[\sqrt{B}]$ (cf. [145]).

Now we turn to Thue's equation

$$f(x,y) = 1, \tag{1.5}$$

with an integral irreducible form $f(x,y)$ of degree $n \geq 3$. By Theorem 4.1 Ch.IV all the solutions to this equation satisfy

$$X = \max(|x|, |y|) < \exp\{c_1 R \ln R^* T \ln T\}, \tag{1.6}$$

where R is the regulator of the field $\mathbb{K}$ obtained by adjoining a root of the form to the field of rational numbers. Here $T = \ln H_f + R^*$, $R^* = \max(R, e)$, and H_f is the height of the form; c_1 depends only on n. If we replace the regulator on the right-hand side of (1.6) by the value $D^{1/2+\varepsilon}/h$ in accordance with the formula (1.2), where D and h are the absolute value of the discriminant and the class number of the field $\mathbb{K}$ respectively, then the resulting inequality will be of the same type as those we have seen in the special cases discussed above. Turning now to the inequality (1.4) we see that, at any rate as regards the equation (1.3) (in fact in all other cases as well), the upper bound (1.6) cannot be improved in any essential way with respect to R and H_f.

As a consequence of the estimate (1.6) we observed that

$$X < \exp\left(c_2 H_f^{3(n-1)+\delta}\right), \tag{1.7}$$

where $\delta > 0$ is arbitrary and $c_2 = c_2(n,\delta)$ (cf. §4 Ch.IV). To derive this estimate we used the Landau inequality (Lemma 2.4 Ch.II).

$$hR < c_3 D^{1/2}(\ln D)^{n-1}, \tag{1.8}$$

and, having no information on the value of class number h, estimated it from below by 1. It is plain that if we could bound h from below by some power of D, then in view of (1.8) and (1.6) this would lead to a strengthening of (1.7): it would be possible to replace the exponent $3(n-1) + \delta$ by a smaller quantity. In what follows we shall see that there exist algebraic number fields of every degree with regulators bounded above by $c_4 D^\varepsilon$ with arbitrary $\varepsilon > 0$. Such fields even constitute some kind of 'majority' among all fields of bounded degree. Consequently, the inequality (1.7) for the corresponding binary forms can be replaced by the stronger inequality

$$X < \exp(c_5 H_f^{\epsilon'}) \tag{1.9}$$

with any $\epsilon' > 0$, $c_5 = c_5(n, \epsilon')$, and in some cases even by an inequality of the type

$$X < \exp(c_6 (\ln H_f)^{n-1}) \tag{1.10}$$

(see below, §§4-6).

On looking through the proofs described in the previous chapters giving bounds for other Diophantine equations, it is easy to observe the same influence of the regulators (and class numbers) on the magnitude of the bounds, though such influence does not show itself as directly, as in the case of Thue equation, it is seen in the intermediate stage of the arguments. Indeed, we used Lemma 3.1 Ch.VI in Ch.VI and VII, the assertion of which contains the regulator of the corresponding field as the main parameter; for applications of this lemma to concrete equations the regulator estimates were needed, and we carried out such in the estimates using the inequality (1.8) where h was replaced by 1. It is clear, that if we could replace h by a large enough value, the results would be better in those cases too.

The opposite influence, that is the influence of good enough estimates of the solutions of diophantine equations on the values of class-numbers, shows itself in some form or other no less strongly. We describe in the next three paragraphs the families of algebraic number fields with large class number, relying on the results obtained above on the bounds for solutions of hyperelliptic diophantine equations and equations of hyperelliptic type. Our arguments are the inversion of the influence discussed above of class numbers on the bounds for the solutions. More hidden ties exist as well, for which results of Chowla [45] give examples:

Let $p > 3$ be a prime such that the imaginary quadratic field $\mathbb{Q}(\sqrt{-p})$ has class number 1. Then the equation

$$x^3 - py^2 = -1728 \tag{1.11}$$

has an integer solution x, y in which $x = [e^{\pi \sqrt{p}/3}]$, where $[z]$ is the nearest integer to z. If $p > 19$ with the same 'class number 1' property, then the equation

$$x^3 - py^2 = -8 \tag{1.12}$$

has an integer solution with $x = \frac{1}{6}[e^{\pi \sqrt{p}/3}]$.

We see that supposing that the field $\mathbb{Q}(\sqrt{-p})$ has class number 1 leads to the conclusion that there exist 'quite large' solutions to (1.11), (1.12). Though the finiteness of the number of imaginary quadratic fields with class number 1 was not proved by this way, the presence of similar ties shows the deep nature of problems concerning the values of ideal class numbers.

2. Real Quadratic Fields

According to the well-known hypothesis called often the hypothesis of Gauss (or of Gauss-Hasse), there exists an infinite number of real quadratic fields $\mathbb{K}$ with class number $h_{\mathbb{K}} = 1$. No approaches to the proof of this hypothesis are known, excluding approaches relying on yet more deep hypotheses. Even the much weaker hypothesis that for some $\delta > 0$ there exists an infinite number of real quadratic fields $\mathbb{K}$ with

$$h_{\mathbb{K}} < D_{\mathbb{K}}^{1/2-\delta} \qquad (2.1)$$

remains unproved; here $D_{\mathbb{K}}$ is the discriminant of $\mathbb{K}$.

There is a variety of results going in the opposite direction (for instance, cf. [92], [147]). These results state that $h_{\mathbb{K}} \to \infty$ as $D_{\mathbb{K}} \to \infty$, with $\mathbb{K}$ running through a special sequence of real quadratic fields. For example, if m takes integral values such that $m^2 \pm 1$ is squarefree, then the fields $\mathbb{Q}(\sqrt{m^2 \pm 1})$ have class numbers h_m satisfying

$$h_m > D_m^{1/2-\delta}, \qquad (2.2)$$

where D_m is the discriminant of the field, $\delta > 0$ is any number, and $m \geq m_0(\delta)$. It can be proved by an application of the sieve method that in fact there exists an infinitude of squarefree $m^2 \pm 1$.

Since for every squarefree integer $D > 0$ the equation

$$m^2 - Dn^2 = 1 \qquad (2.3)$$

has a solution in integers m, n $(n \neq 0)$, every quadratic field $\mathbb{K} = \mathbb{Q}(\sqrt{D})$ is contained in the sequence of fields $\mathbb{Q}(\sqrt{m^2 - 1})$, $(m = 2, 3, \ldots)$. But (2.3) has infinitely many solutions for every D, and therefore every given field $\mathbb{K}$ occurs in the sequence of fields $\mathbb{Q}(\sqrt{m^2 - 1})$ with infinite multiplicity. A natural arithmetic way to reduce the multiplicities to a finite number is by making substitutions $m \to g(m)$ in (2.3) with an integral polynomial $g(x)$ such that the polynomial $G(x) = g^2(x) - 1$ has at least three simple roots. Since the diophantine equation

$$G(x) = Dy^2$$

has only a finite number s of solutions for any D (see Ch. VI), it follows that the multiplicity of each field $\mathbb{Q}(\sqrt{D})$ in the sequence of fields $\mathbb{Q}(\sqrt{g^2(m) - 1})$ so obtained is finite. In the case of special polynomials $g(x) = x^k$ $(k \geq 2)$ it may be even expected that multiplicities are uniformly bounded in D by a value depending only on k.

Each field $\mathbb{Q}(\sqrt{g^2(m) - 1})$ coincides with a field $\mathbb{K}_m = \mathbb{Q}(\sqrt{D_m})$, where D_m is the squarefree kernel of the number $g^2(m) - 1$, because we have

$$g^2(m) - D_m y_m^2 = 1, \qquad y_m \in \mathbb{Z}. \qquad (2.4)$$

Consequently, $g(m) + y_m\sqrt{D_m}$ is a unit of the field $\mathbb{K}_m$ and is a power of the fundamental unit of the field. Therefore the regulator R_m of the field $\mathbb{K}_m$ satisfies

$$R_m < c_7 \ln(m+1) \quad (m = 1, 2, \ldots). \tag{2.5}$$

In the case $g(x) = x^k$ ($k \geq 2$) there are grounds for expecting that R_m is exactly of order $\ln(m+1)$, as this is the same as stating that $g(m) + y_m\sqrt{D_m}$ is a power of the fundamental unit of the field with an exponent bounded by a value depending only on k.

In view of Siegel's theorem we have

$$\ln(h_m R_m) \sim \tfrac{1}{2} \ln D_m \quad (m \to \infty). \tag{2.6}$$

Hence, to decide which of the two possibilities (2.1), (2.2) occurs, we have to determine the order of magnitude of $\ln D_m$ with respect to $\ln\ln(m+1)$ as $m \to \infty$, or to determine the order of D_m with respect to powers of $\ln(m+1)$.

Thus, we come to the problem of estimating the solutions of diophantine equations (2.4) with fixed D_m and unknown m and y_m.

Applying Theorem 4.1 Ch. VI we obtain the following result [218].

Theorem 2.1 *Suppose that $k \geq 2$ is a natural number, and that D_m and h_m are respectively the discriminant and the class number of the field $\mathbb{K}_m = \mathbb{Q}(\sqrt{m^{2k} - 1})$, where $m = 1, 2, \ldots$. Then we have*

$$D_m > c_8 (\ln m)^{\tau(2k)/12 - \delta(k) - \varepsilon}, \tag{2.7}$$

$$h_m > c_9 D_m^{\frac{1}{2} - \frac{12}{tau(2k) - \delta_1(k)} - \varepsilon}, \tag{2.8}$$

where $\tau(2k)$ is the number of divisors of $2k$, and $\delta_1(k) = 12\delta(k)$ with

$$\delta(k) = \begin{cases} 1/6, & \text{if } 2 \nmid k, \; 3 \nmid k, \\ 3/16, & \text{if } 2 \mid k, \; 3 \nmid k, \\ 7/24, & \text{if } 2 \mid k, \; 3 \mid k, \\ 5/24, & \text{if } 2 \nmid k, \; 3 \mid k, \end{cases}$$

$c_8 = c_8(k, \varepsilon)$, $c_9 = c_9(k, \varepsilon)$, *and $\varepsilon > 0$ is arbitrary.*

We see that if $\tau(2k) \geq 28$, then h_m tends to infinity like some power of D_m ($m \to \infty$). Similar theorems open up vast possibilities for investigating fields with certain properties possessed by their class numbers. For instance, one can prove by the application of sieve methods that the number of genera of the field $\mathbb{Q}(\sqrt{m^{2k} - 1})$ may be bounded for infinitely many m by a value depending only on k, while the number of classes in each genus increases without bound as m grows. One can prove also the infinitude of the number of real quadratic fields with class number divisible by a given number, and so on.

Before we proceed with the proof of Theorem 2.1, we prove the following simple lemma.

Lemma 2.1 *Let $f_1(x), \ldots, f_s(x)$ be integral polynomials no two of which have a common zero, and set $f(x) = f_1(x) \cdots f_s(x)$. Then for any integer* *x for which $f(x) \neq 0$, the square-free (positive) kernels $A[f(x)]$, $A[f_1(x)]$, ..., $A[f_s(x)]$ of the numbers $f(x)$, $f_1(x)$, ..., $f_s(x)$ satisfy the inequality*

$$A[f(x)] \geq P^{-s} A[f_1(x)] \dots A[f_s(x)], \qquad (2.9)$$

where $P = \prod_{1 \leq i < j \leq s} R(f_i, f_j)$ with the $R(f_i, f_j)$ the resultants of the polynomials f_i, f_j $(i, j = 1, 2, \dots, s)$.

Proof. For any pair of the polynomials $f_i(x)$, $f_j(x)$ there exists such a pair of integral polynomials $A_{ij}(x)$, $B_{ij}(x)$ that

$$f_i(x)A_{ij}(x) + f_j(x)B_{ij}(x) = R(f_i, f_j).$$

Hence, if for integral x a prime p occurs in both $f_i(x)$ and in $f_j(x)$ then it occurs in $R(f_i, f_j)$ and in P; prime numbers not occurring in P can divide only one of the numbers $f_i(x)$ or $f_j(x)$. Now separate from respectively $A[f(x)], \dots, A[f_s(x)]$ the parts $A^*(x), A_1^*(x), \dots, A_s^*(x)$ constituted from the primes not occurring in P. It is obvious that $A^*(x) = A_1^*(x) \dots A_s^*(x)$, and since both $A[f(x)] \geq |A^*(x)|$ and $|A_i^*(x)| \geq P^{-1}A[f_i(x)]$, we obtain (2.9).

Proof of Theorem 2.1. We start with some general considerations and then use the special conditions of Theorem 2.1.

Let $f(x)$ be an integral polynomial with at least three simple zeros, and let $\alpha, \alpha_1, \alpha_2$ be such a triplet of zeros. Set $\mathbb{F} = \mathbb{Q}(\alpha, \alpha_1, \alpha_2)$, and let d, d_1, d_2 denote the respective degrees of that field over $\mathbb{Q}(\alpha), \mathbb{Q}(\alpha_1)$ and $\mathbb{Q}(\alpha_2)$. Let $\sigma(f)$ be the smallest sum $d + d_1 + d_2$ for all choices of three simple roots of the polynomial $f(x)$. If $f(x)$ does not have three simple zeros we set $\sigma^{-1}(f) = 0$.

By Theorem 4.1 Ch. VI the square-free kernel $A[f(x)]$ of the number $f(x) \neq 0$ with integral $x \neq 0$ satisfies

$$A[f(x)] > c_{10}(\ln|x|)^{\frac{1}{4}\sigma^{-1}(f)-\varepsilon}, \qquad (2.10)$$

where $c_{10} > 0$ depends on ε, and the coefficients and degree of $f(x)$.

If $f(x) = f_1(x) \cdots f_s(x)$, where the $f_i(x)$ are integral pairwise co-prime polynomials, then Lemma 2.1 and inequality (2.10) applied to each polynomial $f_i(x)$ yield

$$A[f(x)] > c_{11}(\ln|x|)^{\frac{1}{4}\sigma^{-1}-\varepsilon}, \qquad (2.11)$$

where

$$\sigma^{-1} = \sum_{i=1}^{s} \sigma^{-1}(f_i). \qquad (2.12)$$

Suppose $e(x), a(x), b(x)$ are integral polynomials related to $f(x)$ by

$$e(x)a^2(x) - f(x)b^2(x) = \pm 1. \qquad (2.13)$$

We take any $b(x)$ and define $e(x)$ and $a(x)$ as the square-free part and the full square of the polynomial $f(x)b^2(x) \pm 1$.

Suppose that $f(x)$ has a positive leading coefficient and that m is a natural number with $f(m) > 0$, $b(m) \neq 0$. Then $e(m) > 0$ and one can suppose that $a(m) > 0$, $b(m) > 0$. We observe from the equality (2.13) that an algebraic number

$$\eta = a(m)\sqrt{e(m)} + b(m)\sqrt{f(m)} > 1$$

is a unit, and that its square is a positive power of the fundamental unit ε_m of the real quadratic field $\mathbb{K}_m = \mathbb{Q}(\sqrt{e(m)f(m)})$. Consequently,

$$\ln \varepsilon_m \leq c_{12} \ln(m+1),$$

where c_{12} is determined in explicit form in terms of the coefficients and degrees of the polynomials $f(x)$ and $b(x)$. It follows from (2.13) that the numbers $e(m)$ and $f(m)$ are co-prime, so the squarefree kernel of the number $e(m)f(m)$ is no less than that of $f(m)$. Therefore we see from (2.11) that the discriminant D_m of the field $\mathbb{K}_m$ satisfies

$$D_m > c_{13}(\ln(m+1))^{\frac{1}{2}\sigma^{-1}-\varepsilon}, \tag{2.14}$$

where σ is determined by the equality (2.12), and $c_{13} > 0$. Hence,

$$R_m = \ln \varepsilon_m < c_{14} D_m^{4\sigma+\varepsilon}.$$

Now we obtain from (2.6) that

$$h_m > c_{15} D_m^{1/2-4\sigma-\varepsilon}, \tag{2.15}$$

which in the case that $\sigma < \frac{1}{8}$ is a pithy inequality showing together with (2.14) that h_m increases without bound as $m \to \infty$.

On setting $f(x) = x^{2k} - 1$ we see that (2.13) holds with $e(x) = 1$, $a(x) = x^k$ and $b(x) = I$. Now decompose $f(x)$ into factors $f_1(x), \ldots, f_s(x)$. We have

$$x^{2k} - 1 = \prod_{d|2k} P_d(x),$$

where the $P_d(x)$ are the cyclotomic polynomials. If $d \neq 1, 2, 3, 4, 6$ the degree of $P_d(x)$, which is $\varphi(d)$, is no less than 4. In the remaining cases it is 1 or 2. Therefore one should take into account the divisibility of k by 2 and 3.

1) $2 \nmid k$, $3 \nmid k$. As 'main' factors $f_i(x)$ we take $P_d(x)$ with $d \neq 1, 2$. Then $\sigma(P_d)$ is equal to 3, and the number of factors is $\tau(2k) - 2$ hence, we obtain from (2.12):

$$\sigma^{-1} = (\tau(2k) - 2)/3. \tag{2.16}$$

2) $2 \mid k$, $3 \nmid k$. For $P_d(x)$ with $d \neq 1, 2, 4$ we have $\sigma(P_d) = 3$, while $\sigma(P_1 P_4) = 4$. Therefore we have

$$\sigma^{-1} = \frac{1}{4} + (\tau(2k) - 3)/3. \tag{2.17}$$

3) $2 \mid k$, $3 \mid k$. For $P_d(x)$ with $d \neq 1, 2, 3, 4, 6$ we find that $\sigma(P_d) = 3$ while $\sigma(P_1 P_3) = \sigma(P_2 P_4) = 4$. Therefore we have

$$\sigma^{-1} = \frac{1}{2} + (\tau(2k) - 5)/3. \tag{2.18}$$

4) $2 \nmid k, 3 \mid k$. Similar to the preceding we have $\sigma(P_d) = 3$ for $d \neq 1, 2, 3, 6$, whilst $\sigma(P_1 P_3) = \sigma(P_2 P_6) = 4$, which gives

$$\sigma^{-1} = \tfrac{1}{2} + (\tau(2k) - 4)/3. \tag{2.19}$$

Now the relations (2.14) – (2.19) give the assertion of the theorem.

3. Fields of Degree 3,4 and 6

The next three theorems may be obtained in a quite similar way to Theorem 2.1; see [219].

Theorem 3.1 *Let $k \geq 1$ be a natural number, and let D_m and h_m denote the discriminant and class number of the field $\mathbb{Q}(\sqrt[3]{m^{3k} - 1})$, where $m = 1, 2, \dots$. Then*

$$|D_m| > c_{16}(\ln m)^{(\tau(3k)-1)/24-\varepsilon}, \tag{3.1}$$

$$h_m > c_{17}|D_m|^{\frac{1}{2} - \frac{24}{\tau(3k)-1} - \varepsilon}, \tag{3.2}$$

where $c_{16} > 0$ and $c_{17} > 0$ are expressible in terms of k and ε, with $\varepsilon > 0$ arbitrary.

Theorem 3.2 *Let $k \geq 1$ be a natural number, and let D_m and h_m denote the discriminant and class number of the field $\mathbb{Q}(\sqrt[4]{m^{4k} - 1})$, where $m = 1, 2, \dots$. Then the inequalities*

$$|D_m| > c_{18}(\ln m)^{\frac{1}{8}\tau(4k)-\delta_2(k)-\varepsilon}, \tag{3.3}$$

$$h_m > c_{19}|D_m|^{\frac{1}{2} - \frac{12}{\tau(4k)-6\delta_2(k)} - \varepsilon}, \tag{3.4}$$

where $\delta_2(k) = 7/12$ if $3 \mid k$, and $\delta_2(k) = 3/8$ if $3 \nmid k$; and $c_{18} > 0$, $c_{19} > 0$ are expressible in terms of k and ε. As usual $\varepsilon > 0$ is arbitrary.

Theorem 3.3 *Let $k \geq 1$ be a natural number, and let D_m and h_m be the discriminant and the class number of the field $\mathbb{Q}(\sqrt[6]{m^{6k} - 1})$, where $m = 1, 2, \dots$. Then we have the inequalities*

$$|D_m| > c_{20}(\ln m)^{(\tau(6k)/4-7/8-\varepsilon)}, \tag{3.5}$$

$$h_m > c_{21}|D_m|^{\frac{1}{2} - \frac{12}{\tau(6k)-3.5} - \varepsilon}, \tag{3.6}$$

where $c_{20} > 0$, $c_{21} > 0$ may be expressed in terms of k and ε, with $\varepsilon > 0$ arbitrary.

To prove Theorem 3.1 we need an estimate for the cube-free divisor of $q^n - 1$, which we denote by $A_3[q^n - 1]$. On denoting the cyclotomic polynomials by $P_d(x)$ we have

$$x^n - 1 = \prod_{d|n} P_d(x). \tag{3.7}$$

Since $P_d(x)$ is a normal polynomial of degree $\varphi(d)$, by Theorem 1.1 Ch.VII we have

$$A_3[P_d(q)] > c_{22}(\ln q)^{1/24-\varepsilon}, \tag{3.8}$$

when $d \geq 3$, where $c_{22} = c_{22}(d,\varepsilon) > 0$. The decomposition (3.7) and an analogue of Lemma 2.1 for cube-free divisors of polynomials now gives: for odd n

$$A_3[q^n - 1] > c_{23}(\ln q)^{(\tau(n)-1)/24-\varepsilon}, \tag{3.9}$$

while for even n

$$A_3[q^n - 1] > c_{24}A_3[q^2 - 1](\ln q)^{(\tau(n)-2)/24-\varepsilon},$$

which again leads to (3.9).

Now set $D = d^3 - 1$, where $d > 0$ is an integer, and write $\alpha = \sqrt[3]{D}$. Then $\eta = d - \alpha$ is a unit of the field $\mathbb{K} = \mathbb{Q}(\alpha)$, and

$$1 < |\eta| < d + |\sqrt[3]{D}| < 2d.$$

Consequently, the regulator $R_{\mathbb{K}} < \ln(2d)$. Setting $A_3[D] = D_1 D_2^2$, where D_1, D_2 are squarefree numbers, we have one of two representations for the discriminant $D_{\mathbb{K}}$:

$$D_{\mathbb{K}} = -3(D_1 D_2)^2, \quad \text{if} \quad A_3[D] \equiv 1 \pmod{9},$$
$$D_{\mathbb{K}} = -27(D_1 D_2)^2, \quad \text{if} \quad A_3[D] \not\equiv 1 \pmod{9}.$$

At any rate $D_{\mathbb{K}}$ is divisible by $A_3[D]$, and then

$$|D_{\mathbb{K}}| > A_3[D] \tag{3.10}$$

holds.

If we now take, in accordance with Theorem 3.1, $d = m^k$ and $n = 3k$, $q = m$, we obtain that

$$R_{\mathbb{K}} < \ln(2m^k) < 2k \ln m \quad (m \geq 2),$$
$$|D_{\mathbb{K}}| > c_{25}(\ln m)^{(\tau(3k)-1)/24-\varepsilon},$$

the latter in view of (3.9), (3.10). This gives (3.1), and together with the Siegel-Brauer theorem it gives (3.2) as well. Theorems 3.2 and 3.3 are proved by similar arguments.

Setting $D = d^4 - 1$, $\alpha = \sqrt[4]{D}$ we observe that the numbers $\eta_1 = d - \alpha$, and $\eta_2 = d^2 - \alpha^2$ are units of $\mathbb{K} = \mathbb{Q}(\alpha)$. The regulator of this system of units is

$$R[\eta_1, \eta_2] = \pm \begin{vmatrix} \ln|d - \alpha| & \ln|d^2 - \alpha^2| \\ \ln|d + \alpha| & \ln|d^2 - \alpha^2| \end{vmatrix},$$

from which we find

$$R[\eta_1, \eta_2] \sim 20(\ln d)^2 \qquad (d \to \infty),$$

taking into account that

$$\alpha = d(1 - d^{-4})^{1/4} = d(1 - \tfrac{1}{4}d^{-4} + O(d^{-8})),$$
$$\alpha^2 = d^2(1 - \tfrac{1}{2}d^{-4} + O(d^{-8})).$$

Therefore $R_{\mathbb{K}} < 21(\ln d)^2$ holds for sufficiently large d.

It follows from the theory of differents of algebraic number fields (that is, multiplicativity of differents in relative extensions) that the discriminant of the field $\mathbb{K}$ is divisible by the square of the discriminant of the field $\mathbb{Q}(\sqrt{d^4 - 1})$. Hence, we have

$$|D_{\mathbb{K}}| \geq A_2^2[d^4 - 1], \qquad (3.11)$$

where $A_2[d^4 - 1]$ is the square-free part of $d^4 - 1$. If we now take $d = m^k$, then in view of (2.17) and (2.18) we obtain

$$A_2[m^{4k} - 1] > c_{26}(\ln m)^{\frac{1}{12}\tau(4k) - \frac{1}{2}\delta_2(k) - \varepsilon},$$

where $\delta_2(k)$ is determined as in Theorem 3.2. We find from (3.11) that

$$|D_{\mathbb{K}}| > c_{26}^2(\ln m)^{\frac{1}{6}\tau(4k) - \delta_2(k) - 2\varepsilon},$$

which in fact coincides with (3.3). Since we have $R_{\mathbb{K}} < 21(k \ln m)^2$ for large enough m, by the Siegel-Brauer theorem we obtain (3.4).

Finally, in the case $D = d^6 - 1$ we set $\alpha = \sqrt[6]{D}$. Then $\eta_1 = d - \alpha$, $\eta_2 = d^2 - \alpha^2$ and $\eta_3 = d^3 - \alpha^3$ are units of the field $\mathbb{K} = \mathbb{Q}(\alpha)$. The regulator $R[\eta_1, \eta_2, \eta_3]$ of this system of units is equal to

$$\pm \begin{vmatrix} \ln|d - \alpha| & \ln|d^2 - \alpha^2| & \ln|d^3 - \alpha^3| \\ \ln|d + \alpha| & \ln|d^2 - \alpha^2| & \ln|d^3 + \alpha^3| \\ \ln|d - \zeta\alpha| & \ln|d^2 - \zeta^2\alpha^2| & \ln|d^3 - \zeta^3\alpha^3| \end{vmatrix},$$

where $\zeta = e^{\pi i/3}$. To compute this determinant, we add the second row to the first and subtract it from the last. Then we find

$$R[\eta_1, \eta_2, \eta_3] = \pm \ln|d^3 + \alpha^3| \left(\ln|d^2 - \alpha^2| \cdot \ln\left|\frac{d^2 - \zeta^2\alpha^2}{d^2 - \alpha^2}\right| - 2\ln|d^2 - \alpha^2| \cdot \ln\left|\frac{d - \zeta\alpha}{d + \alpha}\right| \right).$$

Taking into account that

$$\alpha^2 = d^2(1 - \tfrac{1}{3}d^{-6} + O(d^{-12}))$$

we obtain for $d \to \infty$,

$$R[\eta_1, \eta_2, \eta_3] \sim \pm 3\ln d \left(-4\ln d \cdot \ln\left|\frac{1 - \zeta^2}{2}\right| - 24(\ln d)^2 + 8\ln d \cdot \ln\left|\frac{1 - \zeta}{2}\right| \right),$$

which is $\sim 72(\ln d)^3$. Once again from the theory of differents it follows that the discriminant $D_{\mathbb{K}}$ is divisible by the cube of the discriminant of the quadratic field $\mathbb{Q}(\sqrt{d^6 - 1})$. Consequently,

$$|D_{\mathbb{K}}| \geq A_2^3[d^6 - 1].$$

Setting $d = m^k$, we obtain from the equality (2.18) that

$$|D_{\mathbb{K}}| > c_{27}(\ln m)^{\frac{1}{4}\tau(6k)-7/8-\varepsilon}$$

holds, which gives (3.5). As we have $R_{\mathbb{K}} < 73(k \ln m)^3$ for large m, the Siegel-Brauer theorem gives (3.6).

The inequalities for class numbers indicated in Theorems 2.1, and 3.1–3.3 were obtained on the basis of the 'principal part' of the Siegel-Brauer theorem

$$hR > c(n,\varepsilon)|D|^{1/2-\varepsilon}, \qquad (3.12)$$

where h, R, D and n are respectively the class number, the regulator, the discriminant and the degree of the algebraic number field $\mathbb{K}$; $\varepsilon > 0$ is arbitrary. The value $c(n,\varepsilon)$ in this inequality is ineffective, therefore the inequalities for h_m obtained in this way are ineffective. To obtain effective inequalities one may use the results of Stark [228]. In particular, if $\mathbb{K}$ does not contain a quadratic subfield, then an effective strengthening of (3.12) holds, and in the general case for $\varepsilon \geq 1/n$ and $n \geq 4$ the quantity $c(n,\varepsilon)$ may be represented in effective form. This makes effective the inequality for h_m in Theorem 3.1 and in Theorems 3.2 and 3.3 but gives effective inequalities that are a little weaker.

The inequalities (3.1), (3.2) show that h_m increases without bound together with m when $\tau(3k) \geq 50$. Similarly, an analogous assertion follows from the inequalities (3.3), (3.4) if $\tau(4k) \geq 28$, when $3 \mid k$; and if $\tau(4k) \geq 27$, when $3 \nmid k$. The inequalities (3.5), (3.6) lead to a need for the condition $\tau(6k) \geq 28$. It is obvious that if we have stronger inequalities for squarefree and cube-free divisors of polynomial values than those obtained from Theorem 4.1 Ch.VI and Theorem 1.1 Ch.VII, then it would be possible to weaken or to exclude all (to replace by trivial ones) the restrictions on the number of divisors arising in Theorems 2.1, 3.1–3.3 and those we discussed above.

This shows clearly that a strengthening of the bounds for solutions of diophantine equations allows one to extend one's knowledge of the totality of algebraic number fields with large class numbers. From the other side, as we have seen in §1, the improvements on the bounds to the solutions depend directly on our knowledge about the magnitude of class numbers of the corresponding fields.

4. Superposition of Polynomials

Development of the idea described in two previous paragraphs concerning the substitutions $x \to x^k$ in the general case requires investigation of algebraic properties of integral polynomials obtained by repeated substitutions $x \to g(x)$, where $g(x)$ is an integral polynomial. We introduce here the main result we need for further applications (Lemma 4.4)

We denote by $f * g(x)$ the superposition $f(g(x))$ of two polynomials $f(x)$ and $g(x)$. If $h(x)$ is a third polynomial then we have

$$f * g * h(x) = f * (g * h(x)),$$

and similarly for any number of polynomials. In what follows the polynomials under consideration are different from constants.

Lemma 4.1 Let $f(x), g(x)$ be polynomials with complex coefficients, such that $f(x)$ has no multiple root and $\deg f(x) \geq 2$. Then the squarefree kernel of the polynomial $f * g(x)$ (in the ring $\mathbb{C}[x]$) has degree at least

$$\deg f(x) \deg g(x) - 2 \deg g(x) + 2.$$

Proof. Set $f * g(x) = u(x)v^2(x)$, where $u(x)$ is squarefree in $\mathbb{C}[x]$. Then we have

$$\frac{d}{dx} f * g(x) = (f' * g(x))g'(x) = v(x)w(x),$$

where $w(x) = u'(x)v(x) + 2u(x)v'(x)$. Since $(f(x), f'(x)) = 1$, we have

$$(f * g(x), f' * g(x)) = 1$$

and then $v(x) \mid g'(x)$. Consequently, $\deg v(x) \leq \deg g(x) - 1$. Finally we obtain

$$\deg f * g(x) = \deg f(x) \deg g(x) = \deg u(x) + 2 \deg v(x) \leq$$
$$\leq \deg u(x) + 2 \deg g(x) - 2,$$

which gives the assertion of lemma.

Lemma 4.2 Suppose $f(x)$ is an integral polynomial without multiple roots. Then there exist integral polynomials $g(x)$, $p_1(x)$, $f_1(x)$ effectively determined by $f(x)$ and such that

$$f * g(x) = p_1(x)f_1(x),$$

where $p_1(x)$ is a normal polynomial of degree at least 3, and

$$(p_1(x), f_1(x)) = 1, \quad \deg p_1(x) = \deg g(x).$$

Proof. Let α be a zero of the polynomial $f(x)$, and $\mathbb{F}$ a normal field of degree $k \geq 3$, containing α. Let θ be an integral generating element of the field $\mathbb{F}$. Since the numbers $\theta, \theta^2, \ldots, \theta^k$ are linearly independent over $\mathbb{Q}$ we have

$$\alpha = r_1\theta + r_2\theta^2 + \ldots + r_k\theta^k, \quad r_i \in \mathbb{Q} \quad (i = 1, 2, \ldots, k).$$

From this follows a relation for the traces:

$$S(\alpha\theta^j) = r_1 S(\theta^{1+j}) + \ldots + r_k S(\theta^{k+j}) \qquad (j = 0, 1, \ldots, k - 1).$$

The determinant

$$d = \det(S(\theta^{i+j})) \qquad (i = 1, 2, \ldots, k; j = 0, 1, \ldots, k - 1)$$

differs from the discriminant $D(\theta)$ of the number θ by an integral rational factor $\mathrm{Nm}(\theta)$. Hence it is a rational integer different from zero. Therefore the r_i are integral linear combinations of the numbers

$$S(\alpha\theta^j)d^{-1}, \qquad (j = 0, 1, \ldots, k - 1).$$

If a is the leading coefficient of $f(x)$ then the numbers adr_i all are integers. Consequently, the polynomial $r(x) = r_1 x + r_2 x^2 + \ldots + r_k x^k$ has rational coefficients, the denominators of which divide ad; $r_0(x) = r(adx)$ is an integral polynomial.

Let $p(x)$ be the minimal polynomial of θ. Since we have $f(r(\theta)) = f(\alpha) = 0$, the polynomial $f * r(x)$ is divisible by the irreducible polynomial $p(x)$, and the integral polynomial $f_0(x) = f * r_0(x)$ is divisible by $p_1(x) = p(adx)$. If $p_1(x)$ occurs in $f_0(x)$ to the first power we set $g(x) = r_0(x)$, otherwise we take

$$g(x) = r_0(x) \pm p_1(x),$$

where the sign is chosen so that $g(x)$ has degree k. Then we find that

$$f * (r_0(x) \pm p_1(x)) \equiv f * r_0(x) \pm (f' * r_0(x))p_1(x) \equiv$$
$$\equiv \pm(f' * r_0(x))p_1(x) \not\equiv 0 \quad (\mathrm{mod}\ p_1^2(x)),$$

since we have $(f' * r_0(x), p_1(x)) \equiv 1$.

It is clear that $p_1(x)$ is a normal polynomial and that the degrees of $p_1(x)$ and $g(x)$ coincide.

Finally, we observe that construction of the polynomials $g(x)$, $p_1(x)$ may be done effectively. Indeed, if the degree of α is no less than 3, then the field $\mathbb{F}$ is obtained by successive adjunction of the roots of the minimal polynomial of α. On replacing the successive extensions by a single extension, we find a generating element of $\mathbb{F}$. If the degree of α is 1 or 2, the construction of $\mathbb{F}$ is trivial.

Lemma 4.3 *Suppose $f(x)$ is an integral polynomial of degree no less than 2 and without multiple roots; let $r > 0$ be an arbitrary integer. Then there exist effectively determinable integral polynomials*

$$g, g_1, \ldots, g_{r-1}; p_1, p_2, \ldots, p_r; f_r, q_r,$$

such that $p_1, p_2, \ldots, p_r$ are normal polynomials of degree no less than 3, and so that setting

$$G = g * g_1 * \ldots * g_{r-1},$$
$$P_i = p_i * g_i * g_{i+1} * \ldots * g_{r-1} \quad (i = 1, 2, \ldots, r-1),$$
$$P_r = p_r,$$

we have

$$f * G = P_1 P_2 \ldots P_r f_r q_r^2 \tag{4.1}$$
$$(P_i, P_j) = 1, \quad (P_i, f_r) = 1 \quad (i, j = 1, 2, \ldots, r; i \neq j).$$

Proof. By Lemma 4.2 there exist polynomials g, p_1, f_1 for which $f * g = p_1 f_1$, with p_1 a normal polynomial of degree $k \geq 3$, $(p_1, f_1) = 1$, and with the degrees of g and p_1 coinciding. By Lemma 4.1 the squarefree divisor of $f * g(x)$ is of degree no less than $nk - 2k + 2$, where n is the degree of $f(x)$. If $n \geq 2$, then the square-free part f_{10} of the polynomial $f_1(x)$ has degree no less than $(n-2)k + 2 \geq 2$. Applying Lemma 4.2 to $f_{10}(x)$ we find polynomials g_1, p_2, f_2, for which $f_{10} * g_1 = p_2 f_2$ and where p_2 is a normal polynomial of degree no less than 3, $(p_2, f_2) = 1$, and $\deg g_1 = \deg p_2$. Then we have

$$f * g * g_1 = (p_1 * g_1)(f_{10} * g_1)g_2^2 = (p_1 * g_1)p_2 f_2 q_2^2.$$

Since $(p_1, f_1) = 1$, then $(p_1, f_{10}) = 1$, and then

$$(p_1 * g_1, f_{10} * g_1) = 1.$$

Consequently, we find

$$(p_1 * g_1, p_2) = 1, \quad (p_1 * g_1, f_2) = 1.$$

Next, we separate from $f_2(x)$ its squarefree part $f_{20}(x)$ and apply Lemma 4.2 to f_{20}, and so on. As such the arguments may be continued and we obtain the assertion of the lemma.

Lemma 4.4 *Let $f(x)$ be an integral polynomial of degree no less than 2 and without multiple roots, and take $B > 0$ arbitrary. Then there exists an integral polynomial $G(x)$ effectively determined by $f(x)$ and B, such that for any integer $m \neq 0$ with $f * G(m) \neq 0$ we have*

$$A_2[f * G(m)] > c_{28}(\ln |m|)^B, \tag{4.2}$$

where $c_{28} > 0$ is expressible in terms of the coefficients of the polynomials $f(x)$, $G(x)$, and the number B.

Proof. Taking an integer $r > 12B$, we apply Lemma 4.3. It follows from the equality (4.1) and Lemma 2.1 that

$$A_2[f * G(m)] > c_{29} A_2[P_1(m)] \ldots A_2[P_r(m)],$$

where $c_{29} > 0$ is expressed in terms of the resultants of the polynomials P_i and P_j, and P_i and f_r, $(i, j = 1, 2, \ldots, r; i \neq j)$. As we have

$$P_i(m) = p_i(m_i), \quad m_i = g_i * \cdots * g_{r-1}(m) \quad (i = 1, 2, \ldots, r - 1),$$

the inequality (4.2) follows from the estimate

$$A_2[p(m)] > c_{30}(\ln(|m| + 1))^{1/12-\varepsilon},$$

which is true for any normal polynomial of degree no less than 3 (this is a corollary of Theorem 4.1 Ch. VI; see the discussion at the close of §4 Ch. VI).

Now we can supplement the results of two previous paragraphs by the following assertion.

Theorem 4.1 *Suppose that $t = 2, 3, 4$ or 6, and that $f(x)$ is an integral polynomial of degree at least 2 for which the equation*

$$a^t(x) - f(x)b^t(x) = 1$$

has a solution in integral polynomials $a(x)$, $b(x)$, with $(b(x)$ not identically zero. Then for any real $B > 0$ there exists an integral polynomials $g(x)$, effectively determined by the polynomials $a(x)$, $b(x)$, $f(x)$ and the number B such that the fields $\mathbb{Q}(\sqrt[t]{f * g(m)})$ with $m \geq m_0$ have discriminant D_m and class numbers h_m satisfying

$$|D_m| > c_{31}(\ln m)^B, \qquad h_m > c_{32}|D_m|^{1/2-e/B-\varepsilon}$$

where $e = 1$, if $t = 2$ and $t = 3$; $e = 2$, if $t = 4$; $e = 3$ if $t = 6$; and $c_{31} > 0$, $c_{32} > 0$ are values depending only on the coefficients of the polynomials $a(x)$, $b(x)$, and $f(x)$ and the numbers B, ε, where $\varepsilon > 0$ is arbitrary.

Proof. In the case of $t \neq 3$ one uses Lemma 4.4 and the arguments described in the two previous paragraphs. In the case $t = 3$ we take as a basis the arguments of §3, but Lemmas 4.1, 4.3 and 4.4 should be replaced by the analogous lemmas concerning the cube-free divisors of polynomials, by inserting obvious changes into the arguments described above.

5. The Ankeny-Brauer-Chowla Fields

In 1956 Ankeny, Brauer and Chowla [4] proved the existence of an infinite sequence of algebraic number fields $\mathbb{K}_m$ $(m = 1, 2, \ldots)$ having any prescribed degree $n \geq 2$ and extremely large class numbers:

$$h_m > |D_m|^{1/2-\varepsilon}, \tag{5.1}$$

where h_m and D_m are respectively the class number and the discriminant of the field $\mathbb{K}_m$; $\varepsilon > 0$ is arbitrary. For $n \geq 3$ it turned out to be possible to find such a sequence in the set of the fields generated by the roots of the polynomials

$$f_N(x) = (x - a_1) \ldots (x - a_{n-1})(x - N) + 1, \tag{5.2}$$

where $a_1, \ldots, a_{n-1}$ are any fixed distinct integers, and N runs through all natural numbers starting with some appropriate one (it is easy to see that $f_N(x)$ is then irreducible). Fields generated by the roots of the polynomials (5.2) are purely real, but if n_1, n_2 are any natural numbers with $n_1 + 2n_2 = n$, and the integers $a_1, a_2, \ldots, a_{n_1+n_2-1}$ are such that

$$a_i \neq a_j \quad (i \neq j), \quad a_{n_1+1} > 0, \ldots, a_{n_1+n_2-1} > 0,$$

then the fields generated by the roots of the polynomials

$$g_N(x) = (x - a_1) \ldots (x - a_{n-1})(x^2 + a_{n_1+1}) \cdots$$
$$\cdots (x^2 + a_{n_1+n_2-1})(x^2 + N) + 1 \quad (5.3)$$

for sufficiently large N are irreducible and have exactly n_1 real and n_2 complex conjugate fields. In the set of these fields one can also find a sequence with condition (5.1).

We prove a few assertions on the sequence of fields generated by the roots of the polynomials (5.2) and satisfying the inequality (5.1). First of all, we prove the possibility of an effective choice from the set of all fields of those which satisfy (5.1) and then we prove that such fields constitute the 'vast majority' of the set. The ideas we apply also suffice to deal with, in a similar way, the fields generated by the roots of the polynomials of the form (5.3), as well as more complicated families of the fields generated by the roots of the polynomials

$$(x - a_1(m)) \ldots (x - a_n(m)) \pm 1,$$

where the $a_i(y)$ are integral polynomials with

$$\max_{1 \leq i \leq n-1} \deg a_i(y) < \deg a_n(y).$$

As a preliminary, we set up two main lemmas.

Lemma 5.1 *Let $a_1, \ldots, a_{n-1}$ be distinct integers, an N an integer satisfying $N > \max |a_i|$, $(i = 1, 2, \ldots, n - 1)$. Then the regulator of the field generated by the root of the polynomial (5.2) does not exceed $c_{33}(\ln N)^{n-1}$, where c_{33} depends only on $a_1, \ldots, a_{n-1}$, and n.*

Proof. Denote by θ_N a root of the polynomial $f_N(x)$. Then $\theta_N - a_1, \ldots, \theta_N - a_{n-1}$ are units of the field $\mathbb{K} = \mathbb{Q}(\theta_N)$. The roots $\theta_N^{(i)}$ of the polynomial $f_N(x)$ may be enumerated in such a way that as $N \to \infty$

$$\theta_N^{(i)} \to a_i \quad (i = 1, 2, \ldots, n - 1).$$

Consequently, for $j \neq i$ we have

$$|\theta_N^{(i)} - a_j| \to |a_i - a_j| \neq 0 \quad (N \to \infty),$$

and from the equality

$$(\theta_N^{(i)} - a_1) \cdots (\theta_N^{(i)} - a_i) \cdots (\theta_N^{(i)} - a_{n-1})(\theta_N^{(i)} - N) = -1$$

we observe that

$$|\theta_N^{(i)} - a_i| \sim \frac{1}{N} \prod_{\substack{j=1 \\ j \neq i}}^{n-1} |a_i - a_j|^{-1}.$$

Thus, as $N \to \infty$ we find that

$$\ln |\theta_N^{(i)} - a_j| \sim \begin{cases} \ln |a_i - a_j|, & i \neq j, \\ -\ln N, & i = j. \end{cases}$$

Hence, it follows that

$$\det(\ln |\theta_N^{(i)} - a_j|)_{i,j=1,2,\dots,n-1} \sim (-\ln N)^{n-1}.$$

Since the regulator of $\mathbb{K}$ does not exceed the absolute value of the latter determinant, we obtain the assertion of the lemma.

Lemma 5.2 *Let* $\alpha_1, \dots, \alpha_{n-1}$ *be distinct complex numbers such that*

$$|\alpha_i - \alpha_j| > 2 \qquad (i \neq j; \ i, j = 1, 2, \dots, n). \tag{5.4}$$

Set $f(x,y) = (x - \alpha_1) \dots (x - \alpha_{n-1})(x - y) + 1$, *and let* $D(y)$ *denote the discriminant of the polynomial* $f(x,y)$ *with* y *fixed. Then* $D(y)$ *is a polynomial in* y *of degree* $2(n - 1)$ *and without multiple roots.*

Proof. The equation $f(x,y) = 0$ defines n functions $\tilde{x}_1(y), \dots, \tilde{x}_n(y)$ analytic in a neighborhood of $y = \infty$ for which as $y \to \infty$ we have

$$\tilde{x}_i(y) \to \alpha_i \quad (i = 1, 2, \dots, n-1), \quad \tilde{x}_n(y) - y \to 0. \tag{5.5}$$

Since $D(y)$ may be represented in the form

$$D(y) = \prod_{1 \leq i < j \leq n} (\tilde{x}_i(y) - \tilde{x}_j(y))^2,$$

it follows from (5.5) that as $y \to \infty$,

$$D(y) \sim \prod_{1 \leq i < j \leq n-1} (\alpha_i - \alpha_j)^2 \prod_{1 \leq i < j \leq n-1} (\alpha_i - y)^2 \sim \alpha y^{2(n-1)},$$

where $\alpha \neq 0$. Consequently, $D(y)$, being a polynomial in y, has degree $2(n-1)$.

Let y_0 be a complex number, and let $x_1, \dots, x_n$ be the roots of $f(x, y_0)$ so enumerated that

$$|x_i - \alpha_i| = \min_{i \leq j \leq n} |x_j - \alpha_i| \qquad (i = 1, 2, \dots, n-1).$$

That is, x_1 is the root closest to α_1; x_2 is the root closest to α_2 from the set of remaining roots, and so on. Then we have

$$|x_i - \alpha_i| \leq 1 \qquad (i = 1, 2, \ldots, n-1).$$

Indeed, since $f(\alpha_1, y_0) = 1$, we have

$$|x_1 - \alpha_1|^n \leq \prod_{j=1}^{n} |\alpha_1 - x_j| = 1,$$

which gives $|x_1 - \alpha_1| \leq 1$. Now by the condition (5.4) we obtain

$$|x_1 - \alpha_2| = |x_1 - \alpha_1 + (\alpha_1 - \alpha_2)| > 1.$$

Therefore, it follows from $f(\alpha_2, y_0) = 1$ that

$$|x_1 - \alpha_2||x_2 - \alpha_2|^{n-1} \leq \prod_{j=1}^{n} |\alpha_2 - x_j| = 1,$$

which gives $|x_2 - \alpha_2| \leq 1$, etc.

All the numbers $x_1, \ldots, x_{n-1}$ are different, since if, say, $x_1 = x_2$, then

$$|\alpha_1 - \alpha_2| = |x - \alpha_1 - (x_2 - \alpha_2)| \leq 2,$$

which is excluded by the condition (5.4). Hence, if the polynomial $f(x, y_0)$ has a multiple root, that can be only be x_n coinciding with one of the x_i $(i = 1, 2, \ldots, n-1)$, and the polynomial $f(x, y_0)$ has no roots of multiplicity greater than 2.

Thus, if

$$f(x, y_0) = 0, \quad \frac{\partial}{\partial x} f(x, y_0) = 0, \quad \text{then} \quad \frac{\partial^2}{\partial x^2} f(x, y_0) \neq 0. \qquad (5.6)$$

If y_0 is a root of discriminant $D(y)$, then the equation $f(x, y)$ defining the algebraic function $x(y)$ is satisfied by n Puiseux power series representing this function in the neighborhood of $y = y_0$. The equality $D(y_0) = 0$ is equivalent to the assertion that the pair of equations

$$f(x, y_0) = 0, \qquad \frac{\partial}{\partial x} f(x, y_0) = 0 \qquad (5.7)$$

has a solution $x = x_0$. In view of (5.6) this means that $f(x, y_0)$ has a root x_0 of multiplicity 2. For example,

$$x_0 = x_1(y_0) = x_n(y_0), \quad x_i(y_0) \neq x_j(y_0) \qquad ((i, j) \neq (1, n)). \qquad (5.8)$$

Since $D(y)$ has a representation

$$D(y) = \prod_{1 \leq i < j \leq n} (x_i(y) - x_j(y))^2, \qquad (5.9)$$

it is sufficient to determine the order of $(x_1(y) - x_n(y))^2$ with respect to $y - y_0$ to determine the order of the root y_0 of the polynomial $D(y)$.

Taking the Taylor expansion of the polynomial $f(x,y)$ at the point (x_0, y_0), and applying it to x and y connected by the relation $f(x,y) = 0$, and taking into account that x_0 satisfies (5.7), we obtain for y in a neighborhood of y_0 that

$$\frac{\partial}{\partial y} f(x_0, y_0)(y - y_0) + o(|y - y_0|) + \frac{\partial^2}{\partial x^2} f(x_0, y_0) \frac{(x - x_0)^2}{2} + o(|x - x_0|^2) = 0.$$

Since $\dfrac{\partial^2}{\partial x^2} f(x_0, y_0) \neq 0$, in view of (5.6), we have

$$(x - x_0)^2 = \beta(y - y_0)(1 + o(1)), \quad \beta = -2 \frac{\partial f(x_0, y_0)}{\partial y} \Big/ \frac{\partial^2 f(x_0, y_0)}{\partial x^2}.$$

Now we note that

$$\frac{\partial}{\partial y} f(x, y) = -(x - \alpha_1) \ldots (x - \alpha_{n-1}),$$

hence, $\dfrac{\partial}{\partial y} f(x_0, y_0) \neq 0$, since $x_0 \neq \alpha_i$, which is seen from the equalities

$$\frac{\partial}{\partial y} f(x_0, y_0) = 0, \quad f(\alpha_i, y_0) = 1 \qquad (i = 1, 2, \ldots, n - 1).$$

Therefore, as $y \to y_0$ we have

$$x - x_0 = (1 + o(1)) \sqrt{\beta(y - y_0)} \quad \text{or} \quad -(1 + o(1)) \sqrt{\beta(y - y_0)} \qquad (\beta \neq 0).$$

This shows that

$$x_1(y) - x_0 = (1 + o(1)) \sqrt{\beta(y - y_0)},$$
$$x_n(y) - x_0 = -(1 + o(1)) \sqrt{\beta(y - y_0)}$$

or a similar relation is true with $x_1(y)$ replaced by $x_n(y)$, and vice versa. Hence, as $y \to y_0$, we have

$$(x_1(y) - x_n(y))^2 \sim 4\beta(y - y_0),$$

and then it follows from (5.9) that

$$D(y) \sim 4\beta(y - y_0) \prod_{\substack{1 \leq i < j \leq n \\ (i,j) \neq (1,n)}} (x_i(y) - x_j(y))^2.$$

In view of (5.8) we conclude now that y_0 is a root of multiplicity 1, which completes the proof of the lemma.

Now we can easily prove the following theorem:

Theorem 5.1 *Let $n \geq 3$, and let $a_1, a_2, \ldots, a_{n-1}$ be integers with*

$$|a_i - a_j| > 2 \qquad (i, j = 1, 2, \ldots, n - 1; \; i \neq j).$$

Then for any $B > 2(n-1)$ there exists an integral polynomial $g(x)$ such that the algebraic number field generated by a root of the polynomial

$$(x - a_1)(x - a_2)\ldots(x - a_{n-1})(x - g(m)) + 1 \qquad (m = 1, 2, \ldots)$$

has discriminant D_m, and class number h_m, satisfying

$$|D_m| > c_{34}(\ln m)^B, \qquad h_m > c_{35}|D_m|^{1/2 - (n-1)/B - \varepsilon}.$$

Here $c_{34} > 0$, $c_{35} > 0$ are expressible in terms of $a_1, a_2, \ldots, a_{n-1}$, B and ε, where $\varepsilon > 0$ is arbitrary.

Proof. We apply Lemma 4.4 to the discriminant of the polynomial $D(N)$ defined by the equality (5.2); this can be done in virtue of Lemma 5.2. Then one can find an integral polynomial $g(x)$ such that

$$A_2[D * g(m)] > c_{34}(\ln m)^B \qquad (m = 1, 2, \ldots),$$

where $c_{34} > 0$ depends on $a_1, a_2, \ldots, a_{n-1}$, n and B. The discriminant of the field generated by the root of the polynomial $f_N(x)$ with $N = g(m)$ is divisible by the squarefree part of $D*g(m)$, hence we obtain the assertion of the theorem concerning the discriminant D_m. Lemma 5.1 and the Siegel-Brauer theorem now give the inequality for the class number h_m.

In the next paragraph we pass to the statistical description of the algebraic number fields with large class numbers. Now we present a theorem which shows that fields with the condition (2.1) occur quite rarely in the sequence of fields generated by the polynomials (5.2).

Theorem 5.2 *Suppose n, and $a_1, a_2, \ldots, a_{n-1}$ are as in Theorem 5.1, let $M > 1$ be an integer, and let δ be any number in the interval $0 < \delta < 1/2$. Then the number of N in the range $1 \le N \le M$ for which the discriminants D_N and the class numbers h_N of the field generated by the root of the polynomial (5.2) satisfy the inequality*

$$h_N \le |D_N|^\delta, \tag{5.10}$$

is estimated by a value of the order $O((\ln M)^\lambda)$, where

$$\lambda = 2(2^{11}(n-1)^5 + 1)(n-1)(1 - 2\delta')^{-1}.$$

Here δ' is any number in the range $\delta < \delta' < 1/2$, and the symbol O implies a dependence on $a_1, a_2, \ldots, a_{n-1}$, n, and $\delta' - \delta$.

Proof. By the inequalities (5.10) and (3.12) we find that

$$R_N|D_N|^\delta \ge h_N R_N > c(n, \varepsilon)|D_N|^{1/2 - \varepsilon}.$$

Since by Lemma 5.1 the regulator R_N is estimated by the value $c_{33}(\ln N)^{n-1}$, we have

$$|D_N| < c_{36}(\ln N)^{2(n-1)/(1-2\delta')}, \qquad \delta' = \delta + \varepsilon. \tag{5.11}$$

The discriminant D_N is divisible by the square-free part of the number $D(N)$, i.e., the discriminant of the polynomial $f_N(x)$. Therefore, the number of those $N \leq M$ for which (5.11) holds does not exceed the number of those N for which

$$A_2[D(N)] < c_{36}(\ln N)^{2(n-1)/(1-2\delta')}. \tag{5.12}$$

The latter number may be estimated by summation over all natural numbers D not exceeding the right-hand side of (5.12), the number of those N for which

$$A_2[D(N)] = D \tag{5.13}$$

holds. The inequality (5.2) Ch.VI, obtained as a corollary of the Theorem 5.1 Ch.VI, shows that the number of N with (5.13) does not exceed $c_{37} D^{64(2(n-1))^5}$. This gives the proof of the theorem.

One may derive from the work of Ankeny, Brauer and Chowla [4] that the number of those N in the range $M \leq N \leq 2M$ for which (5.10) does not hold, is not less than

$$c_{38} M (\ln \ln M)^{-2(n-1)},$$

where c_{38} depends on δ. It follows from Theorem 5.2 that this number is $M + O((\ln M)^\lambda)$. Thus, we see that the 'vast majority' of fields does not satisfy (5.10) with any δ close to $1/2$ but less than $1/2$.

We notice that both Theorems 5.1 and 5.2 are based on an analysis of the magnitude of the discriminants of algebraic number fields. One can try, of course, to reach more by strengthening the estimate for the regulator given in Lemma 5.1. But to reach an essential progress by this path is impossible since it may be derived from Lemma 3.1 Ch.VI that

$$R_N > c_{39}(\ln N)^{1-\varepsilon}$$

where $c_{39} > 0$ depends only on $a_1, a_2, \ldots, a_{n-1}$ and n, with $\varepsilon > 0$ arbitrary.

In the next paragraph we discuss the distribution of fields with large class number among all the fields of fixed degree; consideration of the Ankeny-Brauer-Chowla fields will again play an essential role.

6. A Statistical Approach

We now discuss the distribution of algebraic number fields $\mathbb{K}$ with class numbers satisfying the inequality

$$h_{\mathbb{K}} \leq |D_{\mathbb{K}}|^\delta, \tag{6.1}$$

where δ is a fixed number in the range $0 \leq \delta < 1/2$, from a purely statistical viewpoint. The results of the previous paragraphs suggest that it is reasonable to estimate the number of different fields of fixed degree with a bounded regulator and then to compare that with the number of such fields subjected to the additional condition (6.1) (cf. [212], [213]).

We start with real quadratic fields.

Theorem 6.1 *For any $x \geq e$ let $E(x)$ be the number of fundamental units $\varepsilon \leq x$ of all real quadratic fields, and for a fixed δ in the range $0 \leq \delta < 1/2$ let $E_\delta(x)$ be the number of such units in the fields satisfying (6.1). Then we have*

$$E(x) = 2 \sum_{k=1}^{[\ln x]} \mu(k) x^{1/k} + O(\ln x), \qquad (6.2)$$

where $\mu(k)$ is the Möbius function, whilst

$$E_\delta(x) < c_{40}(\ln x)^{2/(1-2\delta')}, \qquad (6.3)$$

where δ' is any number in the interval $\delta < \delta' < 1/2$, and c_{40} depends only on $\delta' - \delta$.

Proof. Any quadratic unit $\eta > 1$ and its conjugate satisfies an equation of the form $z^2 - mz = \pm 1$ with positive integer $m = \eta + \eta'$. The number of those η which lie in the interval $1 < \eta \leq x$ is equal to $2x + O(1)$ and is also $E(x) + E(x^{1/2}) + \ldots + E(x^{1/k}) + \ldots$. Consequently,

$$F(x) = \sum_{k=1}^{\infty} E(x^{1/k}) = 2x + O(1). \qquad (6.4)$$

Since the Möbius function $\mu(l)$ satisfies the equations

$$\sum_{l|n} \mu(l) = \begin{cases} 1, & n = 1, \\ 0, & n \neq 1, \end{cases}$$

we find that

$$\sum_{l=1}^{\infty}{}' \mu(l) F(x^{1/l}) = \sum_{l=1}^{\infty} \mu(l) \sum_{k=1}^{\infty} E(x^{1/kl}) = \sum_{n=1}^{\infty} E(x^{1/n}) \sum_{l|n} \mu(l) = E(x).$$

Thus we obtain

$$E(x) = \sum_{l=1}^{\infty} \mu(l) F(x^{1/l}) = \sum_{1 \leq l \leq \ln x / \ln 2} \mu(l) F(x^{1/l}),$$

as we have $E(x^{1/kl}) = 0$ for $k \geq 1$, $l > \ln x / \ln 2$. It now follows from the right-hand side of (6.4) that

$$E(x) = 2 \sum_{l=1}^{[\ln x / \ln 2]} \mu(l) x^{1/l} + O(\ln x),$$

which yields (6.2).

Inequality (6.3) follows directly from Siegel's theorem. Indeed, if ε_0 is a fundamental unit of the field $\mathbb{K}$, $\varepsilon_0 \leq x$ and (6.1) holds, then we find from (1.2) that for any $\tau > 0$ with $\delta + \tau < 1/2$ we have the following inequalities.

$$c(\tau)D^{1/2-\tau} < h \ln \varepsilon_0 \leq D^{\delta} \ln x,$$
$$D < c_{41}(\ln x)^{2/(1-2\delta')} \quad (\delta' = \delta + \tau).$$

Because each real quadratic field is determined by its discriminant, we find (6.3).

Formulae of the type (6.2) also hold for the number of fundamental units $\varepsilon \leq x$ with the condition $\mathrm{Nm}(\varepsilon) = +1$, respectively with the condition $\mathrm{Nm}(\varepsilon) = -1$.

Denote these numbers by $E^{+}(x)$ and $E^{-}(x)$ respectively. Then we easily find similarly to (6.4) that

$$F^{-}(x) = \sum_{k \equiv 1 \pmod 2} E^{-}(x^{1/k}) = x + O(1),$$

from which we obtain

$$E^{-}(x) = \sum_{\substack{k \equiv 1 \pmod 2 \\ 1 \leq k \leq [\ln x]}} \mu(k)x^{1/k} + O(\ln x). \tag{6.5}$$

Comparing this with (6.2) we find

$$E^{+}(x) = \sum_{\substack{k \equiv 1 \pmod 2 \\ 1 \leq k \leq [\ln x]}} \mu(k)x^{1/k} + 2 \sum_{\substack{k \equiv 0 \pmod 2 \\ 1 \leq k \leq [\ln x]}} \mu(k)x^{1/k} + O(\ln x). \tag{6.6}$$

In particular, we observe that

$$E^{+}(x) = E^{-}(x) - 2x^{1/2} + 2x^{1/6} + \ldots + O(\ln x),$$

Thus the magnitude of $E^{-}(x)$ exceeds that of $E^{+}(x)$ for sufficiently large x.

A detailed analysis shows that the formulae (6.2), (6.5), (6.6) may be given in more precise form, in particular,

$$E(x) = 2 \sum_{k=1}^{[\ln x]} \mu(k)[x^{1/k}] + O\left(\frac{\ln x}{(\ln \ln x)^{\gamma}}\right),$$

if one supposes that

$$\sum_{k \leq x} \mu(k) = O(xe^{-(\ln x)^{\gamma}})$$

holds for $\gamma > 0$ uniformly in x. Nowadays such an estimate is known for $\gamma = 3/5 - \delta$, where $\delta > 0$ is arbitrary.

Now we turn to a comparison of the formulae (6.2), (6.5), (6.6) with the estimate (6.3). It follows from the first three formulae that

$$E(x) \sim 2x, \quad E^+(x) \sim E^-(x) \sim x \quad (x \to \infty).$$

Hence, the fundamental units of real quadratic fields are distributed on the real axis as the members of arithmetic progressions. Considering any quadratic field as generated by its quadratic unit, we see from (6.3) that in the sequence of all real quadratic fields placed according to the size of their fundamental units, the fields with class numbers a 'little less' than the extremal upper bound, i.e. satisfying (2.1), occur quite rarely: in the set of all fields with $\varepsilon \leq x$ the fields with class numbers lower than the extremal bound constitute a set of volume a logarithmic order of the volume of all the fields with $\varepsilon \leq x$. The same phenomenon persists on intervals $x \leq \varepsilon \leq x+t$ of relatively small length: one can take $t = (\ln x)^\alpha$ with any $\alpha > 1$ and again observe the domination of fields with large class number.

On similarly considering cubic fields of negative discriminant, it is easy to see that the number of fundamental units (taken greater than 1) of such fields, not exceeding a given bound x, is equal to

$$\tfrac{8}{3}x^{3/2} + O(x). \tag{6.7}$$

Since any unit uniquely defines its field, the number of fields with fundamental unit not exceeding x, is also given by the asymptotic equality (6.7). At the same time the number of fields with fundamental unit not exceeding x and simultaneously satisfying (6.1), is estimated by the value $O((\ln x)^{5/(1-2\delta')})$. We again come across the phenomenon peculiar to quadratic fields. With inessential changes a similar phenomenon is seen in the set of biquadratic fields with two pairs of complex conjugate fields.

We now pass to considerations from a general point of view and prove the following assertion refining the previously obtained result [213].

Theorem 6.2 *Given integers $n \geq 3$ and t, with $0 \leq t < n/2$, and reals $Z > 0$ and δ, with $0 \leq \delta < 1/2$, let $N_n^{(t)}(Z)$ denote the number of distinct (non-isomorphic) algebraic number fields $\mathbb{K}$ of degree n with regulator $R_\mathbb{K} \leq Z$ and having exactly t pairs of complex conjugate fields; $N_{n,\delta}^{(t)}(Z)$ denotes the number of such fields satisfying the additional condition (6.1). Then we have*

$$N_n^{(t)}(Z) < c_{42}\exp(c_{43}Z), \tag{6.8}$$

$$N_n^{(t)}(Z) > \exp(c_{44}Z^{1/(n-t-1)}), \tag{6.9}$$

where c_{42}, c_{43} and $c_{44} > 0$ depend only on n, and

$$N_{n,\delta}^{(t)}(Z) < c_{45}Z^{n/(1-2\delta')}, \tag{6.10}$$

where δ' is any number in the range $\delta < \delta' < 1/2$, and c_{45} depends only on n and $\delta' - \delta$.

Proof. The proof of the theorem splits into three parts corresponding to the inequalities (6.8), (6.9), and (6.10). We shall prove these successively, relying on the following lemmas.

Lemma 6.1 *Suppose that η is an algebraic integer of degree $m \geq 3$ and*

$$\overline{|\eta|} \leq 1 + \frac{1}{30m^2 \ln(6m)};$$

then η is a root of unity.

Proof. See [31].

Lemma 6.2 *Suppose $\mathbb{K}$ is an algebraic number field of degree $n \geq 3$ and t is the number of pairs of complex conjugate isomorphisms of $\mathbb{K}$, so $0 \leq t < n/2$. Then $\mathbb{K}$ contains a unit η of degree n and of height*

$$h(\eta) < c_{46} \exp(c_{47} R_{\mathbb{K}}), \tag{6.11}$$

where $c_{46} = 2^n e^{n-1)\lambda}$, $c_{47} = \lambda^{-n}$, and the value of λ is determined by (6.13).

Proof. Let s be the number of real isomorphisms of $\mathbb{K}$, so $n = s + 2t$. The number of fundamental units of $\mathbb{K}$ is equal to $r = s + t - 1$. We may suppose that $r \geq 2$, since if $r = 1$ the fundamental unit of the field satisfies (6.11), and its degree is $n = 3$.

We employ the usual enumeration of the conjugate fields $\mathbb{K}^{(i)}$. The indices $(1), (2), \ldots, (s)$ correspond to real fields; and the indices $(s+1), \ldots, (s+2t)$ to complex fields, with $\mathbb{K}^{(s+i)}$ and $\mathbb{K}^{(s+t+i)}$ $(i = 1, 2, \ldots, t)$ complex conjugate fields. We set $e_i = 1$ for real and $e_i = 2$ for complex isomorphisms.

Suppose $\varepsilon_1, \ldots, \varepsilon_r$ is a system of fundamental units of the field $\mathbb{K}$. Consider the system of linear inequalities

$$|x_1 \ln |\varepsilon_1^{(i)}|^{e_i} + \ldots + |x_r \ln |\varepsilon_r^{(i)}|^{e_i}| \leq \lambda_i \qquad (i = 1, 2, \ldots, r), \tag{6.12}$$

where $\lambda_1 = \lambda^{-r+1} R$, $\lambda_i = \lambda$, $(i = 2, 3, \ldots, r)$, and

$$\lambda = \frac{2}{n-2} \ln(1 + \frac{1}{7,5n^2 \ln(3n)}). \tag{6.13}$$

Since the absolute value of the determinant of the system (6.12) is equal to R and the product of the right-hand sides is also R, by Minkowski's theorem there exists a non-trivial integral solution of the system. We set $\eta = \varepsilon_1^{x_1} \ldots \varepsilon_r^{x_r}$, where $(x_1, \ldots, x_r) \neq (0)$ is an integral solution of (6.12), and we shall prove that degree of η is equal to n.

First we observe that $s \geq 1$, $e_1 = 1$, and that it follows from (6.12) that

$$\exp\{-\lambda^{-r+1} R\} \leq |\eta^{(1)}| \leq \exp\{\lambda^{-r+1} R\}, \tag{6.14}$$

$$\exp\{-\lambda\} \leq |\eta^{(i)}|^{e_i} \leq \exp\{\lambda\}, \qquad (i = 1, \ldots, r). \tag{6.15}$$

Because we have

$$\prod_{i=1}^{r+1} |\eta^{(i)}|^{e_i} = |\mathrm{Nm}(\eta)| = 1,$$

it follows from (6.14), (6.15) that

$$|\eta^{(r+1)}|^{e_{r+1}} = \prod_{i=1}^{r} |\eta^{(i)}|^{-e_i} \le e^{(r-1)\lambda} e^{\lambda^{-r+1}R}, \qquad (6.16)$$

and as $|\eta^{(s+i)}| = |\eta^{(s+t+i)}|$, the inequalities (6.14)-(6.16) yield upper bounds for the absolute values of all the numbers $\eta^{(i)}$ $(i = 1, 2, \ldots, n)$.

Suppose η has degree $m < n$. Then the set of numbers $\eta^{(1)}, \eta^{(2)}, \ldots, \eta^{(n)}$ falls into n/m identical groups with m numbers in each group, namely the set of roots of the minimal polynomial for η. If $t \ne 0$, then the numbers $\eta^{(r+1)}$ and $\eta^{(r+t+1)}$ are either equal real numbers or non-real complex conjugate numbers, because they lie in complex conjugate fields $\mathbb{K}^{(r+1)}$ and $\mathbb{K}^{(r+t+1)}$. In the first case at least one group contains no more than one of the numbers $\eta^{(1)}$, $\eta^{(r+1)}$ and $\eta^{(r+t+1)}$. Because the product of the absolute values of the numbers from any one group is equal to 1, the absolute value of this number does not exceed $e^{(m-1)\lambda}$ (a consequence of (6.15)). In the second case $\eta^{(r+1)}$ and $\eta^{(r+t+1)}$ lie in the same group, while $\eta^{(1)}$ as a real number does not coincide either with $\eta^{(r+1)}$, nor with $\eta^{(r+t+1)}$. Since the groups are identical, these numbers coincide with certain numbers of other groups for which (6.15) holds. Thus, in virtue of (6.13) we obtain

$$|\overline{\eta}| \le e^{(\frac{n}{2}-1)\lambda} = 1 + \frac{1}{7.5n^2 \ln(3n)}.$$

Then Lemma 6.1 shows that η is a root of 1, but this is excluded, since $\eta = \varepsilon_1^{x_1} \cdots \varepsilon_r^{x_r}$ and $(x_1, \ldots, x_r) \ne (0)$. We conclude that the degree of η is equal to n.

We observe from (6.14), (6.15) that

$$h(\eta) \le 2^n e^{\lambda^{-n}R + (n-1)\lambda}$$

holds, and in view of (6.13) we obtain (6.11).

Lemma 6.3 *The suppositions of Theorem 6.2 imply the inequality (6.8).*

Proof. Obviously $N_n^{(t)}(Z)$ does not exceed the number of integral polynomials of degree n with leading coefficient 1 and of height not exceeding the right-hand side of (6.11), since Lemma 6.2 guarantees the existence in each field under consideration of a generating element which is the root of such a polynomial. Consequently,

$$N_n^{(t)}(Z) < (2^{n+1} e^{\lambda^{-n}R + (n-1)\lambda} + 1)^n,$$

which gives (6.8).

Lemma 6.4 *Let s and t be natural numbers with $s + 2t = n \ge 3$, and let $a_1, a_2, \ldots, a_{s+t-1}$ be distinct integers, with $a_{s+1} > 0, \ldots, a_{s+t-1} > 0$. Then for* *sufficiently large integral M the number of distinct (non-isomorphic) algebraic number fields generated by the roots of the polynomials*

$$g_N(x) = (x - a_1)\ldots(x - a_s)(x^2 + a_{s+1})\ldots(x^2 + a_{s+t-1})(x^2 + N) + 1,$$

where N takes all integral values in the interval $M \leq N \leq 2M$, is no less than $M/n!$.

Proof. We suppose that $t = 0$ since the general case may be dealt with in a quite similar way.

We can order the roots $\alpha_1, \ldots, \alpha_n$ of the polynomial $g_N(x)$, where N is sufficiently large, so that we have

$$|\alpha_j - a_j| = \min_{1 \leq i \leq n} |\alpha_i - a_j| \quad (1 \leq j \leq n - 1),$$

$$|\alpha_n - N| = \min_{1 \leq i \leq n} |\alpha_i - N|. \tag{6.17}$$

It is then easy to see that

$$|\alpha_j - a_j| < c_{47} N^{-1} \quad (1 \leq j \leq n - 1),$$

$$|\alpha_n - N| < c_{47} N^{-n+1}, \tag{6.18}$$

where c_{47} depends only on n and $a_1, \ldots, a_{n-1}$. We write $\mathbb{K}_i = \mathbb{Q}(\alpha_i)$ where $(i = 1, 2, \ldots, n)$. If the polynomials $g_N(x)$ and $g_{N'}(x)$ define the same field $\mathbb{K}$, and $\alpha'_1, \ldots, \alpha'_N$ are the roots of $g_{N'}(x)$ arranged as for (6.17) and $\mathbb{K}'_i = \mathbb{Q}(\alpha'_i)$, then $\mathbb{K}'_1, \ldots, \mathbb{K}'_n$ is a permutation of $\mathbb{K}_1, \ldots, \mathbb{K}_n$. We can see that in the limit for N such that $M \leq N \leq 2M$ with M sufficiently large, there are no more than $n! - 1$ polynomials $g_{N'}(x)$ defining the field $\mathbb{K}$. Indeed, if there were $n!$ such polynomials, then one of two possibilities hold: $\mathbb{K}'_i = \mathbb{K}_i$, for $(i = 1, 2, \ldots, n)$ with some $N' \neq N$, $M \leq N \leq 2M$, or $\mathbb{K}'_i = \mathbb{K}''_i$ $(i = 1, 2, \ldots, n)$ for a pair of distinct N', N''. In the first case we observe that the product

$$\prod_{i=1}^n (\alpha_i - \alpha'_i) = \mathrm{Nm}(\alpha - \alpha')$$

is a rational integer distinct from zero, and we obtain from (6.18) that

$$1 \leq \prod_{i=1}^n |\alpha_i - \alpha'_i| \leq c_{48}(M^{-1})^{n-1} M = c_{48} M^{-n+2},$$

which is impossible for large enough M. In the second case the same is true for the number

$$\prod_{i=1}^n (\alpha'_i - \alpha''_i) = \mathrm{Nm}(\alpha' - \alpha'').$$

Thus we see that the polynomials $g_N(x)$ define no fewer than $M/n!$ different fields $\mathbb{K}$, which is what is asserted by the lemma.

Lemma 6.5 *Under the conditions of Theorem 6.2 the inequality (6.9) holds.*

Proof. It is easy to see that the regulator of the field $\mathbb{K}$ generated by the root of the polynomial $g_N(x)$ is bounded above by $c_{49}(\ln N)^{s+t-1}$ (see the proof of Lemma 5.1, where the case of $t = 0$ is considered). Taking $M = \exp(c_{50}Z^{1/(s+t-1)})$, by Lemma 6.4 we obtain (6.9).

Lemma 6.6 *Under the conditions of Theorem 6.2 the inequality (6.10) holds.*

Proof. We note that every field $\mathbb{K}$ of degree n contains an integral generating element of height no greater than $2^n(|D_k| + 1)^{1/2}$. Inequality (3.12) and the suppositions (6.1) and $R_{\mathbb{K}} \le Z$ yield

$$|D_k|^{1/2-\varepsilon} < c(n,\varepsilon)h_{\mathbb{K}}R_{\mathbb{K}} < c(n,\varepsilon)|D_k|^\delta Z,$$

so that $|D_k| < c_{51}Z^{2/(1-2\delta')}$, $\delta' = \delta + \varepsilon$. Consequently, any field of degree n with restrictions (6.1) and $R_{\mathbb{K}} \le Z$ is generated by a root of an integral polynomial having degree n, leading coefficient 1 and height no greater than $c_{52}Z^{1/(1-2\delta')}$. On estimating the number of such polynomials we obtain (6.10). Lemmas 6.3, 6.5 and 6.6 yield Theorem 6.2.

A comparison of the inequalities (6.9) and (6.10) shows that

$$N_{n,\delta}^{(t)}(Z) < c_{52}(\ln N_n^{(t)}(Z))^{n(n-t-1)/(1-2\delta')}, \tag{6.19}$$

that is, in the set of fields of degree n with t pairs of complex conjugate fields and with regulator not exceeding a given bound Z, the fields with small class numbers (the inequality (6.1)) constitute a relatively small part.

We have not dealt with the case $t = n/2$. Indeed, then Lemma 6.2 is false and the number of the fields with bounded regulators is infinite. Nevertheless, an assertion similar to the preceding ones holds in this case too, if one makes the necessary adjustments. The assertion of Lemma 6.2 becomes true (with other values for c_{46} and c_{47}), if one excludes the fields $\mathbb{K}$ which are imaginary quadratic extensions of totally real fields; for the remaining fields an estimate of the form (6.8) still holds, so that the corresponding number $N_n^{(t)}(Z)$ is at any rate finite. The inequalities (6.9), (6.10) were obtained independently from the hypothesis $t < n/2$ so we obtain (6.19) again, where $N_{n,\delta}^{(t)}(Z)$ and $N_n^{(t)}(Z)$ now signify the numbers of corresponding fields, excluding imaginary quadratic extensions, of totally real fields.

Concerning the latter fields we note that their class numbers are bounded by the following effective inequality due to Stark [228]:

$$h_{\mathbb{K}} > c_{53}|d|^{\frac{1}{2}-\frac{1}{n}-\varepsilon}|f|^{\frac{1}{2}-\frac{1}{2n}},$$

where d is the discriminant of the totally real subfield $\mathbb{K}$ of degree $n/2$, and $D_{\mathbb{K}} = d^2f$, with $\varepsilon > 0$ arbitrary, and c_{53} depending on n and ε. We see that in the case of $n \ge 4$ the magnitude of $h_{\mathbb{K}}$ increases without bound together with $|D_{\mathbb{K}}|$.

It is interesting to note that for all fields, excepting imaginary quadratic extensions of totally real fields, Remak ([163], [164]) proved the inequality

$$\ln|D_{\mathbb{K}}| \le n\ln n + \frac{R_{\mathbb{K}}}{R_{\min}}\ln 2\frac{2n(n-1)}{r+1}g^{(r-1)/2},$$

where r is the number of fundamental units of the field $\mathbb{K}$, and

$$g = \tfrac{1}{2}(r+5+\sqrt{r^2+2r+17}),$$

whilst $R_{\min}$ is an absolute lower bound for the regulators of all fields of degree n. On considering fields with $R_{\mathbb{K}} \le Z$ we obtain an estimate of the form (6.8) again. It is also proved in Remak's papers that over a given totally real field one can construct infinitely many imaginary quadratic extensions with uniformly bounded regulators.

Recently Zimmert [247] proved that $R_{\min} \ge 0.056$.

7. Conjectures and Perspectives

The following hypothesis [210] gives a possibility for essential improvements on the bounds for solutions of all the diophantine equations discussed in the present volume.

(H) *Given an algebraic number field $\mathbb{K}$ of degree n and some $\varepsilon > 0$, there exists an extension $\mathbb{L}$ of the field $\mathbb{K}$ such that*

$$[\mathbb{L}:\mathbb{K}] < c_{54}, \qquad R_{\mathbb{L}} < c_{55}|D_{\mathbb{K}}|^{\varepsilon},$$

where $R_{\mathbb{L}}$ is the regulator of $\mathbb{L}$, $D_{\mathbb{K}}$ is the discriminant of $\mathbb{K}$, and c_{54} depends only on n; here c_{55} depends on n and ε.

All the basic bounds for solutions of the diophantine equations we have considered depend on the regulators of the corresponding fields, so the influence of hypothesis (H) on the magnitude of these bounds is quite obvious. For example, we would have the bound

$$\max(|x|,|y|) < \exp\{c_{56}(|A|H_f)^{\varepsilon}\},$$

where c_{56} depends only on the degree of the form and on ε, for the solutions of Thue equation (1.1) Ch.IV. A similar estimate would hold for solutions of the elliptic and hyperelliptic equations, and so on. The reason for there being such a strong influence of hypothesis (H) on the bounds for the solutions is explained by Lemma 2.3 Ch.II: the hypothesis implies that the regulator of any algebraic number field $\mathbb{K}$ is approximated by $c_{57}|D_{\mathbb{K}}|^{\varepsilon}$. Consequently, in the Siegel-Brauer formula the regulator may be removed:

$$\ln h_{\mathbb{K}} \sim \frac{1}{2}\ln|D_{\mathbb{K}}| \qquad (|D_{\mathbb{K}}| \to \infty). \tag{7.1}$$

In particular, there should exist only finitely many algebraic number fields of of bounded degree with bounded class number, since for all fields the 'extremal' inequality

$$h_{\mathbb{K}} > c_{58}|D_{\mathbb{K}}|^{1/2-\varepsilon}$$

would hold ($\varepsilon > 0$ is arbitrary, and $c_{58} > 0$ depends only on the degree of $\mathbb{K}$ and ε).

It is apparent that either a justification for or a disproof of hypothesis (H) is of exceptional importance for further development of our present subject.

It is interesting to note that the asymptotic formula (7.1) is equivalent to the following assertion:

Every algebraic unit η of infinite order satisfies

$$\overline{|\eta|} > c_{59} R_\eta, \tag{7.2}$$

where R_η is the regulator of the field $\mathbb{Q}(\eta)$, and $c_{59} > 0$ depends only on the degree of η.

Indeed, given $\varepsilon > 0$, let m be an integer no less than $1/\varepsilon$, and set $d = \deg \eta$. From (7.2) we obtain by Lemma 2.3 Ch.II that

$$\overline{|\eta|}^{1/m} = \overline{|\eta^{1/m}|} > c_{60}(d,m) R_{\eta^{1/m}} > c_{61}(d,m) R_\eta,$$

whence,

$$\overline{|\eta|} > c_{62}(d,\varepsilon) R_\eta^{1/\varepsilon}, \qquad c_{62}(d,\varepsilon) > 0. \tag{7.3}$$

Now let $\mathbb{K}$ be an algebraic number field of degree n. It is well known that one can find an integer $\alpha \in \mathbb{K}$ of degree n and satisfying

$$\overline{|\alpha|} \leq (|D_{\mathbb{K}}| + 1)^{1/2}. \tag{7.4}$$

A root η of the polynomial $x^2 + \alpha x + 1$ is an algebraic unit which may be supposed different from a root of unity, and we have $\overline{|\eta|} \leq \overline{|\alpha|}$. Again by Lemma 2.3 Ch.II we obtain from (7.3), (7.4) that

$$(|D_{\mathbb{K}}| + 1)^{1/2} \geq \overline{|\alpha|} \geq \overline{|\eta|} > c_{62}(d,\varepsilon) R_\eta^{1/\varepsilon} > c_{63}(n,\varepsilon) R_{\mathbb{K}}^{1/\varepsilon},$$

from which it follows that $R_{\mathbb{K}} < c_{64} |D_{\mathbb{K}}|^{\varepsilon/2}$. Thus, from (7.2) we obtain (7.1).

To demonstrate that the converse of (7.2) leads to the negation of the asymptotic formula (7.1), we observe that the discriminant D_η of the field $\mathbb{Q}(\eta)$, as a divisor of the discriminant $D(\eta)$ of the number η, may be estimated by way of

$$|D_\eta| \leq |D(\eta)| = \prod_{1 \leq i < j \leq d} |\eta^{(i)} - \eta^{(j)}|^2 < (2\overline{|\eta|})^{(d-1)d}.$$

Hence, for those units for which (7.2) does not hold (with some definite function $c_{59} = c_{59}(d)$), we find that

$$R_\eta > c_{65} \overline{|\eta|} > c_{66} |D_\eta|^{1/d(d-1)}.$$

Now Landau's inequality (Lemma 2.4 Ch.II) yields

$$h_\eta < c_{67} |D_\eta|^{1/2} (\ln |D_\eta|)^{d-1} R_\eta^{-1} < c_{68} |D_\eta|^{1/2 - 1/d(d-1)} (\ln |D_\eta|)^{d-1},$$

and, if the inequality (7.2) does not hold for infinitely many units of bounded degree, the asymptotic relation (7.1) cannot hold.

These questions on the lower bounds for the sizes or heights of algebraic units have scarcely been investigated, even for the case of real quadratic fields. In the latter case the best result known to the author is that of Yamamoto [246], who proved the following theorem:

Suppose that p_i $(i = 1, 2, \ldots, n)$ are distinct rational prime numbers such that there are an infinite number of real quadratic fields $\mathbb{K}$ in which these numbers decompose into the product of two principal prime ideals. Then the inequality

$$R_{\mathbb{K}} > c_{69}(\ln |D_{\mathbb{K}}|)^{n+1}$$

holds, where $c_{69} > 0$ depends only on n, and the primes $p_1, \ldots, p_n$.

Yamamoto notes that the fields $\mathbb{K}_m = \mathbb{Q}(\sqrt{D_m})$, where

$$D_m = (p^{2m}q + p + 1)^2 - 4p \qquad (m = 1, 2, \ldots) \tag{7.5}$$

with p, q distinct prime numbers, satisfy his theorem, so that one obtains an infinite sequence of real quadratic fields $\mathbb{K}$, the fundamental units ε_0 of which satisfy

$$\varepsilon_0 > e^{c_{70}}(\ln |D_{\mathbb{K}}|)^3.$$

The discriminants of the fields $\mathbb{K}_m$ may be estimated from below as the square-free divisors of the numbers D_m, using the following equality

$$D_m = q^2 x^4 + 2q(p+1)x^2 + (p-1)^2, \quad x = p^m,$$

obtained from (7.5), and Theorem 4.1 Ch.VI.

It is not known whether the hypotheses of Yamamoto's theorem hold when $n \geq 3$.

It is not too hard to see that for a proof of existence of an infinite sequence of real quadratic fields with relatively small class numbers it suffices to prove the existence of infinitely many not too large intervals of natural numbers $(N, N + M)$ containing natural numbers m with $m^2 - 1$ having a relatively small square-free kernel. Such an approach leads to the following question.

Let M_N be the smallest M for which

$$\prod_{p | N(N+1)\ldots(N+M)} p < M^M$$

holds. Is it true that $M_N < (\ln N)^{c_{71}}$ holds infinitely often, where $c_{71} \geq 1$ is an absolute constant?

If the answer is affirmative, then there exists an infinite number of real quadratic fields $\mathbb{K}$, the fundamental units ε_0 of which satisfy

$$\varepsilon_0 > \exp\{c_{72}D_{\mathbb{K}}^{1/3c_{71}}(\ln D_{\mathbb{K}})^{-1}\},$$

hence, the class numbers $h_{\mathbb{K}}$ are less than

$$c_{73} = D_{\mathbb{K}}^{1/2 - 1/(3c_{71})} \ln D_{\mathbb{K}},$$

where $c_{72} > 0$ and $c_{73} > 0$ are absolute constants. Probably, we do not have enough information on the distribution of prime numbers to give the answer to the above question.

Many problems concerning the structure of the products of consecutive natural numbers show a deep connection with the problems of the distribution of prime numbers. For instance, there is a conjecture due to Grimm [80]: if the numbers $N + 1, \ldots, N + M$ are composite, then the number of distinct prime divisors of the product, $(N + 1) \ldots (N + M)$ is no less than M. It follows from this that the consecutive prime numbers p_n satisfy

$$p_{n+1} - p_n < \left(\frac{p_n}{\ln p_n} \right)^{1/2}$$

for sufficiently large n.

Recently Ramachandra, Shorey and Tijdeman [162] have made some progress on Grimm's conjecture, while Dobrowolski [57] has obtained an improvement of Lemma 6.1, close to the hypothesised assertion.

IX. Reducibility of Polynomials and Diophantine Equations

In this final chapter, we deal with a 'meta'-topic, one that lies 'over' the theory of diophantine equations, as it were; namely arithmetic specialisation of polynomials. The main result asserts that under such specialisations the multiplicative structure of the numbers obtained goes some considerable way towards determining the multiplicative structure of the original polynomials. This allows one to give effective versions of Hilbert's irreducibility theorem and to describe all abelian points on algebraic curves. The methods used are quite independent of the theory of linear forms in the logarithms of algebraic numbers. and rely on the study of the arithmetic structure of sums of algebraic power series in all metrics of the field of rational numbers.

1. An Irreducibility Theorem of Hilbert's Type

Let $F(x,y)$ be an integral polynomial irreducible in the polynomial ring $\mathbb{Q}[x,y]$. There are two major problems on the arithmetic properties of such polynomials that have attracted attention for a long time. The first problem is to determine or to describe all rational points on the curve

$$F(x,y) = 0. \tag{1.1}$$

Obviously, this problem is equivalent to the one of determining or describing all rational x_0 for which $F(x_0,y)$ has a linear factor in $\mathbb{Q}[y]$. However, even for the case of special polynomials $F(x,y)$, unsurmountable difficulties sometimes arise in solving this problem, and it has been considered as one of the most difficult problems ever since the time of Fermat and Euler. The second problem is much younger (some 90 years old), but not any easier. This is the problem of determining all rational x_0 for which $F(x_0,y)$ is irreducible in $\mathbb{Q}[y]$.

Let $\mathfrak{D}_F$ and $\mathfrak{H}_F$ be the sets of rational x_0 which constitute the solution of the first and to the second problems respectively. If $\deg_y F(x,y) \geq 2$, these sets have no point in common:

$$\mathfrak{D}_F \cap h_F = \emptyset.$$

At first sight it seems very difficult to predict any relation between the multiplicative structure of a number $x_0 \in \mathbb{Q}$ and the multiplicative structure of a corresponding polynomial $f(x_0,y) \in \mathbb{Q}[y]$. However, such a relation exists, as we shall see below, and that will allow us to characterise $\mathfrak{H}_F$ as an 'extensive' set and will consequently allow us to characterise $\mathfrak{D}_F$ as well.

In 1892 Hilbert [98] proved that $\mathfrak{H}_F$ is infinite for an irreducible polynomial $F(x,y)$. This is the celebrated 'Hilbert Irreducibility Theorem'. Hilbert himself, and later Emmy Noether [148], gave applications of this theorem to the solution of the problem of constructing algebraic equations with given Galois group. In our time Hilbert's theorem remains a corner stone in investigations on the inverse problem of Galois theory.

Hilbert's proof of the irreducibility theorem is ineffective and does not make it possible to determine elements of the set $\mathfrak{H}_F$. However, Hilbert noticed that his reasoning can be supplemented by appropriate estimates to meet this failing.

The theorem has been generalised repeatedly and several new proofs of it have been suggested. In particular, one now knows generalisations to systems of polynomials in many unknowns over a finite extension of the field of rational numbers: these generalisations are of a purely technical nature and the case of polynomials in two unknowns over the field of rational numbers remains the basic one (see [120], [171]). Siegel [193] observed that finiteness theorems for the sets $\mathfrak{D}_G$ for certain polynomials $G \in \mathbb{Q}[x,y]$ connected directly with the initial polynomial F allow one to prove Hilbert's theorem. In that way he obtained a proof of Hilbert's theorem from his theorem on the finiteness of integral points on curves of genus greater than zero. This shows that there is a close connection between the two problems mentioned above on the sets $\mathfrak{D}_F$ and $\mathfrak{H}_F$. In whatever form, such types of connections constitute the basis for many investigations on Hilbert's theorem.

In many papers the set $\mathfrak{H}_F$ is described as an 'extensive' set from various points of view (for example, $\mathfrak{H}_F$ contains 'almost all' natural numbers and its elements lie densely in $\mathbb{Q}$ and in $\mathbb{Q}_p$ for all p). Schinzel [169] proved that $\mathfrak{H}_F$ contains an arithmetic progression, and Fried [74] recently demonstrated by another method that in any concrete case such a progression may be found by a finite number of computations (the progression, of course, depends on the polynomial). An interesting approach is taken in the work of Cohen [50], who gives a statistical characterisation of the set of those natural numbers x_0 for which the polynomial $F(x_0, y)$ has the same Galois group over $\mathbb{Q}$ as does the polynomial $F(x, y)$ over $\mathbb{Q}(x)$.

In this chapter we describe a new approach to assertions similar to Hilbert's theorem, which allows us to go further both in respect of obtaining effective results and in permitting a deeper analysis of the set $\mathfrak{H}_F$. More than that, we describe a phenomenon new in principle: the influence of the multiplicative structure of the number x_0 on the multiplicative structure of the polynomial $F(x_0, y)$. Our arguments rely on the ideas of the theory of diophantine approximation. As a first application of these notions we consider the following theorem [220].

Theorem 1.1 *Let $F(x,y)$ be an integral absolutely irreducible polynomial, with*

$$n = \deg_y F(x,y) \geq 2, \quad F(0,0) = 0, \quad \frac{\partial}{\partial y}F(0,0) \neq 0. \qquad (1.2)$$

Suppose that a and b are rational integers, with $(a,b) = 1$, and let a_p denote the maximal power of some prime p in a (the p-component of a). Let

$$\max(|a|, |b|) < C a_p^{\frac{n}{n-1}-\delta} \qquad (0 < \delta \leq \frac{n}{n-1}). \qquad (1.3)$$

Then, if $a_p > c_1$, the polynomial $F(a/b, y)$ is irreducible in $\mathbb{Q}[y]$, where c_1 is effectively determined in terms of C, δ and $F(x,y)$.

For example, the polynomial $F(p^t, y)$ is irreducible in $\mathbb{Q}[y]$ if $p^t > c_1$. Here p is a prime number and $t > 0$ is an integer, and we may fix p and vary t, or fix t and vary p.

Another corollary deals with the arithmetic structure of rational points on the curve (1.1). Suppose $x_0 \in \mathcal{D}_F$, with $x_0 = u/v$, $(u,v) = 1$ and $|v| \leq C'|u|$. Let p be a prime divisor of u, and u_p the p-component of u. Then

$$u_p < c_2 |u|^{1-(1/n)+\varepsilon}, \qquad (1.4)$$

where $\varepsilon > 0$ is arbitrary and c_2 is effectively expressible in terms of C, ε and the parameters of the polynomial, but does not depend on p.

We see that elements x_0 of the set $\mathcal{D}_F$ for polynomials F satisfying conditions of the Theorem 1.1 have multiplicative structure: the p-component of the numerator of x_0 can not be 'too large' compared with the magnitude of the numerator. Hence, the diophantine equation

$$F(p^t, y) = 0$$

in which p^t and y are unknowns, has only a finite number of solutions, and an effective bound for the size of the solutions may be given. In fact, the bound takes the form

$$\max(p^t, |y|) < c_3 H^{c_4}, \qquad (1.5)$$

where H is the height of the polynomial $F(x,y)$ and c_3, c_4 depend on the degree of F only (later, we obtain such an estimate as a consequence of an improved version of the Theorem 1.1; cf. Theorem 3.1 below).

The exponent $1 - (1/n) + \varepsilon$ in (1.4) is best possible up to ε, as can be seen by considering the polynomial $F(x,y) = x - y(y+1)^{n-1}$.

We now explain the basic idea of the proof of Theorem 1.1. The equation $F(x,y) = 0$ is satisfied by a formal power series

$$f(x) = \sum_{v=1}^{\infty} f_v x^v, \quad f_v \in \mathbb{Q}, \qquad (1.6)$$

determined by the analytic function $y(x)$ satisfying this equation with initial conditions $x = 0$, $y = 0$. By a well known theorem of Eisenstein (see below, Lemma 3.2) there exists a natural number $E = E_f$ such that

$$E^v f_v \in \mathbb{Z} \qquad (v = 1, 2, \ldots).$$

If p is a prime not dividing E, then the series (1.6) converges in the p-adic metric $\lceil \ \rceil_p$ and defines an analytic p-adic function $f(w)_p$ in the disc $|w|_p < 1$, $w \in \mathbb{Q}_p$. We prove that the p-adic number $f(a/b)_p$, where a and b satisfy the conditions of Theorem 1.1, is algebraic of degree n over $\mathbb{Q}$, that is

$$[\mathbb{Q}(f(a/b)_p) : \mathbb{Q}] = n. \tag{1.7}$$

Since the function $f(w)_p$, $(|w|_p < 1, w \in \Omega_p)$ satisfies the equation

$$F(w, f(w)_p) = 0$$

identically w, the polynomial $F(a/b, y)$ of degree n in y has a root $f(a/b)_p$ of degree n, hence, the polynomial $F(a/b, y)$ is irreducible.

To prove (1.7) we consider an auxiliary power series

$$V(x) = A_0(x) + A_1(x)f(x) + \cdots + A_{n-1}(x)f^{n-1}(x), \tag{1.8}$$

in which $A_j(x) \in \mathbb{Z}[x]$, $(j = 0, 1, \ldots, n-1)$ are chosen in such a way that $V(x)$ has a zero of large order at $x = 0$. Our intention is to obtain a polynomial with integer coefficients which takes a 'small' value in the p-adic metric at the point $f(a/b)_p$ but is not equal to zero at this point. Unfortunately, this cannot be done by direct substitution of $x = a/b$ into (1.8) and then considering the expressions in the p-adic metric: $V(a/b)_p$ may be equal to zero (if not, we have achieved our goal!). Hence, it is necessary to differentiate (1.8) many times, until we obtain a value not zero at $x = a/b$. This again leads to an expression of the form (1.8), but we now have polynomials $A_j(x)$ with too large a height threatening the construction.

Nevertheless, this approach can be saved by a purely algebraic idea as in the proof of Theorem 1.1 below.

2. Main argument

In the immediate sequel we assume that the polynomial $F(x, y)$ is reduced to the form

$$F(x, y) = y^n - a_1(x)y^{n-1} - \ldots - a_n(x), \qquad a_i(x) \in \mathbb{Z}[x],$$

which, obviously, loses no generality.

It is apparent that the resultant $D(x)$ of the polynomial $F(x, y)$ and of $(\partial/\partial y)F(x, y)$ in respect of y is not identically zero, so there are only a finite number of x' such that $F(x', y)$ and $(\partial/\partial y)F(x', y)$ have a common zero in respect to y. All such x' satisfy the equation $D(x') = 0$, and hence may be explicitly determined (in particular, their heights are bounded by a power of the height of F). In subsequent discussion when we substitute $x = x_0$, $(x_0 = a/b)$ we may demand that x_0 does not coincide with any of the numbers x'; thus, for instance,

$$\frac{\partial}{\partial y} F(x_0, f(x_0)) \neq 0,$$

where $f(x_0)$ is defined as a sum of the series (1.6) with $x = x_0$, and in the corresponding metric.

We give a short demonstration of the Theorem 1.1 below. The further discussion detailed in §§3-5 will give more details and provides improvements to our arguments, while the introduction of new ideas gives deeper results (see Theorems 3.1, 4.1 and 4.2).

Lemma 2.1 *If the integral polynomial $F(x, y)$ is absolutely irreducible, then for all primes $q > c_5$ the reduced polynomial*

$$F(x, y) \pmod{q} = \overline{F}(x, y)$$

is irreducible $\mathbb{F}_q[x, y]$, where $\mathbb{F}_q$ is the residue field $\pmod{q}$ and c_5 is effectively determined by the coefficients and degree of $F(x, y)$.

This is a well-known theorem of Ostrowski [153] (see also [184] p. 193).

Lemma 2.2 *Let q be a prime, $q > c_5$, with $q \nmid E_f$. Then the formal power series*

$$\overline{f}(x) = \sum_{v=1}^{\infty} \overline{f}_v x^v,$$

obtained from the series (1.6) and the embedding

$$f_v \to \overline{f}_v \pmod{q} \qquad (v = 1, 2, \ldots),$$

is algebraic one over $\mathbb{F}_q[x]$ and

$$[\mathbb{F}[x, \overline{f}(x)] : \mathbb{F}_q[x]] = n.$$

Proof. To see this we need only the remark that $\overline{f}(x)$ satisfies the equation $\overline{F}(x, \overline{f}(x)) = 0$, and the irreducibility of $\overline{F}(x, y)$ in $\mathbb{F}_q[x, y]$ provided by Lemma 1.

Lemma 2.3 *Suppose that the coefficients of the series (1.6) satisfy the inequalities*

$$|f_v| \leq A^v \qquad (v = 1, 2, \ldots), \qquad A \geq 1, \tag{2.1}$$

where $h > 0$ is an arbitrary integer and $0 < \varepsilon < 1$, with $\varepsilon n h \geq 1$. Then there exist polynomials

$$A_j(x) \in \mathbb{Z}[x] \qquad (j = 0, 1, \ldots, n - 1)$$

of degrees no greater than h, and heights no greater than $H_0 = c_6 (2 E_f A)^{nh/\varepsilon}$, and not all equal to zero, such that the power series (1.8) in x has order

$$\operatorname{ord} V(x) \geq (1 - \varepsilon) n h. \tag{2.2}$$

The value c_6 can be expressed effectively in terms of $E_f A$, n, ε and does not depend on h.

Proof. We set

$$A_j(x) = \sum_{i=0}^{h} a_{ij} x^i \qquad (0 \le j \le n-1), \tag{2.3}$$

where the a_{ij} are unknown integers. We find that

$$f^j(x) = \sum_{k=j}^{\infty} d_{kj} x^k,$$

$$d_{kj} = \sum_{v_1 + \ldots + v_j = k} f_{v_1} f_{v_2} \cdots f_{v_j}, \quad E_j^k d_{kj} \in \mathbb{Z}.$$

Then by virtue of (2.1)

$$E_j^k |d_{kj}| \le k^n (E_f A)^k \qquad (1 \le j \le n-1; k = j, j+1, \ldots).$$

Further, for polynomials (2.3), $V(x)$ takes the form

$$V(x) = \sum_{l=0}^{\infty} b_l x^l, \qquad b_l = a'_{l0} + \sum_{j=1}^{n-1} \sum_{\substack{i+k=l \\ 0 \le i \le h, k \ge j}} a_{ij} d_{kj},$$

where $a'_{l0} = a_{l0}$ for $l = 0, 1, \ldots, h$ and $a'_{l0} = 0$ for $l \ge h+1$.

Now set $l_0 = [(1-\varepsilon)(h+1)n]$ and consider the system of equations $b_l = 0$, $(l = 0, 1, \ldots, l_0 - 1)$ linear in a_{ij}. The number of equations is l_0, and the number of unknowns is $(h+1)n > l_0$. Multiplying each equation by $E_f^{l_0}$ we obtain equations with rational integer coefficients. Applying the well known lemma on linear diophantine equations (see [40], Ch.VI, Lemma 3), we find that this system of equations is satisfied by a collection of rational integers a_{ij} not all zero, and with absolute value not exceeding

$$[(h+1)nl_0^n (E_f A)^{l_0}]^{l_0/(h+1)n - l_0}. \tag{2.4}$$

Since we have

$$\frac{l_0}{(h+1)n - l_0} \le \frac{(1-\varepsilon)(h+1)n}{(h+1)n - (1-\varepsilon)(h+1)n} = \frac{1-\varepsilon}{\varepsilon},$$

$$(h+1)nl_0^n (E_f A)^{l_0} \le$$

$$\le ((h+1)n)^{n+1} (E_f A)^{(1-\varepsilon)(h+1)n} < n^{n+1} (2E_f A)^{(h+1)n},$$

the value (2.4) does not exceed

$$H_0 = (n^{n+1} (2E_f A)^{(h+1)n})^{1/\varepsilon},$$

which gives the assertion of the lemma with $c_6 = (n^{n+1}(2E_f A)^n)^{1/\varepsilon}$.

Lemma 2.4 *Suppose that a power series B is defined by*

$$B = B(x, f(x)) = \frac{\partial}{\partial y} F(x, y)|_{y=f(x)},$$

let $D = d/dx$ denote differentiation with respect to x, and suppose that $V = V(x)$ is defined by Lemma 2.3. Then for any integer $s \geq 1$ we have

$$D^s(B^{2s}V) = A_{0s}(x) + A_{1s}(x)f(x) + \cdots + A_{n-1s}(x)f^{n-1}(x), \qquad (2.5)$$

where the A_{is} are integral polynomials of degree at most $h+c_7s$ and with height at most

$$|A_{is}(x)| < c_8^s H_0 \prod_{k=0}^{s-1} (s + h + c_7 k),$$

where c_7 and c_8 are effectively determined by $F(x,y)$.

Proof. (cf. also [184], p. 105). Let

$$P(x, y) = A_0(x) + A_1(x)y + \ldots + A_{n-1}(x)y^{n-1},$$

where the $A_i(x)$ are defined as Lemma 2.3, and $0 \leq l \leq s$. Then

$$D^l(B^{2s}P(x, f(x))) = B^{2s-2l}P_l(x, f(x)), \qquad (2.6)$$

where the

$$P_l(x, y) = A_{0l}(x) + A_{1l}(x)y + \cdots + A_{n-1l}(x)y^{n-1}, \qquad (2.7)$$

are integral polynomials. Indeed, we may prove (2.6) by induction on l. For $l = 0$ the assertion is obvious. We suppose its truth for a given $l < s$ and prove it for $l+1$. After everywhere substituting $y = f(x)$ and partial differentiation we have

$$D^{l+1}(B^{2s}P) = D(D^l B^{2s}P) = D(B^{2s-2l}P_l) =$$

$$= (2s - 2l)B^{2s-2l-1}\left(\frac{\partial}{\partial x}B + \frac{\partial}{\partial y}B \cdot f'(x)\right)P_l +$$

$$+ B^{2s-2l}\left(\frac{\partial}{\partial x}P_l + \frac{\partial}{\partial y}P_l \cdot f'(x)\right).$$

Since

$$f'(x) = -\frac{\partial}{\partial x}F(x, f(x))/\frac{\partial}{\partial y}F(x, f(x)) = \frac{A}{B},$$

where

$$A = a_1'(x)f^{n-1}(x) + \cdots + a_n'(x),$$
$$B = nf^{n-1}(x) - (n - 1)a_1(x)f^{n-2}(x) - \cdots - a_{n-1}(x),$$

we obtain

$$D^{l+1}(B^{2s}P) = B^{2s-2(l+1)}P_{l+1},$$

$$P_{l+1} = 2(s-l)\left(B\frac{\partial}{\partial x}B + A\frac{\partial B}{\partial y}\right)P_l + B\left(B\frac{\partial}{\partial x}P_l + A\frac{\partial}{\partial y}P_l\right).$$

Since $f(x)$ satisfies the equation

$$f^n(x) = a_1(x)f^{n-1}(x) + \cdots + a_n(x),$$

we see that $D^{l+1}(B^{2s}P)$ can be expressed linearly over $\mathbb{Z}[x]$ in terms of $1, f(x), \ldots, f^{n-1}(x)$. Hence, we obtain (2.6), (2.7).

It is not hard to compute that the degrees and heights of the polynomials $A_{il}(x)$ in the decomposition (2.7) satisfy

$$h_{l+1} = \max_{(i)} \deg A_{il+1}(x) \le h_l + c_7$$

$$H_{l+1} = \max_{(i)} \overline{|A_{il+1}(x)|} \le c_8 H_l(s + h_l).$$

The lemma then follows.

Lemma 2.5 *Suppose that the hypotheses of Lemmas 2.3 and 2.4 hold, where $F(x,y)$ is an absolutely irreducible polynomial, and let π_s be the maximal divisor of $s!$ which is made up of powers of primes greater than c_5 and not occurring in E_f. Then all the coefficients of all the polynomials $A_{is}(x)$ determined by (2.5) are divisible by π_s.*

Proof. All the coefficients of the formal power series for $D^s(B^{2s}V)$ are divisible by π_s, since $B^{2s}V(E_f x)$ is an integral power series, and the operator D^s provides each term of the series with a factor of the form $k(k-1)\ldots(k-s+1)$ divisible by $s!$. If q is a prime divisor of π_s, then from (2.7) we obtain the following equality over $\mathbb{F}_q$

$$\overline{A}_{0s}(x) + \overline{A}_{1s}(x)\overline{f}(x) + \cdots + \overline{A}_{n-1s}(x)\overline{f}^{n-1}(x) = 0,$$

where A_{is} and $\overline{f}(x)$ are obtained from $A_{is}(x)$ and $f(x)$ by embedding their coefficients into $\mathbb{F}_q$. By Lemma 2.2 the degree of $\overline{f}(x)$ over $\mathbb{F}_q[x]$ equals n, so that all the polynomials $A_{is}(x)$ must be zero. If q occurs in π_s to more than the first power, then after dividing all the coefficients of all the polynomials $A_{is}(x)$ by q, we may recommence the argument. In this way we find that each of the coefficients of $A_{is}(x)$ is divisible by $q^{\mathrm{ord}_q \pi_s}$, hence, each is divisible by the product of these factors, that is, by π_s.

Lemma 2.6 *Suppose that p is a prime and $f(w)$ is the p-adic function defined by the series (1.6) for $w \in \mathbb{Q}_p$ with $|w|_p < |E_f|_p$. Then the, p-adic function*

$$V(w) = A_0(w) + A_1(w)f(w) + \ldots + A_{n-1}(w)f^{n-1}(w) \qquad (2.8)$$

where $A_0(x), \ldots, A_{n-1}(x)$ are integral polynomials of degrees at most h and not all equal to zero, has no more than $nh + c_9$ zeros in the disc $|w|_p < |E_f|_p$. The value c_9 can be determined effectively in terms of the degree of $F(x, y)$.

Proof. The function $f(w)$ satisfies the irreducible equation $F(w, f(w)) = 0$, hence, it is an algebraic element over $\mathbb{Q}[w]$ (we consider w as a transcendental element over $\mathbb{Q}$),

$$[\mathbb{Q}[w, f(w)] : \mathbb{Q}[w]] = n.$$

We see from the representation (2.8) that $V(w)$ is not identically zero. All the roots of $V(w)$ are roots of its norm from $\mathbb{Q}[w, f(w)]$ to $\mathbb{Q}[w]$, hence, they are the roots of a nonzero polynomial of degree at most $nh + c_9$. This implies the lemma

Lemma 2.7 *Suppose that a and b are integers, with $(a, b) = 1$, and let θ be a p-adic algebraic number satisfying the equation $F(a/b, \theta) = 0$, where $d = [\mathbb{Q}(\theta) : \mathbb{Q}]$. Let $\lambda = B_0 + B_1\theta + \ldots + B_{n-1}\theta^{n-1} \neq 0$, with $B_i \in \mathbb{Z}$ $(i = 0, 1, \ldots, n - 1)$. Then we have*

$$|\lambda|_p > c_{10}(\max(|a|, |b|))^{-nn_1 d} B^{-d}, \tag{2.9}$$

where $B = \max_{(i)} |B_i|$, $n_1 = \deg_x F(x, y)$, and the value $c_{10} > 0$ can be determined effectively in terms of $F(x, y)$.

Proof. Setting $\lambda' = b^{(n-1)n_1}\lambda$, we see that λ' is an algebraic integer, and by the 'product formula' we find that

$$|\mathrm{Nm}(\lambda')||\lambda'|_p \geq 1.$$

Estimating $|\mathrm{Nm}(\lambda')|$ from above we obtain (2.9) (cf. Lemma 3.8).

We now proceed to the proof of Theorem 1.1. Suppose a, b are integers, with $(a, b) = 1$, and let p be a prime dividing a but not dividing E_f. Suppose that $f(w)$, $V(w)$ are the p-adic functions defined for $w \in \mathbb{Q}_p$, $|w|_p < 1$ by the power series (1.6) and the equality (2.8) respectively, where the polynomials $A_0(x), \ldots, A_{n-1}(x)$ are chosen by Lemma 2.3.

We suppose that

$$V^{(i)}(a/b) = 0 \quad (0 \leq i \leq s - 1), \qquad V^{(s)}(a/b) \neq 0.$$

By Lemma 2.6 the number of zeros of $V(w)$ in the disc $|w|_p < 1$ does not exceed $nh + c_9$ while by Lemma 2.3 the function $V(w)$ has a zero at $w = 0$ of multiplicity no less than $(1 - \varepsilon)nh$ since the power series (1.8) has order no less than $(1 - \varepsilon)nh$ in x. Consequently, $s \leq \varepsilon nh + c_9 < 2\varepsilon nh$, if $h > h_0(\varepsilon, F)$.

We set $V_s(x) = D^s(B^{2s}V)$ in accordance with Lemma 2.4. Then

$$\mathrm{ord}V_s(x) \geq (1 - \varepsilon)nh - s \geq (1 - 3\varepsilon)nh.$$

Now write $W(x) = V_s(x)/\pi_s$, where the π_s were defined in Lemma 2.5. Then

$$W(x) = B_0(x) + B_1(x)f(x) + \ldots + B_{n-1}(x)f^{n-1}(x),$$

$$B_i(x) = \frac{1}{\pi_s}A_{is}(x) \in \mathbb{Z}[x], \quad g = \max \deg B_i(x) \leq h + c_7 s.$$

Since $\pi_s \geq s! c_{10}^{-s}$, we find that

$$\max \overline{|B_i(x)|} < (c_8 c_{10})^s H_0 \frac{1}{s!} \prod_{k=0}^{s-1} (s + h + c_7 k) <$$

$$< c_{11}^s H_0 (\frac{h}{s})^s < c_{11}^s c_{12}^{h/\varepsilon} (2\varepsilon n)^{-2\varepsilon n h} < c_{13}^{h/\varepsilon},$$

where none of the values $c_{10}, \ldots, c_{13}$ depend on h. Further, we find that

$$V_s(a/b) = B^{2s}(a/b, f(a/b)) V^{(s)}(a/b) \neq 0,$$

$$W(a/b) = \frac{1}{\pi_s} V_s(a/b) \neq 0, \quad B_i = b^g B_i(a/b) \in \mathbb{Z} \qquad (0 \leq i \leq n-1).$$

Consequently

$$\lambda = b^g W(a/b) = B_0 + B_1 \theta + \ldots + B_{n-1} \theta^{n-1} \neq 0,$$

where $\theta = f(a/b)$ is the p-adic number defined by the sum of the series (1.8) with $x = a/b$ in the metric $|\ |_p$, and

$$B = \max_{(i)} |B_i| \leq (h + c_7 s + 1) c_{13}^{h/\varepsilon} (\max(|a|, |b|))^{h + c_7 s} < c_{14}^{h/\varepsilon} (\max |a|, |b|)^{h(1 + \varepsilon_1)}),$$

where $\varepsilon_1 = 2n\varepsilon c_7$.

We observe now that $f(E_f x)$ is an integral series, hence the coefficients of the series $W(x)$ are p-integral, and since $\text{ord} W(x) \geq (1 - 3\varepsilon) nh$, we find

$$|\lambda|_p \leq p^{-\text{ord}_p a \cdot \text{ord} W(x)} \leq p^{-m(1-3\varepsilon)nh}, \qquad m = \text{ord}_p a.$$

Comparing this inequality with (2.9), taking (1.3) into account and letting h increase without bound, we obtain

$$(1 - 3\varepsilon)n \leq \frac{d \ln c_{14}}{\varepsilon m \ln p} + d(1 + \varepsilon_1) \left(\frac{n}{n-1} - \delta \right).$$

If we suppose that $m \ln p > n\varepsilon^{-2} \ln c_{14}$ and take $\varepsilon = \frac{1}{6} \delta (nc_7 + 2)^{-1}$, then we obtain $d > n - 1$, and hence $d = n$. This completes the proof of the theorem for $p \nmid E_f$. If $p \mid E_f$, we consider $V(w)$ in the disc $|w|_p < |E_f|_p$ $(w \in \mathbb{Q}_p)$ and argue as before.

3. Details and Sharpenings

To obtain further results we have to know how the height of the polynomial $F(x, y)$ influences all the auxiliary values involved in the above discussion. In this paragraph we carry out the necessary additional work. As a first corollary we obtain an explicit form of the dependence of c_1 on H_F, the height of $F(x, y)$. Apart from that, our arguments will give the reader a more complete understanding of the method described briefly in the previous paragraph.

Lemma 3.1 *Under the conditions of Lemma 2.1 one may take $c_5 = (4H_F)^c$, where $c = k^{2^k}$, $k = \frac{1}{2}(n + n_1 - 1)(n + n_1)$, and $n_1 = \deg_x F$.*

Proof. See [184], p. 193.

Lemma 3.2 *Suppose that $F(x, y)$ is integral polynomial, with $F(0, 0) = 0$, and $(\partial/\partial y)F(0, 0) \neq 0$, and that $f(x)$ is a power series, defined by (1.6), with E_f the Eisenstein number of this series. Then*

$$E_f \mid \left(\frac{\partial}{\partial y}F(0, 0)\right)^2.$$

Proof. (cf. [97], 19th lecture). On making the substitution $y = au$, $x = a^2 v$, where $a = (\partial/\partial y)F(0, 0) \neq 0$, we find that the power series $g(v) = a^{-1}f(a^2 v)$ satisfies the equation $G(v, g(v)) = 0$, where $G(v, u)$ is an integral polynomial of the form

$$G(v, u) = b_0(v)u^n + \cdots + b_{n-1}(v)u + b_n(v),$$

with $b_{n-1}(0) = 1$.

We have the formal expansion

$$(b_{n-1}(v))^{-1} = (1 + b_1 v + \cdots + b_r v^r)^{-1} = \sum_{\nu=0}^{\infty} d_\nu v^\nu, \qquad d_\nu \in \mathbb{Z}.$$

Setting

$$g(v) = \sum_{\nu=1}^{\infty} g_\nu v^\nu \qquad (g_\nu = f_\nu a^{2\nu-1}),$$

and making the substitution $u = g(v)$ in the equation $G(v, u) = 0$ we obtain

$$-u = \frac{b_n(v)}{b_{n-1}(v)} + \frac{b_{n-2}(v)}{b_{n-1}(v)}u^2 + \cdots + \frac{b_0(v)}{b_{n-1}(v)}u^n,$$

$$-g(v) = b_n(v)\sum_{\nu=0}^{\infty} d_\nu v^\nu + b_{n-2}(v)(\sum_{\nu=0}^{\infty} d_\nu v^\nu)(\sum_{\mu=1}^{\infty} g_\mu v^\mu)^2 + \cdots.$$

Consequently, it follows that

$$g(v) = \sum_{\nu=0}^{\infty} e_\nu v^\nu + \sum_{\nu=0}^{\infty} e_\nu' v^\nu \left(\sum_{\mu_1,\mu_2=1}^{\infty} g_{\mu_1} g_{\mu_2} v^{\mu_1 + \mu_2}\right) + \cdots$$

$$\cdots + \sum_{\nu=0}^{\infty} e_\nu^{(n-1)} v^\nu \left(\sum_{\mu_1,\ldots,\mu_n=1}^{\infty} g_{\mu_1} \cdots g_{\mu_n} v^{\mu_1 + \cdots + \mu_n}\right).$$

Here all the numbers $e_\nu^{(j)}$ are integers. Comparing the coefficients of identical powers of v on the left and right sides of the last equality, we obtain the infinite system of equations ($m = 1, 2, \ldots$)

$$g_m = e_m + \sum_{\substack{\nu+\mu_1+\mu_2=m \\ \mu_1 \geq 1, \mu_2 \geq 1}} e'_\nu g_{\mu_1} g_{\mu_2} + \cdots + \sum_{\substack{\nu+\mu_1+\cdots+\mu_n=m \\ \mu_1 \geq 1,\ldots,\mu_n \geq 1}} e_\nu^{(n-1)} g_{\mu_1} \cdots g_{\mu_{n-1}}.$$

Thus $g_1 = e_1$, $g_2 = e_2 + e'_0 g_1^2$, ..., g_m is integral polynomial in $g_1, g_2, \ldots, g_{m-1}$ and we see that all the g_m are integers. Since $g_m = a^{2m-1} f_m$, then at any rate the $a^{2m} f_m$ $(m = 1, 2, \ldots)$ are integers and we obtain the assertion of the lemma (supposing that E_f is the least natural E which satisfies the condition that the $E^\nu f_\nu$ are integers for all $\nu = 1, 2, \ldots$).

Lemma 3.3 *Under the conditions of Lemma 2.3 one may take*

$$A = c_{15} H_F^{8(nn_1)^2},$$

where c_{15} is determined by the degree of $F(x,y)$.

Proof. The power series (1.6) defines a regular function of the complex variable x in the disc $|x| < \rho_0$, where ρ_0 is the distance from the point $x = 0$ to the nearest singular point of the algebraic function $y(x)$ defined by the equation $F(x,y) = 0$. Provided that $0 < \rho < \rho_0$ we find that

$$f_\nu = \frac{1}{2\pi i} \int_{|z|=\rho} f(z) z^{-\nu-1} dz, \quad |f_\nu| \leq \rho^{-\nu} \max_{|z|=\rho} |f(z)| \quad (\nu = 1, 2, \ldots).$$
$$(3.1)$$

From the equation $f^n(z) = a_1(z) f^{n-1}(z) + \ldots + a_n(z)$ we obtain for $|z| = \rho$ either $|f(z)| \leq 1$ or

$$|f(z)| \leq n \max(\max_{|z|=\rho} |a_1(z)|, \ldots, \max_{|z|=\rho} |a_n(z)|) \leq$$
$$\leq n \max(1, \rho)^{n_1} (n_1 + 1) H_F, \qquad n_1 = \deg_x F(x,y).$$

Therefore from (3.1) we obtain

$$|f_\nu| \leq n(n_1 + 1)\rho^{-\nu} H_F \quad (\nu = 1, 2, \ldots), \tag{3.2}$$

if we suppose that $\rho \leq 1$ (obviously if $\rho_0 \geq 1$ we can take $\rho = 1$).

Since the singular points of $y(x)$ are the roots of the discriminant of the polynomial $F(x,y)$ with respect to y, i.e., the roots of the integral polynomial

$$D(x) = D(1, -a_1(x), \ldots, -a_n(x)),$$

to estimate the value ρ_0 it suffices to estimate from below the absolute values of the complex roots of the polynomial $D(x)$. The height of this polynomial does not exceed $c_{16} H_F^{2(n-1)n_1}$, and the absolute values of its roots are bounded from above by a similar quantity (c_{16} depends only on the degree of $F(x,y)$). If $\kappa \neq 0$ is one such root and if d is the smallest natural number such that $d\kappa$ is an integer, then

$$1 \leq |\mathrm{Nm}(d\kappa)| \leq |\kappa| (c'_{16} H_F^{4(n-1)n_1})^{\deg \kappa}.$$

Since the degree of $D(x)$ does not exceed $2(n-1)n_1$ we obtain

$$|\kappa| > (c'_{16}H^{4(n-1)n_1})^{-2(n-1)n_1},$$

which is also a lower bound for ρ_0. From (3.2) we now obtain

$$|f_\nu| \le (c_{17}H^{8(n-1)^2n_1^2})^\nu H_F$$

from which the assertion of the lemma follows.

Lemma 3.4 *Let $f(x)$ be a power series satisfying the equation*

$$f^n(x) = a_1(x)f^{n-1}(x) + \ldots + a_n(x), \qquad a_i(x) \in \mathbb{Z}[x],$$
$$\max \overline{|a_i(x)|} \le H_F, \qquad \max \deg a_i(x) \le n_1. \quad (3.3)$$

For $k = n, n+1, \ldots$ set

$$f^k(x) = a_{1k}(x)f^{n-1}(x) + \cdots + a_{nk}(x), \qquad a_{ik}(x) \in \mathbb{Z}[x].$$

Then we have

$$\max_{(i)} \overline{|a_{ik}(x)|} < c_{18}^{k-n}H_F^{k-n+1}, \qquad \max_{(i)} \deg a_{ik}(x) \le n_1(k-n+1),$$

where c_{18} is determined by n_1 $(k = n, n+1, \ldots)$.

Proof. For $k \ge n+1$ we have

$$f^{k+1} = a_{1k}f^n + a_{2k}f^{n-1} + \cdots + a_{nk}f$$
$$= (a_{1k}a_1 + a_{2k})f^{n-1} + (a_{1k}a_2 + a_{3k})f^{n-2} + \cdots + a_{1k}a_n,$$

if we use (3.3). It follows that

$$d_k = \max_{(i)} \deg a_{ik}(x) \le d_{k-1} + n_1, \qquad d_k \le n_1(k-n+1).$$

Putting

$$H^{(k)} = \max_{(i)} \overline{|a_{ik}(x)|} \quad (k = n+1, n+2, \ldots), \qquad H^{(n)} = H_F,$$

we find

$$H^{(k)} < c_{18}H^{(k-1)}H_F, \qquad H^{(k)} < (c_{18}H_F)^{k-n}H^{(n)},$$

from which the assertion of the lemma follows.

Lemma 3.5 *Take $f(x)$ as in Lemma 3.4, and*

$$S(x) = s_0(x) + s_1(x)f(x) + \ldots + s_{n-1}(x)f^{n-1}(x), \qquad s_i(x) \in \mathbb{Z}[x],$$
$$T(x) = t_0(x) + t_1(x)f(x) + \ldots + t_{n-1}(x)f^{n-1}(x), \qquad t_i(x) \in \mathbb{Z}[x].$$

Then we have

$$S(x)T(x) = u_0(x) + u_1(x)f(x) + \ldots + u_{n-1}(x)f^{n-1}(x), \qquad u_i(x) \in \mathbb{Z}[x],$$

and the heights and degrees of the polynomials $u_i(x)$ satisfy the inequalities

$$\max_{(i)} \lceil u_i(x) \rceil < c_{19}\mu \max \lceil s_i(x) \rceil \cdot \max \lceil t_i(x) \rceil H_F^{n-1},$$

$$\max \deg u_i(x) \leq \max \deg s_i(x) + \max \deg t_i(x) + n_1(n-1),$$

where c_{19} is determined by n and n_1, and

$$\mu = \min(\max \deg s_i(x) + \max \deg t_i(x)).$$

Proof. We have

$$ST = \sum_{k=0}^{2n-2} U_k f^k = \sum_{k=0}^{2n-2} U_k(a_{1k}f^{n-1} + \ldots + a_{nk}),$$

where $U_k = \sum_{i+j=k} s_i t_j$ $(k = 0, 1, \ldots, 2n-2)$ and the polynomials a_{ik} are defined for $k \geq n$ in Lemma 3.4, and for $k = 0, 1, \ldots, n-1$ by the equalities

$$a_{ik} = \begin{cases} 1 & \text{for } i = n - k \\ 0 & \text{for } i \neq n - k. \end{cases}$$

Hence, the product ST has the form which is specified in the lemma, and

$$u_r = \sum_{k=0}^{2n-2} U_k a_{n-r,k} \qquad (r = 0, 1, \ldots, n-1).$$

Applying Lemma 3.4 we obtain

$$\max \deg u_r(x) \leq \max \deg s_i(x) + \max \deg t_i(x) + n_1(n-1),$$

and since

$$\lceil U_k \rceil < c'_{19} \min(\max \deg s_i(x), \max \deg t_i(x)) \max_{i+j=k} \lceil s_i(x) \rceil \lceil t_j(x) \rceil,$$

we obtain the estimate for heights of the polynomials u_r which is specified in the lemma; c_{19} depends only on n.

Lemma 3.6 *Under the conditions of Lemma 2.4 one can take $c_7 = 2nn_1$, and $c_8 = c_{20}H_F^{2n}$, where c_{20} is expressible in terms of n and n_1.*

Proof. Apply the previous lemma and the equalities

$A = a'_n(x) + \cdots + a'_1(x)f^{n-1}(x),$

$B = -a_{n-1}(x) - \cdots - (n-1)a_1(x)f^{n-2}(x) + nf^{n-1}(x),$

$\dfrac{\partial B}{\partial x} = -a'_{n-1}(x) - \cdots - (n-1)a'_1(x)f^{n-2}(x),$

$\dfrac{\partial B}{\partial y} = -a_{n-2}(x) - \cdots - (n-1)(n-2)a_1(x)f^{n-3}(x) + n(n-1)f^{n-2}(x),$

$P_l = A_{0l}(x) + A_{1l}(x)f(x) + \cdots + A_{n-1l}(x)f^{n-1}(x),$

$\dfrac{\partial P_l}{\partial x} = A'_{0l}(x) + A'_{1l}(x)f(x) + \cdots + A'_{n-1l}(x)f^{n-1}(x),$

$\dfrac{\partial P_l}{\partial y} = A_{1l}(x) + \cdots + (n-1)A_{n-1l}(x)f^{n-2}(x).$

Using the notation introduced in the proof of Lemma 2.4, we find

$$h_{l+1} \le h_l + 2nn_1, \qquad c_7 = 2nn_1,$$

$$H_{l+1} < c_{20}\{(s-l)H_F^2 H_l H_F^{2(n-1)} + H_F^2 h_l H_l H_F^{2(n-1)}\} = c_{20}H_F^{2n}H_l(s-l+h_l),$$

where c_{20} is expressible in terms of n and n_1. Since $h_l \le h + 2nn_1 l$, we obtain

$$H_l < (c_{20}H_F^{2n})^l H_0 \prod_{k=0}^{l-1}(s+h+2nn_1 k),$$

which proves the lemma.

Lemma 3.7 *Under the conditions of Lemma 2.6, we have $c_9 = (n-1)n_1$.*

Proof. We give here a more detailed proof of Lemma 2.6 removing some uncertainty connected with the notion of the norm of $V(w)$.

Let $f(x), f_1(x), \ldots, f_{n-1}(x)$ be the n different power series expansions in a neighborhood of $x = 0$ of the algebraic function $y = y(x)$ satisfying the equation

$$y^n = a_1(x)y^{n-1} + \ldots + a_n(x), \qquad a_i(x) \in \mathbb{Z}[x].$$

Then we can suppose that $f(x)$ is the series (1.6), and that $f_1(x), \ldots, f_{n-1}(x)$ are, in general, Puiseux series.

Any symmetric function of $f_1(x), \ldots, f_{n-1}(x)$ is a polynomial in x and $f(x)$. Indeed, if $1, S_1, \ldots, S_{n-1}$ denote the elementary symmetric functions of $f_1(x), \ldots, f_{n-1}(x)$, then

$$S_1 = a_1 - f \in \mathbb{Z}[x, f(x)],$$

$$S_k = (-1)^{k-1}a_k - fS_{k-1} \qquad (1 < k < n-1),$$

$$(-1)^{n-1}S_{n-1} = \frac{a_n}{f} = -a_{n-1} - \cdots - a_1 f^{n-2} + f^{n-1}.$$

Therefore, all of the S_k lie in $\mathbb{Z}[x, f(x)]$.

Set $G(x,y) = A_0(x) + A_1(x)y + \cdots + A_{n-1}(x)y^{n-1}$ where the polynomials $A_i(s)$ are defined in Lemma 2.6, and $F(x,y) = y^n - a_1(x)y^{n-1} - \cdots - a_n(x)$.

The resultant $R_y(G, F)$ of the polynomials G and F with respect to y may be represented in the form

$$R_y(G, F) = G(x, f(x)) \prod_{i=1}^{n-1} G(x, f_i(x)) = G(x, f(x))M(x, f(x)),$$

where, in view of the previous remark, $M(x, f(x))$ is an integral polynomial in x and $f(x)$.

Passing now to p-adic variables, we note that

$$R_y(G, F) = V(w)M(w, f(w)),$$

and since $f(w)$ is regular in the disc $|w|_p < |E_f|_p$, this remains true of the function $M(w, f(w))$. Consequently, any zero of $V(w)$ in the disc $|w|_p < |E_f|_p$ is a root of the polynomial $R_y(G, F)$. Since the degree of this polynomial is at most $nh + (n-1)n_1$ and since it is not identically zero, we obtain the assertion of the lemma.

Lemma 3.8 *Under conditions of Lemma 2.7 we have $c_{10} = c_{21}H_F^{-nd}$, where $c_{21} > 0$ is expressible in terms of n and n_1.*

Proof. All complex roots ξ of the polynomial

$$F(a/b, y) = y^n - a_1(a/b)y^{n-1} - \cdots - a_n(a/b)$$

satisfy the inequality

$$|\xi| \le n(n_1 + 1)H_F \max(1, |a/b|)^{n_1}. \tag{3.4}$$

Since θ is a root of $F(a/b, y)$ the minimal polynomial for θ, say, $T(y)$, divides $F(a/b, y)$ and all its complex roots $\xi^{(1)}, \ldots, \xi^{(d)}$ are contained among the roots of $F(a/b, y)$, that is, they satisfy (3.4). Let $\theta^{(1)}, \ldots, \theta^{(d)}$ be the p-adic roots of $T(y)$. Then we have

$$\mathrm{Nm}(\lambda) = \prod_{i=1}^{d} (B_0 + B_1\theta^{(i)} + \ldots + B_{n-1}(\theta^{(i)})^{n-1}) = R(L, T),$$

where $L = L(y) = B_0 + B_1 y + \ldots + B_{n-1}y^{n-1}$. The same resultant $R(L, T)$ may also be written in terms of the complex roots of the polynomial $T(y)$ as

$$\mathrm{Nm}(\lambda) = \prod_{i=1}^{d} (B_0 + B_1\xi^{(i)} + \ldots + B_{n-1}(\xi^{(i)})^{n-1}) = R(L, T),$$

Now applying the estimate (3.4), we obtain

$$|\mathrm{Nm}(\lambda)| \le B^d \{n^2(n_1 + 1)H_F \max(1, |a/b|^{n_1})\}^{(n-1)d}.$$

Since $b^{n_1}\xi^{(i)}$ is an integer it follows that $b^{n_1(n-1)d}\mathrm{Nm}(\lambda)$ is a rational integer, and since $\lambda \ne 0$, we obtain

$$|\mathrm{Nm}(b^{n_1(n-1)}\lambda)| \leq B^d\{n^2(n_1+1)H_F\max(|a|,|b|)^{n_1}\}^{(n-1)d},$$
$$|\mathrm{Nm}(b^{n_1(n-1)}\lambda)|_p > c_{21}(H_F\max(|a|,|b|)^{n_1})^{-nd}B^{-d}.$$

Finally we have

$$|\lambda|_p \geq |b^{n_1(n-1)}\lambda|_p \geq |\mathrm{Nm}(b^{n_1(n-1)}\lambda)|_p,$$

and the assertion of the lemma follows.

We are now in a position to give a sharpened version of Theorem 1.1 relying on the improved lemmas suggested above.

Theorem 3.1 *The assertion of Theorem 1.1 holds with*

$$c_1 = c_{22}^{1/\delta^2} H_F^{2n^2(2n+c)+(6n)^4 n_1^2(n^2 n_1+1)^2/\delta^2},$$

where c_{22} is defined in terms of C, n and n_1, and the quantity c is as indicated in Lemma 3.1.

Proof. First we estimate π_s, the maximal divisor of $s!$ made up of powers of prime numbers bigger than c_5, and not occurring in E_f. Since E_f does not exceed $(n-1)^2 H_F^2$ (this follows from Lemma 3.2), to estimate π_s it is enough to exclude from $s!$ the factor

$$\prod_{p \leq c_5} p^{[s/p]+[s/p^2]+\cdots} \leq \prod_{p \leq c_5} p^{s/(p-1)} < c_5^s e^{O(s)},$$

where we used the asymptotic formula

$$\sum_{p \leq c_5} \frac{\ln p}{p-1} = \ln c_5 + O(1).$$

Consequently, taking into account Lemma 3.1, we obtain

$$\pi_s \geq s!(c_{23}H_F^c)^{-s}, \tag{3.5}$$

where c_{23} is expressed in terms of n and n_1.

Passing now directly to the proof of Theorem 1.1 we find using Lemma 3.6

$$\max \overline{|B_i(x)|} < (c_{20}H_F^{2n})^s H_0 \frac{(c_{23}H_F^c)^s}{s!} \prod_{k=0}^{s-1}(s+h+2nn_1 k) <$$

$$< (c_{24}H_F^{2n+c})^s H_0 \left(\frac{h}{s}\right)^s, \qquad c_{24} = c_{24}(n,n_1).$$

Keeping in mind the explicit form of H_0 (see the proof of Lemma 2.3) and the expression for A indicated in Lemma 3.3, we obtain

$$H_0 < \left(n^{n+1}\left(2c_{15}H_F^{9(nn_1)^2}\right)^{(h+1)n}\right)^{1/\varepsilon}.$$

Hence, we have

$$\max \overline{|B_i(x)|} < \left(c_{25} H_F^{9n^3 n_1^2 + 2n(2n+c)\varepsilon^2}\right)^{(h+1)/\varepsilon},$$

from which it follows that

$$B = \max |B_i| < \left(c_{26} H_F^{9n^3 n_1^2 + 2n(2n+c)\varepsilon^2}\right)^{(h+1)/\varepsilon} (\max(|a|, |b|))^{h(1 + 2n^2 n_1 \varepsilon)},$$

where c_{25} is determined by n and n_1, and c_{26} by C, n, n_1. The final arguments in the proof of Theorem 1.1 show that with

$$\varepsilon = \delta/(12(n^2 n_1 + 1))$$

and provided that

$$p^m > (c_{26} H_F^{9n^3 n_1^2 + 2n(2n+c)\varepsilon^2})^{n/\varepsilon^2}$$

we have $d = n$. This gives Theorem 3.1.

We note that in the expression obtained for c_1, the only fact essential to us was that c_1 be expressed as a power of H_F (and not, say, as an exponential function of H_F); the particular exponent had no special importance. As a simplest corollary we obtain the estimate of the form (1.5) for solutions of the equation $F(p^t, y) = 0$, just as mentioned in §1.

4. Theorems on Reducibility

Theorem 1.1 admits the following strengthening and generalisation [221].

Suppose that $F(x, y)$, $f(x)$, and E_f are as before, and that a and b are integers, with $(a, b) = 1$, and $|b| < C|a|$. Set $x_0 = a/b$. Denote by $\theta_p = f(x_0)_p$ the sum of the series (1.6) at $x = x_0$ in the p-adic metric, where p is a prime divisors of $a \neq 0, 1$ for which the series converges.

Theorem 4.1 *Let*

$$F(x_0, y) = F_1(y) \cdots F_r(y) \tag{4.1}$$

be the decomposition of $F(x_0, y)$ into irreducible factors in $\mathbb{Q}[y]$ and let the d_j denote the degrees $\deg F_j(y)$ $(1 \leq j \leq r)$ of the factors. Then it follows that

$$\sum_{\theta_p \in F_j} \frac{\ln a_p}{\ln |a|} = \frac{d_j}{n} + O\left(\sqrt{\frac{\ln H_F}{\ln |a|}}\right) \qquad (1 \leq j \leq r), \tag{4.2}$$

where a_p is the p-component of a, H_F is the height of $F(x, y)$, and $\theta_p \in F_j$ means that $F_j(\theta) = 0$. The symbol O involves a value which can be effectively determined in terms of C and the degree of $F(x, y)$.

Thus we see that the multiplicative structure of the number a influences the multiplicative structure of the polynomial $F(x_0, y)$ in an essential way.

The theorem gives the 'ergodic law' of the distribution of the sums of the series (1.6) in p-adic metrics as the roots of the irreducible factors of $F(x_0, y)$. We see, in particular, that for any irreducible factor $F_j(y)$ there exists a prime divisor $p|a$ such that the series (1.6) at $x = x_0$ in the metrics $|\ |_p$ converges to a root of $F_j(y)$ provided that $|a|$ is large enough.

Let us discuss several simple corollaries from the formula (4.2).

First of all we see that the decomposition (4.1) entails a decomposition of $|a|$ into coprime factors such that

$$a_j = |a|^{d_j/n + O(\sqrt{\ln H_F / \ln |a|})} \qquad (1 \le j \le r).$$

Consequently, if $F(x_0, y)$ has a linear factor, there must be a factor of $|a|$, say a_1, with

$$a_1 = |a|^{1/n + O(\sqrt{\ln H_F / \ln |a|})}, \qquad (a_1, |a|/a_1) = 1.$$

The absence of such a factor excludes the possibility of a linear factor in $F(x_0, y)$.

A yet more interesting corollary concerns x_0 of the form $x_0 = g^u/v$, where g is fixed, $u > 0$ and v are integers, $(g, v) = 1$, and $|v| \le C g^u$. Since we have

$$\frac{\ln(g^u)_p}{\ln(g^u)} = \frac{\ln(g)_p}{\ln(g)},$$

it follows that the degrees $d_j = d_j(u)$ of all the irreducible factors of $F(g^u/v, y)$ for all sufficiently large u are connected with the fixed set of values $\ln g_p / \ln g$ for $p \mid g$ by equalities of the form

$$\sum_{p \in P_j} \frac{\ln g_p}{\ln g} = \frac{d_j(u)}{n} + O\left(\frac{1}{\sqrt{u}}\right), \qquad (4.3)$$

where the P_j are subsets of the set of all prime divisors of g. The sum in the left-hand side of (4.3) is of the form $\ln a_j / \ln g$ with coprime a_j, if there is more than one of them. A simple analysis shows that if the numbers $\ln a_j / \ln g$ and d_j/n are different, then

$$\left| \frac{\ln a_j}{\ln g} - \frac{d_j}{n} \right| > g^{-n}.$$

Consequently, taking $u > c_{27} g^{2n}$ we find from (4.3) the equalities $a_j^n = g^{d_j}$ for all j, which is possible only in the case of one number a_j and $d_j = n$. Thus, $F(g^u/v, y)$ turns out to be irreducible.

It follows from the last assertion that the diophantine equation

$$F(r^u, y) = 0 \qquad (4.4)$$

in which $r \in \mathbb{Q}$, $|r| > 1$, has only a finite number of solutions in integers $u > 0$ and rational y, and that all the solutions may be found effectively. In fact the polynomial $F(r^u, y)$ is irreducible in $\mathbb{Q}[y]$ for all $u > c_{28} = c_{28}(r, F)$.

Theorem 4.1 in its turn admits a strengthening in which the condition $|b| \leq C|a|$ is excluded. That leads to a corresponding change in the formulae (4.2).

Now let v signify a prime number or the symbol ∞. We define the v-component $(x_0)_v$ of a rational number $x_0 = a/b$, $(a, b) = 1$, by the equality

$$(x_0)_v = \begin{cases} a_p, & v = p, \\ \max(1, |b/a|), & v = \infty. \end{cases}$$

Then we have

$$h(x_0) = \max(|a|, |b|) = \prod_v (x_0)_v,$$

and Theorem 4.1 changes as follows.

Theorem 4.2 *Under the hypotheses of Theorem 4.1 we have for any rational number x_0 with $h(x_0) > 1$ that*

$$\sum_{\theta_v \in F_j} \frac{\ln(x_0)_v}{\ln h(x_0)} = \frac{d_j}{n} + O\left(\sqrt{\frac{\ln H_F}{\ln h(x_0)}}\right) \qquad (1 \leq j \leq r), \qquad (4.5)$$

where θ_v denotes the sum (which can be seen to be finite) of the series (1.6) at $x = x_0$ in each metric corresponding v, and the symbol O involves a value which is effectively determined by the degree of $F(x, y)$.

It is apparent that this theorem implies both Theorem 1.1 and Theorem 4.1; it also leads to several new corollaries.

For instance, we see that the polynomial $F(x_0, y)$ is irreducible for all $x_0 = a/b$, $(a, b) = 1$, provided that $|a| < |b|^{1/n-\epsilon}$, with $\epsilon > 0$ an arbitrary fixed number and $|b|$ large enough. Conversely, if $F(x_0, y)$ has a linear factor, then there exists a v such that $(x_0)_v < c_{29}(h(x_0))^{1/n+\epsilon}$. We see now that the equation (4.4) has only a finite number of solutions if $0 \neq |r| < 1$. Together with the previous result this shows the finiteness of the number of its solutions and the possibility of their determination for all $r \neq 0, \pm 1$ (if $r = 0$ or $r = \pm 1$, then all the solutions are determined trivially).

An interesting corollary concerns 'Abel points' on the curve $F(x, y) = 0$. It is well known that Abel tried to prove the impossibility of a rational point with $x = p^u/b$ (p prime, $(p, b) = 1$) on the Fermat curve $x^n + y^n = 1$, but overlooked a gap in his argument, as was subsequently noticed by Markoff [139]. Abel's conjecture still remains unproved (its almost up to date state is discussed in [101]). We shall call a rational point on the curve with $x = p^u/b$ an 'Abel's point'; here all the numbers $p, u > 0$ and b are unknowns.

On considering such points on the curve $F(x, y) = 0$ with $F(x, y)$ satisfying the conditions of Theorem 4.2 (that means: $F(x, y)$ is absolutely irreducible, $\deg_y F(x, y) = n \geq 2$, $F(0, 0) = 0$, $(\partial/\partial y)F(0, 0) \neq 0$) we find as a direct consequence of this theorem:

Given any $\epsilon > 0$ there are only finitely many Abel's points on the curve with $|x| \geq \epsilon$ and all of them may be found by effective procedure.

Indeed, for an Abel point on the curve it must be that $p^u < |b|$ (if not, the polynomial $F(p^u/b, y)$ is irreducible), and we have to consider two quotients:

$$\frac{\ln(x_0)_p}{\ln h(x_0)} = \frac{u \ln p}{\ln |b|}, \qquad \frac{\ln(x_0)_\infty}{\ln h(x_0)} = \frac{\ln \max(1, |b/p^u|)}{\ln h(x_0)} = 1 - \frac{u \ln p}{\ln |b|}.$$

Hence, by Theorem 4.2, the only limit points for $\ln p^u / \ln |b|$ (with Abel's point $x_0 = p^u/b$ on the curve) can be $1/n$ and $1 - 1/n$. In both cases the limit points for x_0 are formed only by the point $x = 0$.

It is interesting to note that the curve $x - y(y + 1)^{n-1} = 0$ contains infinitely many Abel points with both limit points $1/n$ and $1 - 1/n$ for the quotients $\ln p^u / \ln |b|$.

5. Proofs of the Reducibility Theorems

We start with Theorem 4.1.

Suppose $(a, E_f) = 1$, $\varepsilon > 0$, and $h > 0$ is an integer such that $\varepsilon n h \geq 1$. For a prime $p \mid a$ we construct the auxiliary power series by the method described in §§2, 3

$$W(x) = B_0(x) + B_1(x)f(x) + \ldots + B_{n-1}(x)f^{n-1}(x), \qquad (5.1)$$

where the $B_j(x)$ are integral polynomials of degrees at most $g \leq h(1+4\varepsilon n^2 n_1)$, and with heights at most

$$c_{29}^{(h+1)/\varepsilon}, \qquad c_{29} = c_{25} H_F^{9n^3 n_1^2 + 2n(2n+c)\varepsilon^2}. \qquad (5.2)$$

Further $\text{ord} W(x) \geq (1 - 3\varepsilon)nh$, $W(x_0) \neq 0$, where on substituting $x = x_0$ into (5.1) the value $f(x_0)$ is determined by the sum of the series (1.6) in the p-adic metric (for the chosen $p \mid a$).

Setting

$$b^g W(x_0) = B_0 + B_1 \theta_p + \ldots + B_{n-1} \theta_p^{n-1}, \qquad B_i = b^g B_i(x_0) \in \mathbb{Z},$$

we obtain an integral polynomial

$$L(y) = B_0 + B_1 y + \ldots + B_{n-1} y^{n-1}$$

for which

$$B = \max |B_i| < c_{30}^{(h+1)/\varepsilon} (C'|a|)^{h(1+\varepsilon_1)},$$
$$\varepsilon_1 = 4\varepsilon n^2 n_1, \qquad 0 \neq |L(\theta_p)|_p < p^{-(1-3\varepsilon)nh \, \text{ord}_p a}, \qquad (5.3)$$

where $\theta_p = f(x_0)_p$, and c_{30} is a value of the type (5.2) with some other numerical value for c_{25}, and $C' = \max(1, C)$.

Equality (5.1) and the condition $\text{ord} W(x) \geq (1 - 3\varepsilon)nh$ now show that when we substitute $\theta_q = f(x_0)_q$ in $L(y)$ for primes $q \mid a$ we obtain estimates like those in the right side of (5.3) (it is not guaranteed that $L(\theta_q) \neq 0$).

Let θ_p be a root of $F_1(y)$. Since $L(\theta_p) \neq 0$, it follows that $F_1(y)$ and $L(y)$ do not have common zeros, so their resultant $R(F_1, L) \neq 0$. Denote the roots of $F_1(y)$ in $\overline{\mathbb{Q}}_p$, the algebraic closure of $\mathbb{Q}_p$ by $\theta_{1p}, \ldots, \theta_{dp}$, where $d = \deg F_1(y)$. Then the θ_{ip} are p-integral, one of them is θ_p, and

$$|R(F_1, L)|_p = \prod_{i=1}^{d} |L(\theta_{ip})|_p \leq |L(\theta_p)|_p .$$

If θ_q for $q \mid a$ is also a root of $F_1(y)$, then analogously

$$|R(F_1, L)|_q \leq |L(\theta_q)|_q .$$

Consequently

$$\prod_{q \mid a} |R(F_1, L)|_q \leq \prod_{\theta_q \in F_1} |L(\theta_q)|_q.$$

If we consider the resultant $R(F, L)$ as a determinant formed from the coefficients of $F_1(y)$ and $L(y)$ we see that it is a rational number whose denominator divides $b^{n_1(n-1)}$. Hence

$$\prod_{q \nmid a} |R(F_1, L)|_q \leq |b|^{n_1(n-1)}.$$

We further note that the absolute values of the complex roots ξ of the polynomial $F_1(y)$ which are roots of $F(x_0, y)$ are bounded from above by the inequality (3.4). Therefore

$$|R(F_1, L)| = \prod_{\xi \in F_1} |L(\xi)| < c_{31} \left(B H_F^{n-1} \max(1, |a/b|)^{n_1(n-1)} \right)^d .$$

Combining these estimates, we now find that

$$1 = |R(F_1, L)| \prod_q |R(F_1, L)|_q \leq |R(F_1, L)| |b|^{n_1(n-1)} \prod_{\theta_q \in F_1} |L(\theta_q)|_q <$$

$$< c_{32} H_F^{n^2} \left(|a|^{nn_1} c_{30}^{(h+1)/\varepsilon} (C'|a|)^{h(1+\varepsilon_1)} \right)^d \prod_{\theta_q \in F_1} q^{-(1-3\varepsilon)n \, \mathrm{hord}_p a},$$

where $c_{32} = c_{32}(n, n_1)$. Since h can be chosen arbitrarily large, we obtain

$$(1 - 3\varepsilon)n \sum_{\theta_q \in F_1} \ln a_q \leq d(1 + \varepsilon_1)(\ln |a| + \ln C') + d\varepsilon^{-1} \ln c_{30}, \qquad (5.4)$$

and since ε is arbitrary, we may take

$$\varepsilon = \frac{1}{2n\sqrt{n_1}} \sqrt{\frac{\ln c_{30}}{\ln |a|}}$$

(which can be done if $|a|$ is large enough). We now obtain from (5.4) that

$$\sum_{\theta_q \in F_1} \frac{\ln a_q}{\ln |a|} \leq \frac{d}{n} + 3n\sqrt{n_1}\sqrt{\frac{\ln c_{30}}{\ln |a|}} + 2\frac{\ln C'}{\ln |a|}.$$

Since the value c_{30} has the form (5.2) we find that

$$\sum_{\theta_q \in F_1} \frac{\ln a_q}{\ln |a|} \leq \frac{d}{n} + O\left(\sqrt{\frac{\ln H_F}{\ln |a|}} + \frac{\ln C'}{\ln |a|}\right). \tag{5.5}$$

In a similar way we obtain analogous inequalities for the other polynomials $F_j(y)$ and recalling that the sum of the left-hand sides of the inequalities taken over all $F_j(y)$ is equal to 1, while the sum of the right-hand sides is equal to $1 + O(\ldots)$, we obtain the asymptotic equalities (4.2).

If $(a, E_f) \neq 1$, then the contribution of primes $q \mid |E_f|$ for which $a_q > (E_f)_q$ is estimated as before, since then $|x_0|_q < |E_f|_q$ and x_0 lies in the domain of convergence of the series (1.6) in the q-adic metric. But, if $a_q \leq (E_f)_q$, then this contribution is at most

$$O\left(\sum_{q|E_f} \frac{\ln(E_f)_q}{\ln |a|}\right) = O\left(\frac{\ln E_f}{\ln |a|}\right) = O\left(\frac{\ln H_F}{\ln |a|}\right),$$

and we obtain (4.2) again.

The proof of Theorem 4.2 follows our general scheme of argument, but both the ordinary absolute value and the p-adic metrics have to be involved, and some extra computations have to be made.

Firstly, we observe that if $|x_0| \geq (4A)^{-1}$, where A is defined as in Lemma 2.3, then the assertion of Theorem 4.2 follows from Theorem 4.1. Indeed, in this case $|b| \leq 4A|a|$ which corresponds to $C = 4A$. The inequality (5.5) and analogous inequalities for the other polynomials $F_j(y)$ show that C influences the remainder term in (4.2) as $O(\ln \max(1, C)/\ln |a|)$, while Lemma 3.3 gives an estimate of $O(\ln H_F/\ln |a|)$ for this value . One takes account of the archimedean component on the left-hand side of (4.5) only in the case of convergence of the series (1.6) at $x = x_0$. The radius of convergence of the series coincides with the distance from $x = 0$ of the nearest root of the discriminant $D(x)$, introduced in the proof of Lemma 3.3. Consequently, it is bounded from above by $c_{33}H_F^{2(n-1)n_1}$, where $c_{33} = c_{33}(n, n_1)$. It follows that $|x_0|$ is bounded by the same value. We see that $|\ln |a/b|| = O(\ln H_F)$, and then we find

$$\frac{\ln H_F}{\ln |a|} = \frac{\ln H_F}{\ln |b| + O(\ln H_F)} = \frac{\ln H_F}{\ln |b|} + O\left(\left(\frac{\ln H_F}{\ln |b|}\right)^2\right).$$

This shows that when $|x_0| \geq (4A)^{-1}$ the contribution of the archimedean component to (4.5) may be moved to the remainder term. The difference between the contributions of the archimedean components in the equalities (4.2) and (4.5) is at most

$$\sum_{p|a} \ln a_p \left(\frac{1}{\ln |a|} - \frac{1}{\ln |b|} \right) = O \left(\frac{\ln H_F}{\ln |b|} \right).$$

Thus, indeed, when $|x_0| \geq (4A)^{-1}$ (4.5) follow from (4.2).

Suppose now that $|x_0| \leq (4A)^{-1}$. After taking suitable numbers ε and h we have to construct an integral polynomial $L(y)$ which will provide an analogue of (5.3) in the archimedean metrics; θ_p is replaced by θ_∞ and $|\ |_p$ by the ordinary absolute value. In the non-archimedean metrics we are to have inequalities of the type on the right side of (5.3). To attain this aim we rely on the construction of formal power series as described above. Passing to the numbers (substituting $x = x_0$) we do the estimates in archimedean metrics.

First of all we estimate the coefficients b_l of the power series $V(x)$ (see the proof of Lemma 2.3). We find that

$$|b_l| \leq H_0 + H_0 l^n A^l \sum_{\substack{1 \leq j \leq n-1 \\ 0 \leq i \leq h}} 1 = H_0(1 + (n-1)(h+1)l^n A^l),$$

$$|b_l| < 2nh H_0 l^n A^l \qquad xs(l = l_0, l_0 + 1, \dots), \tag{5.6}$$

while $b_l = 0$ $(l = 0, 1, \dots, l_0 - 1)$. Using (5.6), we obtain

$$\frac{1}{s!}|V^{(s)}(x_0)| \leq \sum_{l=l_0}^{\infty} \binom{l}{s} |b_l||x_0|^{l-s} < \sum_{l=l_0}^{\infty} 2^{l+1} nh H_0 l^n A^l |x_0|^{l-s} =$$

$$= 2nh H_0 (2A)^{l_0} |x_0|^{l_0-s} \sum_{l=0}^{\infty} (2A|x_0|)^l (l_0 + l)^n.$$

Since we are supposing that $|x_0| \leq (4A)^{-1}$, we have

$$\frac{1}{s!}|V^{(s)}(x_0)| \leq c_{34} hh H_0 l_0^n (2A)^{l_0} |x_0|^{l_0-s},$$

where c_{34} is defined by n.

Observe now that

$$V_s(x_0) = B^{2s}(x_0, f(x_0)) V^{(s)}(x_0), \tag{5.7}$$

where $f(x_0)$ is the sum of the series (1.6) in the metric $|\ |_\infty$ (i.e. in the ordinary absolute value). Since $f(x_0)$ is a root of the polynomial $F(x_0, y)$ it follows that $|f(x_0)|$ satisfies the inequality (3.4):

$$|f(x_0)| \leq n(n_1 + 1) H_F \max(1, |x_0|)^{n_1}.$$

Consequently, we have

$$|B(x_0, f(x_0))| = \left| \frac{\partial}{\partial y} F(x_0, f(x_0)) \right| < c_{35} H_F^n,$$

where $c_{35} = c_{35}(n, n_1)$. Thus we obtain

$$\frac{1}{s!}|V_s(x_0)| < c_{36}^s H_F^{2ns} h H_0 l_0^n (2A)^{l_0} |x_0|^{l_0-s},$$

from which using (3.5) we obtain

$$|W(x_0)| = \frac{1}{\pi_s}|V_s(x_0)| < c_{37}^s H_F^{2ns+cs} h H_0 l_0^n (2A)^{l_0} |x_0|^{l_0-s}, \qquad (5.8)$$

where c_{36}, c_{37} are defined by n and n_1. In view of (5.7) we have $W(x_0) \neq 0$ and the polynomial $L(y)$ defined for the metric $| \ |_\infty$ at the point $\theta_\infty = f(x_0)_\infty$ satisfies the inequalities

$$0 \neq |L(\theta_\infty)| < c_{37}^s H_F^{2ns+cs} h H_0 l_0^n (2A)^{l_0} |x_0|^{l_0-s} |b|^g <$$
$$< c_{37}^s H_F^{n+c} c_{38}^{(h+1)n/\varepsilon} |x_0|^{l_0-s} |b|^g,$$

where c_{38} is of the form

$$c_{38} = (2nc_{15})^2 H_F^{9(nn_1)^2}, \qquad h \geq h_0(n,\varepsilon) \qquad (5.9)$$

(we used the estimates for H_0, A and l_0). Since $l_0 - s \geq (1 - 3\varepsilon)nh$ and $g \leq h(1 + \varepsilon_1)$, $\varepsilon_1 = 4n^2 n_1 \varepsilon$, we finally obtain

$$0 \neq |L(\theta_\infty)| < c_{37}^s H_F^{n+c} c_{38}^{(h+1)n/\varepsilon} |x_0|^{(1-3\varepsilon nh)} |b|^{h(1+\varepsilon_1)}. \qquad (5.10)$$

Note that the height B of the polynomial $L(y)$ is estimated as before, but it should be taken into account that

$$h(x_0) = \max(|a|, |b|) = |b|, \qquad B = \max|B_i| < c_{30}^{h/\varepsilon}|b|^{h(1+\varepsilon_1)}. \qquad (5.11)$$

Now let $F_1(y)$ be an irreducible divisor of $F(x_0, y)$, with θ_∞ as a zero. Since $L(\theta_\infty) \neq 0$, it follows that $F_1(y)$ and $L(y)$ are coprime and we find from (5.10) and from the estimate (3.4) for the complex roots of $F_1(y)$ that

$$0 \neq |R(F_1, L)| = |L(\theta_\infty)| \prod_{\substack{\xi \in F_1 \\ \xi \neq \theta_\infty}} |L(\xi)| <$$
$$< c_{39} H_F^{n^2+c} c_{38}^{(h+1)n/\varepsilon} |x_0|^{(1-3\varepsilon)nh} |b|^{h(1+\varepsilon_1)} B^{d-1},$$

where c_{39} is determined by n and n_1. Using the estimates obtained above, we find

$$1 = |R(F_1, L)| \prod_q |R(F_1, L)|_q \leq |R(F_1, L)||b|^{n_1(n-1)} \prod_{\theta_q \in F_1} |L(\theta_q)|_q <$$
$$< c_{39} H_F^{n^2+c} c_{38}^{(h+1)n/\varepsilon} |x_0|^{(1-3\varepsilon)nh} |b|^{h(1+\varepsilon_1)+nn_1} B^{d-1} \prod_{\theta_q \in F_1} q^{-(1-3\varepsilon)nh \operatorname{ord}_q a}.$$

Recalling the inequality (5.11) for the estimate of B and that we may take h unboundedly large, we obtain

$$\left(\left|\frac{a}{b}\right| \prod_{\theta_q \in F_1} q^{\operatorname{ord}_q a}\right)^{(1-3\varepsilon)n} \le c_{38}^{n/\varepsilon} |b|^{d(1+\varepsilon_1)} c_{30}^{(d-1)/\varepsilon}.$$

Consequently

$$(1-3\varepsilon)n \sum_{\theta_v \in F_1} \ln(x_0)_v \le d(1+\varepsilon_1)\ln h(x_0) + n\varepsilon^{-1}\ln(c_{38}c_{30}).$$

We remain free to choose ε and the values c_{30} and c_{38} are estimated by (5.2) and (5.9). Hence finally we obtain

$$\sum_{\theta_v \in F_1} \frac{\ln(x_0)_v}{\ln h(x_0)} \le \frac{d}{n} + O\left(\sqrt{\frac{\ln H_F}{\ln h(x_0)}}\right).$$

Similar inequalities hold for all of the polynomials $F_j(y)$. That, together with the multiplicative representation of the height $h(x_0)$ in terms of its local components $(x_0)_v$, yields a proof of the equalities (4.5). This completes the proof of the Theorem 4.2.

6. Further Results and Remarks

Our general supposition that $F(x,y)$ is absolutely irreducible is purely formal since we may suppose that it is irreducible in $\mathbb{Q}[x,y]$ and then its absolute irreducibility will follow from the conditions (1.2). Indeed suppose that over some finite extension of $\mathbb{Q}$ we have the factorisation

$$F(x,y) = G_1(x,y) \cdots G_r(x,y) \tag{6.1}$$

into absolutely irreducible polynomials. Say, $G_1(0,0) = 0$. Then $G_1(x,y)$ is not in $\mathbb{Q}[x,y]$ unless it coincides with $F(x,y)$. Hence, there exists a polynomial $\overline{G}_1(x,y)$ conjugate to $G_1(x,y)$ and different from $G_1(x,y)$, which is absolutely irreducible and satisfies $\overline{G}_1(0,0) = 0$. Since the conjugate equality

$$F(x,y) = \overline{G}_1(x,y) \cdots \overline{G}_r(x,y)$$

follows from (6.1) , we have

$$F(x,y) = G_1(x,y)\overline{G}_1(x,y) \cdots .$$

It then follows that $F(0,y)$ has $y = 0$ as a root of multiplicity at least 2, which is impossible in view of the condition $(\partial/\partial y)F(0,0) \ne 0$.

In practical applications of Theorem 1.1 it is more convenient to justify that $F(0,y)$ has a simple rational root instead of requiring condition (1.2) . Generalising this condition, we may suppose that $F(0,y)$ has a simple root generating a field $\mathbb{K}$ of degree $k < n$. The method described above is applicable in this case too, and leads to the following assertion [223].

Theorem 6.1 *Let $F(x, y)$ be an absolutely irreducible integral polynomial with $\deg_y F(x, y) = n \geq 2$. Suppose that $F(0, y)$ has a simple root of degree $k < n$, and let a and b be rational integers, $(a, b) = 1$, such that for at least one prime p dividing a the p-component a_p satisfies the inequality*

$$a_p > \max(|a|, |b|)^{1 - 1/nk + \delta}, \quad 0 < \delta < 1/nk.$$

Suppose

$$F(a/b, y) = F_1(y) \cdots F_r(y)$$

be a decomposition of the polynomial $F(a/b, y)$ into irreducible factors in $\mathbb{Q}[y]$ and set $d_j = \deg F_j(y)$ $(j = 1, 2, \ldots, r)$. Then all of the numbers

$$k d_j \qquad (j = 1, 2, \ldots, r)$$

are divisible by n whenever

$$\max(|a|, |b|) \geq (H_F + 1)^{c_{40}/\delta^2},$$

where H_F is the height of $F(x, y)$, and c_{40} can be determined explicitly by the degree of $F(x, y)$.

We see, in particular, that if $(k, n) = 1$, then $F(a/b, y)$ is irreducible in $\mathbb{Q}[y]$. In any case $d_j \geq n/k$ $(j = 1, 2, \ldots, r)$, and so $F(a/b, y)$ has no linear factor. If we assume again, as in §1, that $x_0 \in \mathcal{D}_F$ and $x_0 = u/v$, with $(u, v) = 1$ and $|v| \leq |u|$ then we find that all the p-components u_p satisfy the inequality $u_p \leq |u|^{1 - 1/nk + \delta}$, whenever

$$|u| > (H_F + 1)^{c_{40}/\delta^2},$$

where $\delta > 0$ is arbitrary. It then follows that all the solutions of the Diophantine equation

$$F(p^t, y) = 0$$

in unknown primes p, integers $t > 0$, and y satisfy the inequality

$$\max(p^t, |y|) < (H_F + 1)^{c_{41}},$$

where c_{41} can be determined explicitly by the degree of $F(x, y)$.

A more important corollary of Theorem 6.1 is given by the following theorem concerning binary diophantine equations in which one unknown belongs to a special infinite set of natural numbers [223].

Theorem 6.2 *Let $A = \{a_m\}_1^\infty$ be a sequence of natural numbers satisfying the following two conditions:*
a) For each m the number a_m has a prime divisor p such that the p-component $(a_m)_p$ of a_m satisfies the inequality

$$(a_m)_p \geq a_m^{1 - z(m)},$$

where $x(m) > 0$, *and* $x(m) \to 0$ *as* $m \to \infty$.

b) For each prime q all the numbers a_m *with* $m \geq m_0(q)$ *are divisible by q.*

Then for any integral polynomial $F(x, y)$, *irreducible in* $\mathbb{Q}[x, y]$, *of degree at least 2 with respect to y, and such that* $F(0, y)$ *has at least one simple root, the diophantine equation*

$$F(x, y) = 0, \qquad x \in A, \quad y \in \mathbb{Z}, \tag{6.2}$$

has only a finite number of solutions. All the solutions can be determined effectively if the functions $x(m)$ *and* $m_0(q)$ *are defined effectively.*

The proof of this theorem is as follows. We distinguish two cases corresponding to condition a) and b) imposed on the sequence A: the polynomial $F(0, y)$ is reducible or irreducible in $\mathbb{Q}[y]$. If it is reducible, we would obtain the theorem as a corollary of Theorem 6.1 if $F(x, y)$ were absolutely irreducible. By arguments similar to those we described at the beginning of this paragraph we find that if (6.2) has a solution (x_0, y_0) with $(\partial/\partial y)F(x_0, y_0) \neq 0$, then $F(x, y)$ must be absolutely irreducible. Since $F(x, y)$ is irreducible in $\mathbb{Q}[x, y]$ and $(\partial/\partial y)F(x, y)$ has lower degree in y, the equalities

$$F(x_0, y_0) = 0, \qquad \frac{\partial}{\partial y}F(x_0, y_0) = 0$$

are independent and only a finitely many points (x_0, y_0) satisfy them. It is apparent that we may suppose the existence of a solution of (6.2) different from these special points, and hence the absolute irreducibility of $F(x, y)$.

If $F(0, y)$ is an irreducible polynomial of degree at least 2, then by Frobenius' theorem there exists an effectively determinable prime number q such that $F(0, y)$ (mod q) has no linear factor. Then $F(a_m, y)$ (mod q) for $m \geq m_0(q)$ likewise has no linear factors, and the equality $F(a_m, y) = 0$ for integral y is impossible.

Finally, the case of a linear polynomial $F(0, y)$ is covered by Theorem 6.1.

Theorem 6.2, being effective, may replace Siegel's theorem on integer points on algebraic curves to give an effective version of Hilbert's irreducibility theorem in its full generality. The corresponding result is as follows [223].

Theorem 6.3 *Let* $F(x, y)$ *be an integral polynomial, irreducible in* $\mathbb{Q}[x, y]$. *Then for all integers a, except for a finite number of effectively determinable singular numbers a, and for all* $m \geq m_0(a, F)$ *the polynomials* $F(a + a_m, y)$ *are irreducible in* $\mathbb{Q}[y]$. *Here the* a_m *are terms of the sequence A introduced in Theorem 6.2, and* $m_0(a, F)$ *is a quantity which is explicitly determinable from a and the parameters of F.*

In its turn, Theorem 6.3 supplied with some extra estimates allows one to construct in explicit form the universal subsets of Hilbert sets: those sequences of natural numbers b_m ($m = 1, 2, \ldots$) such that for any irreducible polynomial $F(x, y) \in \mathbb{Q}[y]$ the polynomials $F(b_m, y)$ are irreducible in $\mathbb{Q}[y]$, whenever

$m \geq m_0(F)$. Here $m_0(F)$ may be determined in explicit form in terms of the height and the degree of $F(x, y)$. For example, one can take

$$b_m = a(m) + m! P_m \qquad (m = 1, 2, \ldots),$$

where $a(m)$ is an unbounded increasing sequence of natural numbers subject to the condition

$$a(m) \leq \exp\left\{\frac{\ln \ln m}{\ln \ln \ln m}\right\} \qquad m \geq 3^{2^7},$$

and P_m is a power of a prime number with $P_m > (m!)^{\psi(m)}$, where $\psi(m) \to \infty$ as $m \to \infty$. Setting, in particular,

$$a(m) = [\exp(\ln \ln m)^{1/2}], \qquad \psi(m) = \ln \ln m,$$

we find for the value $m_0(F)$ an expression of the form

$$m_0(F) = \exp \exp \max \left(c_{42}, (\ln H_F)^2\right), \tag{6.3}$$

where c_{42} is defined by the degree of $F(x, y)$. Hence,

$$b_m = \left[\exp \sqrt{\ln \ln m}\right] + m! 2^{m^2} \qquad m \geq 9$$

is an example of a universal subset of Hilbert sets.

In 1955 Gilmore and Robinson [79] had proved that universal subsets of Hilbert sets exist, but the sets were thought of as mysterious until Theorem 6.1 became known.

Theorem 4.2 admits a generalisation to polynomials $F(x, y)$ from $I_{\mathbb{K}}[x, y]$, where $\mathbb{K}$ is a field of algebraic numbers of finite degree k over $\mathbb{Q}$. Suppose that S is a full system of nonequivalent canonical valuations (metrics) $|\ |_v$ on the field $\mathbb{K}$, and $\mathbb{K}_v$ is a completion of $\mathbb{K}$ in the metric $|\ |_v$. Denote by $\mathbb{K}^v$ the isomorphic image of $\mathbb{K}$ in $\mathbb{K}_v$. For $0 \neq \kappa \in \mathbb{K}$ we set

$$(\kappa)_v = \max(1, |\kappa|_v^{-k_v}), \qquad k_v = [\mathbb{K}_v : \mathbb{Q}_v], \qquad H_{\mathbb{K}}(\kappa) = \prod_{v \in S} (\kappa)_v .$$

Let $F(x, y)$ be an irreducible polynomial in $I_{\mathbb{K}}[x, y]$, satisfying (1.2) and let $f(x)$ be the power series (1.6) (in which now $f_v \in \mathbb{K}$). For $x_0 \in \mathbb{K}$ we define the sums $f(x_0)_v$ of the series (1.6) in the metric $v \in S$, when the series converges.

Suppose that $F(x, y)$ has a decomposition of the form (4.1) in $\mathbb{K}[y]$. Then in $\mathbb{K}^v[y]$ we have the 'conjugate' decompositions

$$F^v(x_0, y) = F_1^v(y) \cdots F_r^v(y),$$

and $f(x_0)_v = \theta_v$ turns out to be a root of one of the polynomials $F_j^v = F_j^v(y)$, which we signify as $\theta_v \in F_j^v$. Similarly to Theorem 4.2, the next theorem shows the way in which the numbers θ_v are distributed among the factors of $F^v(x_0, y)$ as v changes in S (cf. [224], [225]).

Theorem 6.4 *Suppose that $H_{\mathbb{K}}(x_0) > 1$ and that*

$$d_j = \deg F_j(y) \qquad (j = 1, 2, \ldots, r).$$

Then for any $j = 1, 2, \ldots, r$ we have

$$\sum_{\theta_v \in F_j^v} \ln(x_0)_v / \ln H_{\mathbb{K}}(x_0) = \frac{d_j}{n} + O\left(\left(\frac{\ln \lceil F \rceil}{\ln H_{\mathbb{K}}(x_0)}\right)^{1/2}\right), \qquad (6.4)$$

where the sum is taken over all $v \in S$ for which θ_v is a root of $F_j^v(y)$, where $\lceil F \rceil$ is the size of the polynomial $F(x, y)$, and the symbol O implies a value which can be effectively determined by the degrees of F and of $\mathbb{K}$.

Since both Theorems 4.2 and 6.4 wore proved on the basis of the theory of diophantine approximation it is interesting to notice that in fact these theorems are equivalent to some assertions on diophantine approximation to the values θ_v (for all $v \in S(x_0)$, where $S(x_0)$ is that subset of S consisting of those v for which θ_v is defined). Thus, for instance, Theorem 6.4 both implies and may be derived from the following theorem (*cf.* [225]).

Theorem 6.5 *For $v \in S(x_0)$ set*

$$\lambda_v = \frac{nk}{k_v} \cdot \frac{\ln(x_0)_v}{\ln H_{\mathbb{K}}(x_0)} - e_v + \varepsilon$$

where $e_v = 1$ in the case of an archimedean v, and $e_v = 0$ for a non-archimedean v. Then the system of inequalities

$$|P(\theta_v)|_v < \lceil P \rceil^{-\lambda_v} \qquad (v \in S(x_0)) \qquad (6.5)$$

in nonzero polynomials $P(y) \in I_{\mathbb{K}}[y]$ of degrees at most $n - 1$, has only a finite number of solutions whenever

$$H_{\mathbb{K}}(x_0) \geq (\lceil F \rceil + 1)^{c_{43}/\varepsilon^2}, \qquad (6.6)$$

for $\varepsilon > 0$, where c_{43} is effectively determined by the degrees of $F(x, y)$ and of $\mathbb{K}$. If $\varepsilon < 0$, then the system (6.5) has infinitely many solutions in polynomials $P(y)$.

Instead of inequalities (6.5) it is reasonable to consider inequalities of a different type. Set

$$\mu_v = \frac{n}{k_v} \cdot \frac{\ln(x_0)_v}{\ln H_{\mathbb{K}}(x_0)}(1 + \varepsilon),$$

and for $P(y) = \pi_0 + \pi_1 y + \ldots + \pi_{n-1} y^{n-1} \in \mathbb{K}[y]$ define a height

$$H_{\mathbb{K}}(P) = \prod_{v \in S} H_v, \qquad H_v = \max(|\pi_0|_v, \ldots, |\pi_{n-1}|_v)^{k_v}.$$

Then one can show that the system of inequalities

$$|P(\theta_v)|_v < H_{\mathbb{K}}(P)^{-\mu_v} H_v \qquad (v \in S(x_0))$$

for $\varepsilon > 0$ has only a finite number of solutions in polynomials $P(y) \in \mathbb{K}[y]$ whenever a condition of type (6.6) holds, and has infinitely many solutions if $\varepsilon < 0$.

Recently Bombieri [32] and Dèbes [54] obtained further generalisations and improvements of Theorems 4.2 and 6.4. In particular, Bombieri derived these theorems from Weil's 'Theorème de décomposition' and the theory of heights of points on algebraic curves, so his arguments are based on the notions and principles far beyond the theory of diophantine approximation. Bombieri relies on yet another approach is his recent development of the theory of Siegel's G-functions, while Dèbes demonstrates the possibilities of Gelfond's method in solving these problems. The dissertation of Dèbes [54] contains a detailed discussion of the subject, as well as new results, corollaries and examples.

It is probable, that under the conditions of Theorem 6.1 one cannot remove the assumption that $k < n$; that is, one may not assume that $k = n$ since otherwise it would follow that any nonlinear polynomial represents only a finite number of primes, contrary to the widespread belief that there are infinitely many primes of the form $y^2 + 1$, or of the form $y^2 - 2$ etc. Indeed, it is commonly believed that any irreducible polynomial without a constant divisor represents infinitely many primes — Bouniakowsky's conjecture). Since no non-linear irreducible polynomial with this property is known, this peculiarity of Theorem 6.1 deserves special attention.

The estimate (6.3) seems to be unsatisfactory from both a theoretical and a practical point of view, and in fact it may be improved considerably. But it is more important to find sequences of natural numbers increasing not so rapidly as A in Theorem 6.2 which may replace A in equation (6.2) with the same finiteness result. In particular, it would be interesting to know whether such a sequence may increase more slowly than an exponential function.

The remainder terms in the formulae (4.2), (4.5) or (6.4) are of the same order of magnitude independently of the method used to prove them. Any improvement of these terms is very important, since an understanding of their nature may lead to further results on rational points on algebraic curves, in particular to an effective version of Faltings theorem [65]. The work of Bombieri mentioned above ([32]) sheds some light on this problem.

References

1. Abel, N. H.: Oeuvres **2**. Christiania, 1881.
2. Adams, W. W.: Transcendental numbers in the p-adic domain. Amer. J. Math. **88** (1966) 279–308.
3. Amice, Y.: Les nombres p-adiques. Press Univ. de France, Paris, 1975.
4. Ankeny, N. C., Brauer, R., Chowla, S.: A note on the class- numbers of algebraic number fields. Amer. J. Math. **78** (1956) 51–61.
5. Avanesov, E. T.: On a class of binary biquadratic forms. Acta Arith. **26** (1974) 189–195 [Russian].
6. Baker, A.: Rational approximations to certain algebraic numbers. Proc. London Math. Soc. (3) **14** (1964) 385–398.
7. Baker, A.: Rational approximations to $\sqrt[3]{2}$ and other algebraic numbers. Quart. J. Math. Oxford Ser. (2) **15** (1964) 375–383.
8. Baker, A.: Linear forms in the logarithms of algebraic numbers. Mathematika **12** (1966), 204–216.
9. Baker, A.: Simultaneous rational approximations to certain algebraic numbers. Proc. Camb. Phil. Soc. **69** (1967) 693–703.
10. Baker, A.: Linear forms in the logarithms of algebraic numbers II. Mathematika **14** (1967) 102–107.
11. Baker, A. : Linear forms in the logarithms of algebraic numbers III. Mathematika **14** (1967) 220–224.
12. Baker, A.: Linear forms in the logarithms of algebraic numbers IV. Mathematika **15** (1968) 204–216.
13. Baker, A.: Contributions to the theory of Diophantine equations I. On the representation of integers by binary forms. Phil. Trans. Royal Soc. London, Ser. **A263** (1968) 173–191.
14. Baker, A.: Contributions to the theory of Diophantine equations II. The Diophantine equation $y^2 = x^3 + k$. Phil. Trans. Royal Soc. London, Ser. **A263** (1968) 193–208.
15. Baker, A.: The Diophantine equation $y^2 = ax^3 + bx^2 + cx + d$. J. London Math. Soc. **43** (1968) 1–9.
16. Baker, A.: Bounds for the solutions of the hyperelliptic equation. Proc. Cab. Phil. Soc. **65** (1969) 439–444.
17. Baker, A.: Effective methods in the theory of numbers. In: Proc. Intern. Congress Math., Nice, 1970, **1** pp. 19–26, Paris, Gauthier Villars, 1971.
18. Baker, A.: Effective methods in Diophantine problems I, II. Proc. Symposia Pure Math. Amer. Math. Soc. **20** (1971) 195–200 *ibid.* **24** (1971) 1–7.
19. Baker, A.: Recent advances in transcendence theory. Proc. Intern. Conf. Number Theory, Moscow, 14-18 September 1971. Trudy Math. Inst. Steklov **132** (1973) 67–69.

20. Baker, A.: A sharpening of the bounds for linear forms in the logarithms. Acta Arith. **21** (1972) 117–129.

21. Baker, A.: A sharpening of the bounds for linear forms in the logarithms, II. Acta Arith. **24** (1973) 33–36.

22. Baker, A.: A sharpening of the bounds for linear forms in the logarithms, III. Acta Arith. **27** (1974) 247–252.

23. Baker, A.: A central theorem in transcendence theory. In: Diophantine Approximation and its applications, pp. 1–23. New York: Acad. Press, 1973.

24. Baker, A.: Transcendental Number Theory. Cambridge Univ. Press, 1975.

25. Baker, A.: The theory of linear forms in logarithms. In: Transcendence Theory: Advances and Applications, pp.1–27. London: Acad. press, 1977.

26. Baker, A., Coates, J.: Integer points on curves of genus 1. Proc. Camb. Phil. Soc. **67** (1970) 595–602.

27. Baker, A., Davenport, H.: The equations $3x^2 - 2 = y^2$ and $8x^2 - 7 = z^2$ Quart. J. Math. Oxford Ser. (2) **20** (1969) 129–137.

28. Baker, A., Stark, H.: On a fundamental inequality in number theory. Ann. Math. **94** (1971) 190–199.

29. Bertrand, D.: Problems arithmétiques lies a l'exponentielle p-adique sur les courbes elliptiques. C. R. Acad. Sci. Paris **282** (1976) 1399–1401.

30. Bertrand, D.: Approximations Diophantiennes p-adiques et courbes elliptiques. Theses l'Univ. P. et M. Curie, Paris 1977.

31. Blanksby, P. E., Montgomery, H. L.: Algebraic integers near the unit circle. Acta. Arith. **18** (1971) 355–369.

32. Bombieri, E.: On Weil's "Theorème de décomposition". Amer. J. Math. **105** (1983) 295–308.

33. Borevich, Z. I., Shafarevich, I. R.: Number theory, Moscow, 1964, London: Academic Press 1966.

34. Brauer, R.: On the zeta functions of algebraic number fields. Amer. J. Math. **69** (1947) 243–250.

35. Brindza, B.: On S-integral solutions of the equation $y^m = f(x)$. Acta Math. Hung. **44** (1–2) (1984) 133–139.

36. Brumer, A.: On the units of algebraic number fields. Mathematika **14** (1967) 121–124.

37. Bundschuh, P.: Zur Approximation gewisser p-adischer algebraischer Zahlen durch rationale Zahlen. J. Reine Angew. Math. **265** (1973) 154–159.

38. Cassels, J. W. S.: On the equation $a^x - b^y = 1$. Amer. J. Math. **75** (1953) 159–162.

39. Cassels, J. W. S.: On the equation $a^x - b^y = 1$. II. Proc. Camb. Phil. Soc. **56** (1960) 97-103. Corrigendum: *ibid.* **57** (1961) 187.

40. Cassels, J. W. S.: An introduction to Diophantine approximation. Cambridge Univ. Press 1957.

41. Cassels, J. W. S.: On a class of exponential equations. Arkiv für Math. **4** (1960) 231–233.

42. Cassels, J. W. S.: Integral points on certain elliptic curves. Proc. London Math. Soc. (3) **14a** (1965) 55–57.

43. Cassels, J. W. S.: Diophantine equations with special reference to elliptic curves. J. London Math. Soc. **41** (1966) 195–291.

44. Catalan, E.: Note extraite d'une lettre adressée a l'éditeur. J. Reine Angew. Math. **27** (1844) p.192.

45. Chowla, S.: On a conjecture of Marshall Hall. Proc. Nat. Acad. Sci. U.S.A. **56** (1966) 417–418.

46. Coates, J.: An effective p-adic analogue of a theorem of Thue. Acta Arith. **15** (1969) 279–305.

47. Coates, J.: An effective p-adic analogue of a theorem of Thue, II The greatest prime factor of a binary form. Acta Arith. **16** (1970) 399–412.

48. Coates, J.: An effective p-adic analogue of a theorem of Thue, III. The Diophantine equation $y^2 = x^3 + k$. Acta Arith. **16** (1970) 425–435.

49. Coates, J.: Construction of rational functions on a curve. Proc. Camb. Phil. Soc. **68** (1970) 105–123.

50. Cohen, S. D.: The distribution of the Galois groups of integral polynomials. Illinois J. Math. **23** (1979) 135–152.

51. Davenport, H.: A note on Thue's theorem. Mathematika **15** (1968) 76–87.

52. Dèbes, P.: Specialisations de polynoms. Math. Rep. Acad. Sci., Roy. Soc. Canada 5 (5) (1983).

53. Dèbes, P.: Spécialisations de polynoms. Publ. Math. de l'Univ. P. et M. Curie **58** (1983) III.1–III.6.

54. Dèbes, P.: Valeurs algébriques de functions algébriques et theorème d'irreductibilité de Hilbert. Theses l'Univ. P. et M.Curie: Paris 1984.

55. Delone, B. N., Faddeev, D. K.: The theory of irrationalities of the third degree. AMS Transl. Math. Monographs vol **10**: Providence, R. I., 1964.

56. Dickson, L. E.: History of the theory of numbers. New York: Chelsea, 1952.

57. Dobrowolski, E.: On a question of Lehmer and the number of irreducible factors of a polynomial. Acta Arith. **34** (1979) 391–401.

58. Dörge, R.: Einfacher Beweis des Hilbertschen Irreduzibilitäts Sätze. Math. Ann. **26** (1927) 176–182.

59. Dyson, F. J.: The approximation to algebraic numbers by rationals. Acta Math. **79** (1947) 225–240.

60. Erdös, P., Shorey, T. N.: On the greatest prime factor of $2^p - 1$ for a prime p and other expressions. Acta Arith: **38** (1976) 257–265.

61. Evertse, J.–H.: On the equation $ax^n - by^n = c$ Composito Math. **43** (1982) 289–315.

62. Evertse, J.–H.: On the representation of integers by binary cubic forms of positive discriminant. Invent. Math. **73** (1983) 117–138.

63. Evertse, J.–H.: Upper bounds for the numbers of solutions of diophantine equations. MC-tract, Math. Centrum: Amsterdam 1983.

64. Evertse, J.–H.: On equations in S-units and the Thue-Mahler equation. Invent. Math. **75** (1984) 561–584.

65. Faltings, G.: Endlichkeitssätze für abelsche Varietäten über Zahlkörpern. Invent. Math. **73** (1983) 349–366.

66. Feldman, N. I.: A refinement of two effective inequalities of A. Baker. Mat. Zametki 6 767–769 = Math. Notes **6** (1969) 925–926.

67. Feldman, N. I.: Estimation of an incomplete linear form in certain algebraic numbers. Mat. Zametki 7 569–580 = Math. Notes **7** (1970) 343–349.

68. Feldman, N. I.: Effective bounds for the size of solutions of certain Diophantine equation. Mat. Zametki 8 361–371 Math Notes **8** (1970) 674–679.

69. Feldman, N. I.: An effective power sharpening of a theorem of Liouville. Izv. Akad. Nauk SSSR Ser. Mat. **35** 973–990 = Math. USSR Izvestija 5 (1971) 985–1002.

70. Feldman, N. I., Chudakov, N. G.: On a theorem of Stark. Mat. Zametki **11** (1972) 329–340 [Russian].

71. Franz, W.: Untersuchungen sum Hilbertschen Irreduzibilitätssatz. Math. **33** (1981) 275–293.

72. Fried, M.: Arithmetical properties of value sets of polynomials. Acta Arith. **15** (1969) 91–115.

73. Fried, M.: On a theorem of Ritt and related Diophantine problems. J. Reine Angew. Math. **264** (1973) 40–55.

74. Fried, M.: On Hilbert's Irreducibility Theorem. J.Number Theory **6** (1974), 211–231.

75. Gauss, C. F.: Disquisitiones arithmeticae Göttingen 1801.

76. Gelfond, A. O.: Sur le septième problème de Hilbert, Dokl. Acad. Nauk SSSR 2 (1934) 1–6.
77. Gelfond, A. O.: Transcendental and algebraic numbers. Moscow, 1952. New York: Dover publ. 1960.
78. Gelfond, A. O.: Selected works, Moscow 1973 [Russian].
79. Gilmore, P. C., Robinson, A.: Mathematical consideration of the relative irreducibility of polynomials. Canadian J. Math. 7 (1955), 483–409.
80. Grimm, C. A.: A conjecture on consecutive composite numbers. Amer. Math. Monthly 76 (1969), 1126–1128.
81. Győry, K.: Sur les polynôms à coefficients entiers et de discriminant donné. Acta Arithm. 23, 419–426 (1973).
82. Győry, K.: Sur les polynoms à coefficients entiers et de discriminant donné, II. Publ. Math. (Debrecen) 21, 125–144 (1974).
83. Győry, K.: Sur les polynoms à coefficients entiers et de discriminant donné, III. Publ. Math. (Debrecen) 23 (1976), 141–165.
84. Győry, K.: Polynomials with given discriminant. Coll. Math. Soc. János Bolyai 13: Debrecen 1974. Topics in number theory: Amsterdam 1976.
85. Győry, K.: Représentation des nombres entiers par des formes binaires. Publ. Math. (Debrecen) 24 (1977), 363–375.
86. Győry, K.: On polynomials with integer coefficients and given discriminant. IV.Publ.Math.(Debrecen), 25 (1978), 155–167:
87. Győry, K: On the greatest prime factors of decomposable forms at integer points. Ann. Acad. Sci. Fenn. Ser. A.l. Math. 4 (1978/79), 341–355.
88. Győry, K.: On the number of solutions of linear equations in units of an algebraic number fields. Comment. Math. Helvetici 54 (1979), 583–600.
89. Győry, K.: Effective finiteness theorems for polynomials with given discriminant and integral elements with given discriminant over finitely generated domains. J. Reine Angew. Math. 346 (1984), 54–100.
90. Győry, K., Papp, Z. Z.: Effective estimates for the integer solutions of norm form and discriminant form equations. Publ. Math. (Debrecen) 25 (1978), 311–325.
91. Győry, K., Pethő, A.: Sur la distribution des solutions des equations du type "norm-forme", Acta Math. Acad. Sci., Hung. 26 (1975), 135–142.
92. Hasse, H.: Über mehrklassige, aber eingeschlechtige reellquadratische Zahlkörper, Elemente der Mathematik 20 (1965), 49–59.
93. Hasse, H.: Zahlentheorie, Berlin: Acad. Verlag, 1969.
94. Hasse, H., Bernstein, L.: Einheitenberechnung mittels des Jacobi-Perronschen Algorithmuth. J. Reine Angew. Math. 218 (1965), 51–69.
95. Hasse, H., Bernstein, L: An explicit formula for the units of an algebraic number field of degree $n \geq 2$. Pacific J. Math. 30 (1969), 293–365.
96. Hecke, E.: Vorlesungen über die Theorie der algebraischen Zahlen. Leipzig 1923.
97. Hermite, Ch.: Cours d'Analyse. Moscow, 1936 [Russian]. Cours de M. Ch. Hermite: Paris, 1891.
98. Hilbert, D.: Uber die Irreduzibilität ganzer rationaler Funktionen mit ganzzahligen Koeffizienten. J. Reine Angew. Math. 110 (1892), 104–129.
99. Hilliker, D. L., Straus, E. G.: Determination of bounds for the solutions to those binary diophantine equations that satisfy the hypotheses of Runge's theorem. Trans. Amer. Math. Soc. 280 (1983), 637–657.
100. Inaba, E.: Uber den Hilbertschen Irreduzibilitätssatz. Jap. J. Math. 19 (1944), 1–25.
101. Inkeri, K.: A note on Fermat's conjecture. Acta Arith. 29 (1976), 251–256.
102. Inkeri, K., van der Poorten, A. J.: Some remarks on Fermat's conjecture. Acta Arith. 36 (1980), 107–111.

103. Keates, M.: On the greatest prime factor of a polynomial. Proc. Edinburgh Math. Soc. (2) **16** (1968/69), 301-303.

104. Kleiman, H.: On the Diophantine equation $f(x,y) = 0$. J. Reine Angew. Math. **286/287** (1976), 124-131.

105. Knobloch, H. W.: Zum Hilbertschen Irreduzibilitätssatz. Abh. Math. Sem. Univ. Hamburg **19** (1955), 176-190.

106. Knobloch, H. W.: Die Seltenheit der reduziblen Polynome. Jber. Deutsch. Math. Verein. **59** (1956), Abt.1, 12-19.

107. Koblitz, N.: p-adic Numbers, p-adic Analysis and Zeta-functions. Berlin Heidelberg New York: Springer 1977.

108. Kotov, S. V.: On the norm of ideal divisors of a binary form with algebraic coefficients. Vesci Akad. Nauk BSSR Ser. Fiz.-Mat. Nauk **3** (1972), 14-22 [Russian].

109. Kotov, S. V.: On the greatest prime divisor of a polynomial. Math. Zametki **13** (1973), 515-522 [Russian].

110. Kotov, S. V.: The law of the iterated logarithm for binary forms with algebraic coefficients, Dokl. Akad. Nauk BSSR **17** (1973), 591-594 [Russian].

111. Kotov, S. V.: The equation of Thue-Mahler in relative fields. Acta Arith. **27** (1975), 293-315 [Russian].

112. Kotov, S. V.: Uber die maximale Norm der Idealteiler des Polynoms $\alpha x^m + \beta y^m$ mit algebraiscben Koefficienten. Acta Arith. **31** (1976), 219-230.

113. Kotov, S. V.: On the effectivisation of Mahler's theorem on the rational points on the curves of genus 1. Dokl. Akad. Nauk BSSR **21** (1977), 101-104 [Russian].

114. Kotov, S. V.: Die arithmetische Structur der rationalen Punkte auf Kurven vom Geschlecht Eins. Acta Arith. **35** (1978), 103-115.

115. Kotov, S. V., Sprindžuk, V. G.: An effective analysis of the Thue-Mahler equation. Dokl. Akad. Nauk BSSR **17** (1973), 393-395 [Russian].

116. Kotov, S. V., Sprindžuk, V. G.: The equation of Thue-Mahler in a relative Field and approximation of algebraic numbers by algebraic numbers. Izv. Akad. Nauk SSSR Ser. Mat. **41** (1977), 723-751. Math. USSR Izvestija **11** (1977), 677-707.

117. Kotov, S. V., Trelina, L. A.: S-ganze Punkte auf elliptischen Kurven. J. Reine Angew. Math. **306** (1979), 28-41.

118. Lagarias, J. C., Montgomery, H. L., Odlyzko, A. M.: A bound for the least prime ideal in the Chebotarev density theorem. Invent. Math. **54** (1979), 271-296.

119. Landau, E.: Abschätzungen von Charaktersummen, Einheiten und Klassenzahlen. Nachr. Kgl. Ges. Wiss. Göttingen, Mat. Phys. Kl., 79-97 (1918).

120. Lang, S.: Diophantine Geometry. New York: Int. Publ. 1962.

121. Lang, S.: Algebraic Numbers. Reading, Mass., — Palo Alto London: Addison-Wesley Publ. Co, 1964.

122. Lang, S.: Transcendental numbers and diophantine approximations. Bull. Amer. Math. Soc. **77** (1971), 635-677.

123. Lang, S.: Diophantine approximation on Abelian varieties with complex multiplication. Advances in Math. **17** (1975), 281-336.

124. Lang, S.: Elliptic curves. Diophantine analysis. Berlin Heidelberg New York: Springer 1978.

125. Lavrik, A. F.: A note on a theorem of Siegel-Brauer concerning parameters of the algebraic number Fields. Mat. Zametki **8** (1970), 259-263. [Russian].

126. Lenskoi, D. N.: Functions in non-archimedean normed Fields. Saratov: Izdat. Saratov Univ. 1962 [Russian].

127. LeVeque, W. J.: On the equation $y^m = f(x)$. Acta Arith. **9** (1964), 209-219.

128. Liouville, J.: Sur des classes très étendues de quantités dont la valuer n^e est ni algebrique, ni même réductible á des irrationnelles algebriques. C.R. Acad. Sci. Paris **18** (1844), 883-885, 910-911.

129. Mahler, K.: Zur Approximation algebraischer Zahlen, I. Uber den grössten primteiler binärer formen. Math. Ann. **107** (1933), 691-730.

130. Mahler, K.: Zur Approximation algebraischer Zahlen, II. Über die Anzahl der Darstellungen grosser Zahlen durch binäre formen. Math. Ann. **108** (1933), 37–55.

131. Mahler, K.: Zur Approximation algebraischer Zahlen, III. Über die mittlere Anzahl der Darstellungen grossen Zahlen durch binäre Formen. Acta Math. **62** (1934), 91–166.

132. Mahler, K.: Über die rationalen Punkte auf Kurven vom geschlecht Eins. J. Reine Angew. Math. **170** (1934), 168–178.

133. Mahler, K.: Über transzendente p-adische Zahlen. Compositio Math. **2** (1935), 259–275.

134. Mahler, K.: Über den grössten Primteiler spezieller Polynome zweiten Grades. Archiv für Math. Naturvid. B41 (1935), N 6.

135. Mahler, K.: On the greatest prime Factor of $ax^m + by^m$. Nieuw Arch. Wisk. (3) **1** (1953), 113–122.

136. Mahler, K.: Lectures on Diophantine approximation. Notre Dame Univ. 1961.

137. Mahler, K.: On the approximation of algebraic numbers by algebraic integers. J. Austral. Math. Soc. **3** (1963), 408–434.

138. Mahler, K.: Introduction to p-adic numbers and their functions. Cambridge Univ. Press 1973.

139. Markoff, V.: L'Intermédiaire des math. **2** (1895), 23. Ibid. **3** (1901), 305–306.

140. Masser, D. W.: Elliptic functions and Transcendence. Berlin Heidelberg Hew York: Springer 1975.

141. Mordell, L. J.: The Diophantine equation, $y^2 - k = x^3$. Proc. London Math. Soc. (2) **13** (1913), 60–80.

142. Mordell, L. J.: Indeterminate equations of the third and forth degrees. Quart. J. Pure and Appl. Math. **45** (1914), 170–186.

143. Mordell, L. J.: On the rational solutions of the indeterminate equations of the 3rd and 4th degrees. Proc. Camb. Phil. Soc. **21**, 179–192 (1922).

144. Mordell, L, J.: A chapter in the theory of numbers. Cambridge Univ. Press 1947.

145. Mordell, L. J.: Diophantine Equations. London: Acad. Press 1969.

146. Nagell, T.: Über den grössten Primteiler gewisser Polynome dritten Grades. Math. Ann. **114** (1937), 284–292.

147. Nagell, T.: Bemerkung über die Klassenzahl reellquadratischer Zahlkörper. Det Kongelige Norske Vid. Selsk. **3** (1928), 7–10.

148. Noëther, E.: Gleichungen mit vorgeschriebener Gruppe. Math. Ann. **78** (1917), 221–229.

149. Noëther, E.: Ein algebraisches Kriterium für absolute Irreduzibilität. Math. Ann. **85** (1922), 26–33.

150. Ore, Ö.: Existenzbeweise für algebraische Körper mit vorgeschriebenen Eigenschaften. Math. Z. **25** (1926), 474–489.

151. Osgood, C. F.: The Diophantine approximation of roots of positive integers. J. Res. Nat. Bur. Stand., **74B** (1970), 241–244.

152. Osgood, C. F.: The, simultaneous Diophantine approximation of certain k-th roots. Proc. Camb. Phil. Soc: **67** (1970), 75–86.

153. Ostrowski, A.: Zur arithmetischen Theorie der algebraischen Grössen. Göttinger Nachr. (1919), 279–298.

154. Parry, C. J.: The p-adic generalisation of the Thue-Siegel theorem. J. London Math. Soc. **15** (1940), 293–305.

155. Parry, C. J.: The p-adic generalization of the Thue–Siegel theorem. Acta Math. **83** (1959), 1–100.

156. Polya, G.: Zur arithmetischen Untersuchung der Polynome. Math. Zeitschr. **1** (1918), 143–148.

157. van der Poorten, A. J.: Effectively computable bounds for the solutions of certain Diophantine equations. Acta Arithm. **33** (1977), 195-207.
158. van der Poorten, A. J.: Linear forms in logarithms in the p-adic case. In: Transcendence Theory: Advances and Applications, p. 29–57, London: Acad. Press 1977.
159. van der Poorten, A. J., Loxton, J. H.: Computing the effectively computable bound in Baker's inequality for linear forms in logarithms. Bull. Austral. Math. Soc. **15** (1976), 33–57.
160. van der Poorten, A. J., Loxton, J. H.: Multiplicative relations in number Fields. Bull. Austral. Math. Soc. **16** (1976), 83–98.
161. Preus, G., Schmidt, F. K.: Über den Hilbertschen Irreduzibilitätssatz. Math. Nachr. **4** (1951), 348–365.
162. Ramachandra, K., Shorey, T. N., Tijdeman, R.: On Grimm's problem relating to factorization of a block of consecutive integers. J. Reine Angew. Math. **273** (1975), 109–124. Ibid. **288** (1976),192–201.
163. Remak, R.: Über Grössenbereichungen zwischen Diskriminante und Regulator eines algebraischen Zahlkörpers. Compositio Math. **10** (1952), 245–285.
164. Remak, R.: Über algebraische Zahlkörper mit schwachen Einheitsdefect, Compositio Math. **12** (1954), 35–80.
165. Roth, I.: Rational approximations to algebraic numbers. Mathematika **2** (1955), 1–20. Corrigendum, ibid., p.168.
166. Runge, C.: Über ganzzahlige Lösungen von Gleichungen zwischen zwei Veränderlichen, J. Reine Angew. Math, **100** (1887), 425–435.
167. Schinzel, A.: On two theorems of Gelfond and some of their applications. Acta Arith. **13** (1967/68), 177–236.
168. Schinzel, A.: An improvement of Runge's theorem on Diophantine equations. Comment. Pontific. Acad. Sci. **2** (1969), 1–9.
169. Schinzel, A.: On Hilbert's Irreducibility Theorem. Ann. Polon. Math. **16** (1965), 333–340.
170. Schinzel, A.: Reducibility of polynomials. In: Proc. Intern. Congress Math., Nice, 1970, **1** p.491–496, Paris: Gauthier-Villars 1971.
171. Schinzel, A.: Selected topics on polynomials. Ann. Arbor: The Univ. of Michigan Press 1982.
172. Schinzel, A., Tijdeman, R.: On the equation $y^m = f(x)$. Acta Arith. **31** (1976), 199–204.
173. Schlickewei, H. P.: Linearformen mit algebraischen Koefficienten, Manuscripts Math. **18** (1976), 147–185.
174. Schlickewei, H. P.: Die p-adische Verallgemeinerung des Satzes von Thue-Siegel-Roth-Schmidt. J. Reine Angew. Math. **288** (1976), 85–105.
175. Schlickewei, H. P.: On products of special linear forms with algebraic coefficients. Acts Arith. **31** (1976), 389–398.
176. Schlickewei, H. P.: Über die Diophantische Gleichung $x_1 + x_2 + \cdots + x_n = 0$ Acta Arith. **33** (1977), 183–185.
177. Schmidt, W.: On heights of algebraic subspaces and Diophantine approximation. Ann. Math. (2), **85** (1967), 430–472.
178. Schmidt, W.: Simultaneous approximation to algebraic numbers by rationals. Acts Math: **125** (1970), 189–201.
179. Schmidt, W.: Linearformen mit algebraischen Koefficienten, I, II. J. Number Theory **3** (1971), 253–277.
180. Schmidt, W.: Some recent progress in Diophantine approximations. In: Proc. Intern. Congress Math., Nice, **1** 1970, p. 497–503, Paris: Gauthier-Villars 1971.
181. Schmidt, W.: Norm form equations. Ann, Math, **96** (1972), 526–551.
182. Schmidt, W.: Approximation to algebraic numbers, L'Enseignement Math. Ser. 2, **17** (1971), 187–253.

183. Schmidt, W.: Applications of Thue's method in various branches of number theory. In: Proc. Intern. Congress Math, Vancouver, 1974, **1**, 177–185, Vancouver 1975.

184. Schmidt, W.: Equations over finite fields. An elementary approach. Berlin Heidelberg New York: Springer 1976.

185. Schmidt, W.: Diophantine Approximation. Springer: Berlin, Heidelberg, New York. 1980.

186. Schneider, Th.: Transzendenzuntersuchungen periodischer Functionen, I. Transzendenz von Potenzen. J. Reine Angew, Math. **172** (1934), 65–69.

187. Schnirelman, L, G.: On Functions in normed algebraically closed fields. Izv. Akad. Nauk SSSR Ser. Mat. **23** (1938), 487–496 [Russian].

188. Shorey, T. N., Tijdeman, R.: On the greatest prime factor of polynomials at integer points. Compositio Math. **33** (1976), 187-195.

189. Shorey, T. N., Tijdeman, R.: New application of Diophantine approximations to Diophantine equations. Math. Scand. **39** (1976), 5–18.

190. Shorey, T. N., van der Poorten, A. J., Tijdeman, R., Schinzel, A.: Applications of Gelfond-Baker method to Diophantine equations. In: Transcendence Theory: Advances and Applications, p.59–77, London: Acad. Press, 1977.

191. Siegel, C. L.: Approximation algebraischer Zahlen. Math. Z. **10** (1921), 173–213.

192. Siegel, C. L. (under the pseudonym X), The integer solutions of the equation $y^2 = ax^n + bx^{n-1} + \ldots + k$. J. London Math. Soc. **1** (1926), 66–68.

193. Siegel, C. L.: Über einige Anwendungen Diophantischer Approximationen. Abh. Preuss. Acad. Wiss. Phys.–Mat. Kl.1, (1929), 41–69.

194. Siegel, C. L.: Über die Klassenzahl quadratischer Zahlkörper. Acta Arith. **1** (1935), 83–86.

195. Siegel, C. L.: Abschätzung von Einheiten. Nachr. Akad. Wiss. Göttingen. Mat.-Phys. KI. **9** (1969), 71–86.

196. Skolem, T.: Diophantische Gleichungen. Berlin: Springer 1938.

197. Sprindžuk, V. G.: On the number of solutions of the Diophantine equation $x^3 = y^2 + A$. Dokl. Akad, Nauk BSSR **7** (1963), 9–11 [Russian].

198. Sprindžuk, V. G.: Concerning Baker's theorem on linear forms in logarithms. Dokl. Akad. Nauk BSSR **11** (1967), 767–769 [Russian].

199. Sprindžuk, V. G.: Effectivization in certain problems of Diophantine approximation theory. Dokl. Akad, Nauk BSSR **12**, 293–297 (1968). [Russian].

200. Sprindžuk, V. G.: Estimates of linear forms with p-adic logarithms of algebraic numbers. Vesci Akad. Nauk BSSR Ser. Fiz.-Mat, Nauk **4** (1968), 5–14 [Russian].

201. Sprindžuk, V. G.: Effective estimates in "ternary", exponential Diophantine equations. Dokl, Akad, Nauk BSSR **13** (1969), 777–780 [Russian].

202. Sprindžuk, V. G.: An effective estimate of rational approximations to algebraic numbers. Dokl. Akad, Nauk BSSR **14** (1970), 681–684 [Russian].

203. Sprindžuk, V, G.: A new application of p-adic analysis to representations of numbers by binary forms, Izv. Akad. Nauk SSSR Ser. Mat. **34**, 1038–1063

204. Sprindžuk, V. G.: New applications of analytic and p-adic methods in Diophantine approximations. In: Proc, Intern. Congress Bath., Nice, 1970, **1**, p505–509, Paris; Gauthier-Villars 1971.

205. Sprindžuk, V. G.: An improvement of the estimate of rational approximations to algebraic numbers. Dokl. Akad. Nauk BSSR **15** (1971), 101–104 [Russian].

206. Sprindžuk, V. G.: On the rational, approximations to algebraic numbers. Izv. Akad. Nauk SSSR Ser. Mat. **35** 991–1007 (1971) Math. USSR Izvestija **5** (1971), 1003-1019.

207. Sprindžuk, V. G,: An estimate of the greatest prime divisor of a binary form. Dokl. Akad. Nauk BSSR **15** (1971), 389–391 [Russian].

208. Sprindžuk, V. G.: On bounds for the units of algebraic number fields. Dokl. Akad. Nauk BSSR **15** (1971), 1065–1068 [Russian].

209. Sprindžuk, V. G.: On bounds for the solutions of the Thue equation. Izv. Akad. Nauk SSSR Ser. Mat. **36** (1972), 712–741. Math. USSR Izvestija **6** (1972), 705–734.

210. Sprindžuk, V. G.: Square-free divisors of polynomials and class-numbers of algebraic number fields. Acta Arith. **24** (1973), 143–149 [Russian].

211. Sprindžuk, V. G.: On the structure of numbers representable by the binary forms. Dokl. Akad. Nauk BSSR **17** (1973), 685–688 [Russian].

212. Sprindžuk, V. G.: The distribution of the fundamental units of real quadratic fields. Acta Arith. **25** (1974), 405–409.

213. Sprindžuk, V. G.: "Almost every" algebraic number-field has a large class-number. Acta Arith. **25** (1974), 411–413.

214. Sprindžuk, V. G.: The fields of algebraic numbers with a large class-number. Izv. Akad. Nauk SSSR Ser. Mat. **38** (1974), 971–982. Math. USSR Izvestija **8** (1974), 967–978.

215. Sprindžuk, V. G.: An effective analysis of the Thue and Thue–Mahler equations. In: Current problems of analytic number theory, p.199–222. Minsk 1974 [Russian].

216. Sprindžuk, V. G.: Representation of numbers by the norm forms with two dominating variables. J. Number Theory **6** (1974), 481-486.

217. Sprindžuk, V. G.: Diophantine equations and class–numbers. Mat. Zametki **17** (1975), 161–168.

218. Sprindžuk, V. G.: A hyperelliptic diophantine equation and class-numbers. Acta Arith. **30** (1976), 95–108 [Russian].

219. Sprindžuk, V. G.: Arithmetic structure of integral polynomials and class–numbers. Trudy Math. Inst. Steklov **143** (1977), 152–174. Proc. Steklov Inst. Math., 1980, issue 1, p. 163–186.

220. Sprindžuk, V. G.: Hilbert's irreducibility theorem and rational points on algebraic curves. Dokl. Akad. Nauk SSSR **247**, 285–289, Soviet Math. Dokl. **20** (1979), 701–705.

221. Sprindžuk, V. G.: Reducibility of polynomials and rational points on algebraic curves. Dokl. Akad. Nauk SSSR **250**, 1327–1330 Soviet Math. Dokl. **21** (1980), 331–334.

222. Sprindžuk, V. G.: Achievements and problems of the theory of diophantine approximations. Uspechi Mat. Nauk **35** (1980), 3–68 [Russian].

223. Sprindžuk, V. G.: Diophantine equations involving unknown primes. Trudy Math. Inst. Steklov **158**, 180–196 (1981). Proc. Steklov. Inst. Math, 1983, Issue 4, p. 197–214.

224. Sprindžuk, V. G.: Arithmetic specializations in polynomials. J. Reine Angew. Math. **340** (1983), 26-52.

225. Sprindžuk, V. G.: Diophantine approximations to the values of algebraic functions. Dokl. Akad. Nauk BSSR **29** (1985), 101–103 [Russian].

226. Stark, H.: Further advances in the theory of linear forms in logarithms. In: Diophantine Approximation and its Applications, p.255–293, New York: Acad. Press 1973.

227. Stark, H.: Effective estimates of solutions of some Diophantine equations. Acta Arith. **24**, 251–259.

228. Stark, H.: Some effective cases of the Brauer–Siegel theorem. Invent. Math. **23** (1974), 135–152.

229. Thue, A.: Über Annäherungswerte algebraischer Zahlen. J. Reine Angew. Math. **135** (1909), 284–305.

230. Thue, A.: Berechnung aller Lösungen gewisser Gleichungen von der form $ax^r - by^r = f$. Norske Vid. selskap. Skrifter. Math. Natuw. Kl. **4** (1918), 1–9.

231. Tijdeman, R.: Some applications of Baker's sharpened bounds to Diophantine equations. Sem. Delange-Pisot-Poitou **16** (1975), N 24.
232. Tijdeman, R.: On the equation of Catalan. Acta Arith. **29** (1976), 197–209.
233. Tijdeman, R.: Applications of the Gel'fond-Baker method to rational number theory. In: Coll. Math. Soc. Jànos Bolyai **13**: Debrecen 1974. Topics in number theory: Amsterdam: 1976.
234. Tijdeman, R.: On the Gel'fond-Baker method and its applications. In: Mathematical developments arising From Hilbert problems. Proc. Sympos. Pure Math. Amer. Math. Soc. **28** (1976), 241–286.
235. Tschebotareff, N.: Die Bestimmung der Dichtigkeit einer Menge von Primzahlen welche zu einer gegebenen Substitutionenklasse gehoren. Math. Ann. **95** (1926), 191–228.
236. Turk, J.: Polynomial values and almost powers. Michigan Math. J. **29** (1982), 213–230.
237. Trelina, L. A.: On algebraic integers with discriminants having fixed prime divisors. Math. Zametki **21** (1977), 289–296 [Russian].
238. Trelina, L. A.: On the greatest prime factor of an index form. Dokl. Akad. Nauk BSSR **21** (1977), 975–976 [Russian].
239. Trelina, L. A.: On S-integral solutions of the hyperelliptic equation. Dokl. Akad. Nauk BSSR **22** (1978), 881–884 [Russian].
240. Trelina, L. A.: On the representation of powers by polynomials in the fields of algebraic numbers. Dokl. Akad. Nauk BSSR **29** (1985), 5–8 [Russian].
241. Vinogradov, A. I., Sprindžuk, V. G.: On the representation of numbers by binary forms. Mat. Zametki **3** (1968), 369–376 [Russian].
242. Waldschmidt, M.: Transcendence Methods. Queen's papers in pure and applied mathematics **52**, Kingston 1979.
243. Waldschmidt, M.: A lower bound for linear forms in logarithms. Acta Arith. **37** (1980), 257–283.
244. Weil, A.: Arithmetic on algebraic varieties. Annals of Math. **53** (1951), 412-444.
245. Weyl, H.: Algebraic theory of numbers. Princeton 1940.
246. Yamamoto, Y.: Real quadratic number fields with large fundamental units. Osaka J. Math. **8** (1971), 261–270.
247. Zimmert, R.: Ideale kleiner Norm in Idealklassen und eine Regulatorbeschätzung. Invent. Math. **62** (1981), 367–380.

Lecture Notes in Mathematics

For information about Vols. 1–1384
please contact your bookseller or Springer-Verlag

Vol. 1385: A.M. Anile, Y. Choquet-Bruhat (Eds.), Relativistic Fluid Dynamics. Seminar, 1987. V, 308 pages. 1989.

Vol. 1386: A. Bellen, C.W. Gear, E. Russo (Eds.), Numerical Methods for Ordinary Differential Equations. Proceedings, 1987. VII, 136 pages. 1989.

Vol. 1387: M. Petkovi´c, Iterative Methods for Simultaneous Inclusion of Polynomial Zeros. X, 263 pages. 1989.

Vol. 1388: J. Shinoda, T.A. Slaman, T. Tugué (Eds.), Mathematical Logic and Applications. Proceedings, 1987. V, 223 pages. 1989.

Vol. 1000: Second Edition. H. Hopf, Differential Geometry in the Large. VII, 184 pages. 1989.

Vol. 1389: E. Ballico, C. Ciliberto (Eds.), Algebraic Curves and Projective Geometry. Proceedings, 1988. V, 288 pages. 1989.

Vol. 1390: G. Da Prato, L. Tubaro (Eds.), Stochastic Partial Differential Equations and Applications II. Proceedings, 1988. VI, 258 pages. 1989.

Vol. 1391: S. Cambanis, A. Weron (Eds.), Probability Theory on Vector Spaces IV. Proceedings, 1987. VIII, 424 pages. 1989.

Vol. 1392: R. Silhol, Real Algebraic Surfaces. X, 215 pages. 1989.

Vol. 1393: N. Bouleau, D. Feyel, F. Hirsch, G. Mokobodzki (Eds.), Séminaire de Théorie du Potentiel Paris, No. 9. Proceedings. VI, 265 pages. 1989.

Vol. 1394: T.L. Gill, W.W. Zachary (Eds.), Nonlinear Semigroups, Partial Differential Equations and Attractors. Proceedings, 1987. IX, 233 pages. 1989.

Vol. 1395: K. Alladi (Ed.), Number Theory, Madras 1987. Proceedings. VII, 234 pages. 1989.

Vol. 1396: L. Accardi, W. von Waldenfels (Eds.), Quantum Probability and Applications IV. Proceedings, 1987. VI, 355 pages. 1989.

Vol. 1397: P.R. Turner (Ed.), Numerical Analysis and Parallel Processing. Seminar, 1987. VI, 264 pages. 1989.

Vol. 1398: A.C. Kim, B.H. Neumann (Eds.), Groups – Korea 1988. Proceedings. V, 189 pages. 1989.

Vol. 1399: W.-P. Barth, H. Lange (Eds.), Arithmetic of Complex Manifolds. Proceedings, 1988. V, 171 pages. 1989.

Vol. 1400: U. Jannsen. Mixed Motives and Algebraic K-Theory. XIII, 246 pages. 1990.

Vol. 1401: J. Steprans, S. Watson (Eds.), Set Theory and its Applications. Proceedings, 1987. V, 227 pages. 1989.

Vol. 1402: C. Carasso, P. Charrier, B. Hanouzet, J.-L. Joly (Eds.), Nonlinear Hyperbolic Problems. Proceedings, 1988. V, 249 pages. 1989.

Vol. 1403: B. Simeone (Ed.), Combinatorial Optimization. Seminar, 1986. V, 314 pages. 1989.

Vol. 1404: M.-P. Malliavin (Ed.), Séminaire d´Algèbre Paul Dubreil et Marie-Paul Malliavin. Proceedings, 1987–1988. IV, 410 pages. 1989.

Vol. 1405: S. Dolecki (Ed.), Optimization. Proceedings, 1988. V, 223 pages. 1989. Vol. 1406: L. Jacobsen (Ed.), Analytic Theory of Continued Fractions III. Proceedings, 1988. VI, 142 pages. 1989.

Vol. 1407: W. Pohlers, Proof Theory. VI, 213 pages. 1989.

Vol. 1408: W. Lück, Transformation Groups and Algebraic K-Theory. XII, 443 pages. 1989.

Vol. 1409: E. Hairer, Ch. Lubich, M. Roche. The Numerical Solution of Differential-Algebraic Systems by Runge-Kutta Methods. VII, 139 pages. 1989.

Vol. 1410: F.J. Carreras, O. Gil-Medrano, A.M. Naveira (Eds.), Differential Geometry. Proceedings, 1988. V, 308 pages. 1989.

Vol. 1411: B. Jiang (Ed.), Topological Fixed Point Theory and Applications. Proceedings. 1988. VI, 203 pages. 1989.

Vol. 1412: V.V. Kalashnikov, V.M. Zolotarev (Eds.), Stability Problems for Stochastic Models. Proceedings, 1987. X, 380 pages. 1989.

Vol. 1413: S. Wright, Uniqueness of the Injective III₁ Factor. III, 108 pages. 1989.

Vol. 1414: E. Ramirez de Arellano (Ed.), Algebraic Geometry and Complex Analysis. Proceedings, 1987. VI, 180 pages. 1989.

Vol. 1415: M. Langevin, M. Waldschmidt (Eds.), Cinquante Ans de Polynômes. Fifty Years of Polynomials. Proceedings, 1988. IX, 235 pages.1990.

Vol. 1416: C. Albert (Ed.), Géométric Symplectique et Mécanique. Proceedings, 1988. V, 289 pages. 1990.

Vol. 1417: A.J. Sommese, A. Biancofiore, E.L. Livorni (Eds.), Algebraic Geometry. Proceedings, 1988. V, 320 pages. 1990.

Vol. 1418: M. Mimura (Ed.), Homotopy Theory and Related Topics. Proceedings, 1988. V, 241 pages. 1990.

Vol. 1419: P.S. Bullen, P.Y. Lee, J.L. Mawhin, P. Muldowney, W.F. Pfeffer (Eds.), New Integrals. Proceedings, 1988. V, 202 pages. 1990.

Vol. 1420: M. Galbiati, A. Tognoli (Eds.), Real Analytic Geometry. Proceedings, 1988. IV, 366 pages. 1990.

Vol. 1421: H.A. Biagioni, A Nonlinear Theory of Generalized Functions, XII, 214 pages. 1990.

Vol. 1422: V. Villani (Ed.), Complex Geometry and Analysis. Proceedings, 1988. V, 109 pages. 1990.

Vol. 1423: S.O. Kochman, Stable Homotopy Groups of Spheres: A Computer-Assisted Approach. VIII, 330 pages. 1990.

Vol. 1424: F.E. Burstall, J.H. Rawnsley, Twistor Theory for Riemannian Symmetric Spaces. III, 112 pages. 1990.

Vol. 1425: R.A. Piccinini (Ed.), Groups of Self-Equivalences and Related Topics. Proceedings, 1988. V, 214 pages. 1990.

Vol. 1426: J. Azéma, P.A. Meyer, M. Yor (Eds.), Séminaire de Probabilités XXIV, 1988/89. V, 490 pages. 1990.

Vol. 1427: A. Ancona, D. Geman, N. Ikeda, École d'Eté de Probabilités de Saint Flour XVIII, 1988. Ed.: P.L. Hennequin. VII, 330 pages. 1990.

Vol. 1428: K. Erdmann, Blocks of Tame Representation Type and Related Algebras. XV. 312 pages. 1990.

Vol. 1429: S. Homer, A. Nerode, R.A. Platek, G.E. Sacks, A. Scedrov, Logic and Computer Science. Seminar, 1988. Editor: P. Odifreddi. V, 162 pages. 1990.

Vol. 1430: W. Bruns, A. Simis (Eds.), Commutative Algebra. Proceedings. 1988. V, 160 pages. 1990.

Vol. 1431: J.G. Heywood, K. Masuda, R. Rautmann. V.A. Solonnikov (Eds.), The Navier-Stokes Equations – Theory and Numerical Methods. Proceedings, 1988. VII, 238 pages. 1990.

Vol. 1432: K. Ambos-Spies, G.H. Müller, G.E. Sacks (Eds.), Recursion Theory Week. Proceedings, 1989. VI, 393 pages. 1990.

Vol. 1433: S. Lang, W. Cherry, Topics in Nevanlinna Theory. II, 174 pages.1990.

Vol. 1434: K. Nagasaka, E. Fouvry (Eds.), Analytic Number Theory. Proceedings, 1988. VI, 218 pages. 1990.

Vol. 1435: St. Ruscheweyh, E.B. Saff, L.C. Salinas, R.S. Varga (Eds.), Computational Methods and Function Theory. Proceedings, 1989. VI, 211 pages. 1990.

Vol. 1436: S. Xambó-Descamps (Ed.), Enumerative Geometry. Proceedings, 1987. V, 303 pages. 1990.

Vol. 1437: H. Inassaridze (Ed.), K-theory and Homological Algebra. Seminar, 1987–88. V, 313 pages. 1990.

Vol. 1438: P.G. Lemarié (Ed.) Les Ondelettes en 1989. Seminar. IV, 212 pages. 1990.

Vol. 1439: E. Bujalance, J.J. Etayo, J.M. Gamboa, G. Gromadzki. Automorphism Groups of Compact Bordered Klein Surfaces: A Combinatorial Approach. XIII, 201 pages. 1990.

Vol. 1440: P. Latiolais (Ed.), Topology and Combinatorial Groups Theory. Seminar, 1985–1988. VI, 207 pages. 1990.

Vol. 1441: M. Coornaert, T. Delzant, A. Papadopoulos. Géométrie et théorie des groupes. X, 165 pages. 1990.

Vol. 1442: L. Accardi, M. von Waldenfels (Eds.), Quantum Probability and Applications V. Proceedings, 1988. VI, 413 pages. 1990.

Vol. 1443: K.H. Dovermann, R. Schultz, Equivariant Surgery Theories and Their Periodicity Properties. VI, 227 pages. 1990.

Vol. 1444: H. Korezlioglu, A.S. Ustunel (Eds.), Stochastic Analysis and Related Topics VI. Proceedings, 1988. V, 268 pages. 1990.

Vol. 1445: F. Schulz, Regularity Theory for Quasilinear Elliptic Systems and – Monge Ampère Equations in Two Dimensions. XV, 123 pages. 1990.

Vol. 1446: Methods of Nonconvex Analysis. Seminar, 1989. Editor: A. Cellina. V, 206 pages. 1990.

Vol. 1447: J.-G. Labesse, J. Schwermer (Eds.), Cohomology of Arithmetic Groups and Automorphic Forms. Proceedings, 1989. V, 358 pages. 1990.

Vol. 1448: S.K. Jain, S.R. López-Permouth (Eds.), Non-Commutative Ring Theory. Proceedings, 1989. V, 166 pages. 1990.

Vol. 1449: W. Odyniec, G. Lewicki. Minimal Projections in Banach Spaces. VIII, 168 pages. 1990.

Vol. 1450: H. Fujita, T. Ikebe, S.T. Kuroda (Eds.), Functional-Analytic Methods for Partial Differential Equations. Proceedings, 1989. VII, 252 pages. 1990.

Vol. 1451: I.. Alvarez-Gaumé, E. Arbarello, C. De Concini, N.J. Hitchin, Global Geometry and Mathematical Physics. Montecatini Terme 1988. Seminar. Editors: M. Francaviglia, F. Gherardelli. IX, 197 pages. 1990.

Vol. 1452: E. Hlawka, R.F. Tichy (Eds.), Number-Theoretic Analysis. Seminar, 1988–89. V, 220 pages. 1990.

Vol. 1453: Yu.G. Borisovich, Yu.E. Gliklikh (Eds.), Global Analysis – Studies and Applications IV. V, 320 pages. 1990.

Vol. 1454: F. Baldassari, S. Bosch, B. Dwork (Eds.), p-adic Analysis. Proceedings, 1989. V, 382 pages. 1990.

Vol. 1455: J.-P. Françoise, R. Roussarie (Eds.), Bifurcations of Planar Vector Fields. Proceedings, 1989. VI, 396 pages. 1990.

Vol. 1456: L.G. Kovács (Ed.), Groups – Canberra 1989. Proceedings. XII, 198 pages. 1990.

Vol. 1457: O. Axelsson, L.Yu. Kolotilina (Eds.), Preconditioned Conjugate Gradient Methods. Proceedings, 1989. V, 196 pages. 1990.

Vol. 1458: R. Schaaf, Global Solution Branches of Two Point Boundary Value Problems. XIX, 141 pages. 1990.

Vol. 1459: D. Tiba, Optimal Control of Nonsmooth Distributed Parameter Systems. VII, 159 pages. 1990.

Vol. 1460: G. Toscani, V. Boffi, S. Rionero (Eds.), Mathematical Aspects of Fluid Plasma Dynamics. Proceedings, 1988. V, 221 pages. 1991.

Vol. 1461: R. Gorenflo, S. Vessella, Abel Integral Equations. VII, 215 pages. 1991.

Vol. 1462: D. Mond, J. Montaldi (Eds.), Singularity Theory and its Applications. Warwick 1989, Part I. VIII, 405 pages. 1991.

Vol. 1463: R. Roberts, I. Stewart (Eds.), Singularity Theory and its Applications. Warwick 1989, Part II. VIII, 322 pages. 1991.

Vol. 1464: D. L. Burkholder, E. Pardoux, A. Sznitman, Ecole d'Eté de Probabilités de Saint- Flour XIX-1989. Editor: P. L. Hennequin. VI, 256 pages. 1991.

Vol. 1465: G. David, Wavelets and Singular Integrals on Curves and Surfaces. X, 107 pages. 1991.

Vol. 1466: W. Banaszczyk, Additive Subgroups of Topological Vector Spaces. VII, 178 pages. 1991.

Vol. 1467: W. M. Schmidt, Diophantine Approximations and Diophantine Equations. VIII, 217 pages. 1991.

Vol. 1468: J. Noguchi, T. Ohsawa (Eds.), Prospects in Complex Geometry. Proceedings, 1989. VII, 421 pages. 1991.

Vol. 1469: J. Lindenstrauss, V. D. Milman (Eds.), Geometric Aspects of Functional Analysis. Seminar 1989-90. XI, 191 pages. 1991.

Vol. 1470: E. Odell, II. Rosenthal (Eds.), Functional Analysis. Proceedings, 1987-89. VII, 199 pages. 1991.

Vol. 1471: A. A. Panchishkin, Non-Archimedean L-Functions of Siegel and Hilbert Modular Forms. VII, 157 pages. 1991.

Vol. 1472: T. T. Nielsen, Bose Algebras: The Complex and Real Wave Representations. V, 132 pages. 1991.

Vol. 1473: Y. Hino, S. Murakami, T. Naito, Functional Differential Equations with Infinite Delay. X, 317 pages. 1991.

Vol. 1474: S. Jackowski, B. Oliver, K. Pawałowski (Eds.), Algebraic Topology, Poznań 1989. Proceedings. VIII, 397 pages. 1991.

Vol. 1475: S. Busenberg, M. Martelli (Eds.), Delay Differential Equations and Dynamical Systems. Proceedings, 1990. VIII, 249 pages. 1991.

Vol. 1476: M. Bekkali, Topics in Set Theory. VII, 120 pages. 1991.

Vol. 1477: R. Jajte, Strong Limit Theorems in Noncommutative L_2-Spaces. X, 113 pages. 1991.

Vol. 1478: M.-P. Malliavin (Ed.), Topics in Invariant Theory. Seminar 1989-1990. VI, 272 pages. 1991.

Vol. 1479: S. Bloch, I. Dolgachev, W. Fulton (Eds.), Algebraic Geometry. Proceedings, 1989. VII, 300 pages. 1991.

Vol. 1480: F. Dumortier, R. Roussarie, J. Sotomayor, H. Żoładek, Bifurcations of Planar Vector Fields: Nilpotent Singularities and Abelian Integrals. VIII, 226 pages. 1991.

Vol. 1481: D. Ferus, U. Pinkall, U. Simon, B. Wegner (Eds.), Global Differential Geometry and Global Analysis. Proceedings, 1991. VIII, 283 pages. 1991.

Vol. 1482: J. Chabrowski, The Dirichlet Problem with L^2-Boundary Data for Elliptic Linear Equations. VI, 173 pages. 1991.

Vol. 1483: E. Reithmeier, Periodic Solutions of Nonlinear Dynamical Systems. VI, 171 pages. 1991.

Vol. 1484: H. Delfs, Homology of Locally Semialgebraic Spaces. IX, 136 pages. 1991.

Vol. 1485: J. Azéma, P. A. Meyer, M. Yor (Eds.), Séminaire de Probabilités XXV. VIII, 440 pages. 1991.

Vol. 1486: L. Arnold, H. Crauel, J.-P. Eckmann (Eds.), Lyapunov Exponents. Proceedings, 1990. VIII, 365 pages. 1991.

Vol. 1487: E. Freitag, Singular Modular Forms and Theta Relations. VI, 172 pages. 1991.

Vol. 1488: A. Carboni, M. C. Pedicchio, G. Rosolini (Eds.), Category Theory. Proceedings, 1990. VII, 494 pages. 1991.

Vol. 1489: A. Mielke, Hamiltonian and Lagrangian Flows on Center Manifolds. X, 140 pages. 1991.

Vol. 1490: K. Metsch, Linear Spaces with Few Lines. XIII, 196 pages. 1991.

Vol. 1491: E. Lluis-Puebla, J.-L. Loday, H. Gillet, C. Soulé, V. Snaith, Higher Algebraic K-Theory: an overview. IX, 164 pages. 1992.

Vol. 1492: K. R. Wicks, Fractals and Hyperspaces. VIII, 168 pages. 1991.

Vol. 1493: E. Benoît (Ed.), Dynamic Bifurcations. Proceedings, Luminy 1990. VII, 219 pages. 1991.

Vol. 1494: M.-T. Cheng, X.-W. Zhou, D.-G. Deng (Eds.), Harmonic Analysis. Proceedings, 1988. IX, 226 pages. 1991.

Vol. 1495: J. M. Bony, G. Grubb, L. Hörmander, H. Komatsu, J. Sjöstrand, Microlocal Analysis and Applications. Montecatini Terme, 1989. Editors: L. Cattabriga, L. Rodino. VII, 349 pages. 1991.

Vol. 1496: C. Foias, B. Francis, J. W. Helton, H. Kwakernaak, J. B. Pearson, H_∞-Control Theory. Como, 1990. Editors: E. Mosca, L. Pandolfi. VII, 336 pages. 1991.

Vol. 1497: G. T. Herman, A. K. Louis, F. Natterer (Eds.), Mathematical Methods in Tomography. Proceedings 1990. X, 268 pages. 1991.

Vol. 1498: R. Lang, Spectral Theory of Random Schrödinger Operators. X, 125 pages. 1991.

Vol. 1499: K. Taira, Boundary Value Problems and Markov Processes. IX, 132 pages. 1991.

Vol. 1500: J.-P. Serre, Lie Algebras and Lie Groups. VII, 168 pages. 1992.

Vol. 1501: A. De Masi, E. Presutti, Mathematical Methods for Hydrodynamic Limits. IX, 196 pages. 1991.

Vol. 1502: C. Simpson, Asymptotic Behavior of Monodromy. V, 139 pages. 1991.

Vol. 1503: S. Shokranian, The Selberg-Arthur Trace Formula (Lectures by J. Arthur). VII, 97 pages. 1991.

Vol. 1504: J. Cheeger, M. Gromov, C. Okonek, P. Pansu, Geometric Topology: Recent Developments. Editors: P. de Bartolomeis, F. Tricerri. VII, 197 pages. 1991.

Vol. 1505: K. Kajitani, T. Nishitani, The Hyperbolic Cauchy Problem. VII, 168 pages. 1991.

Vol. 1506: A. Buium, Differential Algebraic Groups of Finite Dimension. XV, 145 pages. 1992.

Vol. 1507: K. Hulek, T. Peternell, M. Schneider, F.-O. Schreyer (Eds.), Complex Algebraic Varieties. Proceedings, 1990. VII, 179 pages. 1992.

Vol. 1508: M. Vuorinen (Ed.), Quasiconformal Space Mappings. A Collection of Surveys 1960-1990. IX, 148 pages. 1992.

Vol. 1509: J. Aguadé, M. Castellet, F. R. Cohen (Eds.), Algebraic Topology - Homotopy and Group Cohomology. Proceedings, 1990. X, 330 pages. 1992.

Vol. 1510: P. P. Kulish (Ed.), Quantum Groups. Proceedings, 1990. XII, 398 pages. 1992.

Vol. 1511: B. S. Yadav, D. Singh (Eds.), Functional Analysis and Operator Theory. Proceedings, 1990. VIII, 223 pages. 1992.

Vol. 1512: L. M. Adleman, M.-D. A. Huang, Primality Testing and Abelian Varieties Over Finite Fields. VII, 142 pages. 1992.

Vol. 1513: L. S. Block, W. A. Coppel, Dynamics in One Dimension. VIII, 249 pages. 1992.

Vol. 1514: U. Krengel, K. Richter, V. Warstat (Eds.), Ergodic Theory and Related Topics III, Proceedings, 1990. VIII, 236 pages. 1992.

Vol. 1515: E. Ballico, F. Catanese, C. Ciliberto (Eds.), Classification of Irregular Varieties. Proceedings, 1990. VII, 149 pages. 1992.

Vol. 1516: R. A. Lorentz, Multivariate Birkhoff Interpolation. IX, 192 pages. 1992.

Vol. 1517: K. Keimel, W. Roth, Ordered Cones and Approximation. VI, 134 pages. 1992.

Vol. 1518: H. Stichtenoth, M. A. Tsfasman (Eds.), Coding Theory and Algebraic Geometry. Proceedings, 1991. VIII, 223 pages. 1992.

Vol. 1519: M. W. Short, The Primitive Soluble Permutation Groups of Degree less than 256. IX, 145 pages. 1992.

Vol. 1520: Yu. G. Borisovich, Yu. E. Gliklikh (Eds.), Global Analysis – Studies and Applications V. VII, 284 pages. 1992.

Vol. 1521: S. Busenberg, B. Forte, H. K. Kuiken, Mathematical Modelling of Industrial Process. Bari, 1990. Editors: V. Capasso, A. Fasano. VII, 162 pages. 1992.

Vol. 1522: J.-M. Delort, F. B. I. Transformation. VII, 101 pages. 1992.

Vol. 1523: W. Xue, Rings with Morita Duality. X, 168 pages. 1992.

Vol. 1524: M. Coste, L. Mahé, M.-F. Roy (Eds.), Real Algebraic Geometry. Proceedings, 1991. VIII, 418 pages. 1992.

Vol. 1525: C. Casacuberta, M. Castellet (Eds.), Mathematical Research Today and Tomorrow. VII, 112 pages. 1992.

Vol. 1526: J. Azéma, P. A. Meyer, M. Yor (Eds.), Séminaire de Probabilités XXVI. X, 633 pages. 1992.

Vol. 1527: M. I. Freidlin, J.-F. Le Gall, Ecole d'Eté de Probabilités de Saint-Flour XX – 1990. Editor: P. L. Hennequin. VIII, 244 pages. 1992.

Vol. 1528: G. Isac, Complementarity Problems. VI, 297 pages. 1992.

Vol. 1529: J. van Neerven, The Adjoint of a Semigroup of Linear Operators. X, 195 pages. 1992.

Vol. 1530: J. G. Heywood, K. Masuda, R. Rautmann, S. A. Solonnikov (Eds.), The Navier-Stokes Equations II – Theory and Numerical Methods. IX, 322 pages. 1992.

Vol. 1531: M. Stoer, Design of Survivable Networks. IV, 206 pages. 1992.

Vol. 1532: J. F. Colombeau, Multiplication of Distributions. X, 184 pages. 1992.

Vol. 1533: P. Jipsen, H. Rose, Varieties of Lattices. X, 162 pages. 1992.

Vol. 1534: C. Greither, Cyclic Galois Extensions of Commutative Rings. X, 145 pages. 1992.

Vol. 1535: A. B. Evans, Orthomorphism Graphs of Groups. VIII, 114 pages. 1992.

Vol. 1536: M. K. Kwong, A. Zettl, Norm Inequalities for Derivatives and Differences. VII, 150 pages. 1992.

Vol. 1537: P. Fitzpatrick, M. Martelli, J. Mawhin, R. Nussbaum, Topological Methods for Ordinary Differential Equations. Montecatini Terme, 1991. Editors: M. Furi, P. Zecca. VII, 218 pages. 1993.

Vol. 1538: P.-A. Meyer, Quantum Probability for Probabilists. X, 287 pages. 1993.

Vol. 1539: M. Coornaert, A. Papadopoulos, Symbolic Dynamics and Hyperbolic Groups. VIII, 138 pages. 1993.

Vol. 1540: H. Komatsu (Ed.), Functional Analysis and Related Topics, 1991. Proceedings. XXI, 413 pages. 1993.

Vol. 1541: D. A. Dawson, B. Maisonneuve, J. Spencer, Ecole d' Eté de Probabilités de Saint-Flour XXI - 1991. Editor: P. L. Hennequin. VIII, 356 pages. 1993.

Vol. 1542: J.Fröhlich, Th.Kerler, Quantum Groups, Quantum Categories and Quantum Field Theory. VII, 431 pages. 1993.

Vol. 1543: A. L. Dontchev, T. Zolezzi, Well-Posed Optimization Problems. XII, 421 pages. 1993.

Vol. 1544: M.Schürmann, White Noise on Bialgebras. VII, 146 pages. 1993.

Vol. 1545: J. Morgan, K. O'Grady, Differential Topology of Complex Surfaces. VIII, 224 pages. 1993.

Vol. 1546: V. V. Kalashnikov, V. M. Zolotarev (Eds.), Stability Problems for Stochastic Models. Proceedings, 1991. VIII, 229 pages. 1993.

Vol. 1547: P. Harmand, D. Werner, W. Werner, M-ideals in Banach Spaces and Banach Algebras. VIII, 387 pages. 1993.

Vol. 1548: T. Urabe, Dynkin Graphs and Quadrilateral Singularities. VI, 233 pages. 1993.

Vol. 1549: G. Vainikko, Multidimensional Weakly Singular Integral Equations. XI, 159 pages. 1993.

Vol. 1550: A. A. Gonchar, E. B. Saff (Eds.), Methods of Approximation Theory in Complex Analysis and Mathematical Physics IV, 222 pages, 1993.

Vol. 1551: L. Arkeryd, P. L. Lions, P.A. Markowich, S.R. S. Varadhan. Nonequilibrium Problems in Many-Particle Systems. Montecatini, 1992. Editors: C. Cercignani, M. Pulvirenti. VII, 158 pages 1993.

Vol. 1552: J. Hilgert, K.-H. Neeb, Lie Semigroups and their Applications. XII, 315 pages. 1993.

Vol. 1553: J.-L- Colliot-Thélène, J. Kato, P. Vojta. Arithmetic Algebraic Geometry. Editor: E. Ballico. VII, 223 pages. 1993.

Vol. 1554: A. K. Lenstra, H. W. Lenstra, Jr. (Eds.), The Development of the Number Field Sieve. VIII, 131 pages. 1993.

Vol. 1555: O. Liess, Conical Refraction and Higher Microlocalization. X, 389 pages. 1993.

Vol. 1556: S. B. Kuksin, Nearly Integrable Infinite-Dimensional Hamiltonian Systems. XXVII, 101 pages. 1993.

Vol. 1557: J. Azéma, P. A. Meyer, M. Yor (Eds.), Séminaire de Probabilités XXVII. VI, 327 pages. 1993.

Vol. 1558: T. J. Bridges, J. E. Furter, Singularity Theory and Equivariant Symplectic Maps. VI, 226 pages. 1993.

Vol. 1559: V. G. Sprindžuk, Classical Diophantine Equations. XII, 228 pages. 1993.